Ethical Traceability and Communicating Food

The International Library of Environmental, Agricultural and Food Ethics

VOLUME 15

For other titles published in this series, go to
www.springer.com/series/6215

Christian Coff • David Barling
Michiel Korthals • Thorkild Nielsen
Editors

Ethical Traceability and Communicating Food

Springer

Ethical traceability

Editors
Christian Coff
Centre for Ethics and Law
Copenhagen
Denmark

David Barling
Centre for Food Policy
City University
London
United Kingdom

Michiel Korthals
Applied Philosophy Group
Wageningen University
The Netherlands

Thorkild Nielsen
Technical University of Denmark
Department of Manufacturing and Management
Lyngby
Denmark

ISBN 978-1-4020-8523-9 e-ISBN 978-1-4020-8524-6

Library of Congress Control Number: 2008927742

Printed on acid-free paper

9 8 7 6 5 4 3 2 1

springer.com

Foreword

The theme of this book evolved from the idea of linking three concepts around food: traceability, ethics and informed choice. We believe that the current development and implementation of traceability in the agri-food sector offers an interesting way not only of handling food safety but also of addressing and communicating ethical issues arising from current food production practices.

Practices in the agri-food sector worry food consumers (as we all are, since we need to eat and drink to stay alive). But how can consumers act upon their concerns? Paradoxically, although consumers are bombarded with information on food – from the media, the food industry, food authorities, NGOs and interest groups – details about how foods are actually produced is often hard to find. Much of the information available is superficial, conflicting or partial, and it is hard for consumers seeking to make informed food choices to know which information to trust. The consumers we interviewed for this project felt that information about food products was withheld and manipulated.

Traceability, which provides a record of the history and journey of a given food, and which is increasingly used in the food sector for legal and commercial reasons, has the potential to communicate a more authentic picture of how food is produced. The idea of 'ethical traceability' adapts the idea of traceability to record and communicate the ethical aspects of a food's production history. It seems to us to offer a mechanism for communicating more comprehensively and reliably the information about food production practices that consumers need in order to be able to make food choices consistent with their values. Used imaginatively, it could also provide an opportunity for two-way communication along food chains, allowing the views of consumer-citizens to be taken into account along the length of the chain.

In order to address the concepts of traceability, ethics and informed food choice, both philosophical and sociological investigations are incorporated in this book. By combining philosophical and sociological research and employing a cross-cutting approach, both theoretical and empirical aspects of ethical traceability and informed choice could be investigated. So even though the sociological investigations and philosophical reflections are presented in separate parts of the book (Parts II and III, respectively), each part is informed by the other. The questions drawn up for the purpose of conducting the empirical case studies were formulated in collaboration with the philosophers. For instance, the ethical concerns around food mapped by the philosopher Michiel Korthals in the initial phase of the project were used as the ethical foci for the empirical case studies of specific national food supply chains. The philosophers also contributed to the final analysis of the empirical results. Equally, the directions that the philosophical work took were influenced by the problems and questions raised by the sociological studies.

The book consists of four parts:

Part I covers the interdisciplinary merging of philosophical, sociological, political science and juridical investigations on the use of traceability in law, politics and advertising.

Part II consists of sociological case studies on the current use of traceability in the food chain, the actors' attitudes to ethical traceability and informed choice, and the way information flows along the chain. The three case studies cover bacon production in Denmark, wheat-bread production in the UK and olive oil production in Greece, representing three mature, complex and economically important production systems.

Part III consists of philosophical examinations of the challenges of ethical traceability and informed food choice in the light of selected philosophical and political theories and discussions. To some extent this part also considers how ethical traceability can be made operational from consumers' and producers' viewpoints, enabling informed food choice.

Part IV, the conclusion of the book, first considers the communicative and participative aspects of ethical traceability, looking at how to develop new ways of communicating ethics in the agri-food sector which, crucially, incorporate the consumers' point of view. Secondly, it presents and reflects upon the main findings of the whole book. Finally, in the annex, it presents the thoughts of two key branches of the European Commission on the future need for and direction of ethical traceability, in speeches given by Mariann Fischer Boel, European Commissioner for Agriculture and Rural Development, and Margaritis Schinas, Head of Cabinet to Markus Kyprianou, Commissioner for Health and Consumer Protection.

Christian Coff
David Barling
Michiel Korthals

Editors' Preface

This book is more of a collective endeavour than many anthologies. From May 2004 to April 2007, the authors met regularly for discussions on the individual contributions and co-authored chapters. Likewise, the links between the chapters and the overall outline of the book were discussed during meetings and in extensive email exchanges.

Chapter 1 sets the scene for traceability, ethics and informed food choice. It describes the history and use of traceability, introduces food ethics and links these two issues to the idea of informed food choice. Chapter 1 has thus served as the common starting point for all authors, from which the juridical and philosophical reflections and empirical sociological surveys presented in the first three parts of the book depart. The last part closes the circle, reprising and harvesting the insights of the intermediate chapters and revisiting the questions raised in Chapter 1.

None of the authors involved was from the outset an 'expert' on the technical and legal aspects of traceability. Hence, it was necessary to establish an initial competence on these issues within the group. Building on these initial studies of the historical, technical, practical and juridical dimensions of traceability, efforts could be put into linking traceability with the disciplinary fields of the academics involved – that is, of ethics and applied social sciences.

We are grateful to the EU and its Sixth Framework Programme[1] for supporting our work on the ethical dimensions of traceability. A one-day conference was held in Brussels on September 20, 2006, with the participation of, among others, Mariann Fischer Boel, European Commissioner for Agriculture and Rural Development, and Margaritis Schinas, Head of Cabinet to Markus Kyprianou, Commissioner for Health and Consumer Protection. We are grateful for their participation in the conference and for allowing us to use their speeches in this book.

Last but not least, we are grateful to Rosalind Sharpe at City University, London, for her editorial help with the preparation of the book.

February 2008

Christian Coff
David Barling
Michiel Korthals
Thorkild Nielsen

[1] FP6-2003-Science and Society-4: Deepening the Understanding of Ethical Issues.

Contents

Part IV Conclusions and Outlook

List of Abbreviations

ABARE	Australian Bureau of Agricultural and Resource Economics
ABIM	Association of Baking Ingredients Manufacturers
ACCS	Assured Combinable Crops Scheme
AFS	Assured Food Standards
AIC	Agricultural Industries Confederation
AoA	Agreement on Agriculture (by WTO)
AOC	Appellation d'Origine Contrôlée
BAT	Best Available Technology
BSE	Bovine Spongiform Encephalopathy – mad-cow disease
BSPB	British Society of Plant Breeders
CAP	Common Agricultural Policy (of the EU)
CCFICS	Codex Committee on Food Import and Export Inspection and Certification Systems
CCGP	Codex Committee on General Principles
CEN	European Committee for Standardization
CIES	Comité International d'Entreprises à Succursales (International Committee of Food Retail Chains)
Codex	Codex Alimentarius
CPA	Crop Protection Association
CSL	Central Sciences Laboratory
CSR	Corporate Social Responsibility
CSO	Civil Society Organizations
DEFRA	Department for the Environment, Food and Rural Affairs
DG	Directorate General
DG SANCO	Health and Consumer Protection Directorate General
DNA	Deoxyribonucleic acid
DUS	Distinct, Uniform and Stable
EC	European Commission
EFSA	European Food Safety Authority
EMAS	Environmental Management and Audit Scheme (EU)
ERM	Ethical Room for Manoeuvre

ETI	Ethical Trading Initiative
EU	European Union
	Eurep Euro Retailer Produce Working Group
	EurepGAP Eurep's Good Agricultural Practice
FAO	Food and Agriculture Organization (of the United Nations)
FSA	Food Standards Agency (UK)
GATT	General Agreement on Tariffs and Trade (administrated by WTO since 1990s)
GI	Geographical Indication
GMO	Genetically modified organisms
HACCP	Hazard Analysis of Critical Control Points
HGCA	Home Grown Cereals Authority
ICT	Information and Communication Technology
IGC	International Grains Council
IGD	Institute of Grocery Distribution
IOOC	International Olive Oil Council
IPPC	International Plant Protection Convention
ISB	In-store Bakery
ISO	International Organization for Standardization
NABIM	National Association of British and Irish Millers
NASA	National Aeronautics and Space Administration
NFU	National Farmers Union
NGOs	Non-Governmental Organization
OJL	Official Journal of the European Communities
PASA	Pennsylvania Association for Sustainable Agriculture
PCR	Polymerase Chain Reaction
PDA	Personal Digital Assistant
PDO	Protected Designations of Origin
PGI	Protected Geographical Indications
PMWS	Post-weaning Multisystemic Wasting Syndrome
PRC	Pesticides Residues Committee
RFID	Radio Frequency IDentification
RSPCA	Royal Society for the Protection of Animals (UK)
SEDEX	Suppliers Ethical Data Exchange
SME	Small and Medium-Sized Enterprise
SPS	Sanitary and Phytosanitary Agreement
SRC	Socially Responsible Corporations
TASCC	Trade Assurance Scheme for Combinable Crops
TBT	Technical barriers to Trade
TRIPS	Trade Related Intellectual Property Rights Agreement

UAA	Utilized Agricultural Area
UK	United Kingdom
UN	United Nations
US	United States
USDA	United States Department of Agriculture
VCU	Value for Cultivation and Use
WHO	World Health Organization
WTO	World Trade Organization
WWF	World Wide Fund for Nature

List of Tables

List of Figures

Contributors

Alessandro Arienzo is a researcher in political philosophy and the history of political thought at the University of Naples 'Federico II'. He has published on early modern theories of reason of state and on contemporary theories of political governance in the European Union. He is currently working on global security governance and the development of emergency powers, and writing on European food politics as a case of security politics. Recent publications include '*Governo, Governamentalità, Governance. Riflessioni sul neo-liberalismo contemporaneo*' [Government, Governamentality, Governance. Reflections on contemporay neo-liberalism], in *Biopolitica e Democrazia*, Milano: Meltemi: pp. 251–278, and '*Il governo delle emergenze e la conservazione politica: ragion di stato democratica e security governance*' [The government of emergencies and political conservatism: democratic reason of state and security governance], pp. 35–57 in V. Dini (ed.) *Eccezioni*. Naples: Dante & Descartes.

David Barling is a senior lecturer at the Centre for Food Policy, City University, London. His main areas of research are around the governance of the food supply and food standards at national and international levels, and the relationship of food policy to sustainability. He has published a number of book chapters, and articles in a range of journals including: *Social Policy and Administration*, *Political Quarterly*, *Journal of Agricultural and Environmental Ethics*, *European Environment*, *European Journal of Public Health*, and *Public Health Nutrition*. He is currently completing a co-authored book on food policy for Oxford University Press.

Volkert Beekman joined the Applied Philosophy Group of Wageningen University in The Netherlands in 2004. He has been occupied with two European research projects, on (1) consumer perceptions of animal welfare, and (2) traceability and informed choice in food ethics. He received his Ph.D. on the basis of a thesis *A Green Third Way? Philosophical Reflections on Government Intervention in Non-Sustainable Lifestyles* at Wageningen University in 2001. Since 2000 he has been employed at the Agricultural Economics Research Institute (LEI) in The Hague as ethicist and sociologist working on consumer concerns about food. He has been co-ordinator of a European research project (*The Development of Ethical Biotechnology Assessment Tools for Agriculture and Food Production*) and co-ordinator of research programmes on food safety and food quality funded by the Dutch Ministry of Agriculture, Nature and Food Quality. He has been newsletter chief-editor and second secretary of the European Society for Agricultural and Food Ethics (EurSafe) and is editor of the *Journal of Agricultural and Environmental Ethics*. He regularly publishes papers in the areas of agricultural, environmental and food ethics in journals like *Journal of Agricultural and Environmental Ethics* and *Ethical Theory and Moral Practice*.

Mariann Fischer Boel has been European Commissioner for Agriculture and Rural Development since 2004.

Marco Castagna is Ph.D. student in 'ethics and anthropology' at Lecce University and lecturer in philosophy and theories of language at the University of Naples. He has published several articles on Paul Ricœur's hermeneutics and is at present researching the ethical and philosophical implications of the act of reading. He contributes to *Alternative*, a bimonthly review of politics and culture, and 'Kainos' an online review of philosophical criticism. He is also member of the SFL (Philosophy of Language Society).

Christian Coff was research director at Centre for Ethics and Law in Copenhagen from 2004 to 2007. He studied agricultural science at the Royal Veterinary and Agricultural University of Copenhagen and has a Ph.D. in food ethics from the Danish Educational University. In 2001 he founded the first consumer-supported agriculture (CSA) in Denmark (www.landbrugslauget.dk). His recent research projects include *Consumers, Ethics and Traceability* (2003–2004) in collaboration with Danish Consumers Co-operative Society (FDB) and *Ethical Traceability and Informed Food Choice* (2004–2007). He is author of the book *The Taste for Ethics. An Ethic of Food Consumption* (2006) and 'Ethical Traceability', pp. 56–61 in K. Matthias and M. Lien (eds.) *Ethics and the Politics of Food* (2006).

Emmanouil Kabourakis, M.Sc., Ph.D., is researcher in ecological production systems at the National Agricultural Research Foundation, Greece. He graduated in plant sciences at the Agricultural University of Athens, and did postgraduate studies on ecological agriculture at Wageningen Agricultural University. He works in research on ecological food production systems and sustainable rural development. He is interested in designing and developing sustainable food systems that empower local communities, link consumers with producers and enhance diversity. He lectures on agroecology and ecological food systems at graduate and postgraduate level. Among other things, he has published 'Learning processes in designing and disseminating ecological olive production systems in Crete', in M. Cerf et al. (eds.) *Cow Up a Tree. Knowing and Learning for Change in Agriculture. Case Studies from Industrialised Countries* (2001).

Michiel Korthals is professor of applied philosophy at Wageningen University. He studied philosophy, sociology, German and anthropology at the University of Amsterdam and the Karl Ruprecht Universität in Heidelberg. His academic interests include bioethics and ethical problems concerning food production and environmental issues, deliberative theories, and American Pragmatism. Main publications include *Philosophy of Development* (with Wouter van Haaften and Thomas Wren Kluwer, 1996); *Pragmatist Ethics for a Technological Culture* (with Jozef Keulartz et al. Kluwer 2002); *Before Dinner. Philosophy and Ethics of Food* (Springer 2004) and *Ethics for Life Sciences* (Springer, 2005). He has written articles in *Science, Technology and Human Values, Development World Bioethics, Journal of Agricultural and Environmental Ethics, Gastronomica, Philosophy and Social Criticism* and many other journals.

Niels Heine Kristensen is associate professor at the Department of Management Engineering at the Technical University of Denmark. He is currently interested in how policies are constructed and implemented, especially in relation to environmental and food issues. He is a frequent advisor to Danish ministries (Agriculture, Food and Fisheries; Environment, Industry), the EU Commission and other bodies. Recent publications include 'Public sector procurement of organic food: school meals in Denmark', in G. Holt and M. Reed (eds.) *Sociological Perspectives of Organic Agriculture: From Pioneer to Policy* (with Astrid Dahl, 2006) and 'The liquidity of the Organic movement – reconversion or reillumination', a paper presented at Working Group 5 at the XXI Congress of the ESRS (2005).

Tim Lang is professor of food policy at the Centre for Food Policy, City University, London, UK. He is interested in how competing policies affect the shape of the food supply chain, what people eat and the social, health and environmental outcomes. Since the 1990s, he has been a frequent advisor to the World Health Organization and other bodies. In 1987 he was special advisor to the European Commissioner for the Environment on food matters, taking a special interest in niche markets. He has been advisor to four UK Parliamentary inquiries on food standards, globalization and obesity. In 2005, he was appointed Land Use and Natural Resources Commissioner on the UK Government's Sustainable Development Commission, and in 2005–2008 led a team reviewing how Government relates with food retailers on sustainable development matters. He is author and co-author of over 100 articles and reports. Recent books include *The Unmanageable Consumer* (with Y. Gabriel, 2006), *Food Wars: The Battle for Mouths, Minds and Markets* (with M. Heasman, 2004) and *The Atlas of Food* (with E. Millstone, 2003).

Guido Nicolosi is researcher at University of Catania (Italy), where he is professor of sociology of cultural and communicative processes at the Faculty of Political Science and scientific coordinator at 'Centre for Technological Innovation in Socio-linguistic and Territorial Systems in the Mediterranean Area' (Braudel Centre). His research programme is 'The body between social practices and cultural changes: food, communication and symbolic integrity'. He published in Italy *Corpi al limite. Linguaggio, natura e pratiche sociali* (2005) and *Lost food. Comunicazione e cibo nella società ortoressica* (2007).

Thorkild Nielsen has an M.Sc. in geography and is senior researcher at the Department of Management Engineering at the Technical University of Denmark. Since 1990 most of his work has focused on strategies and policies for sustainable food systems. Projects have related to: (1) public procurement as a policy tool, (2) ethics and foods and (3) management systems and innovation. Among his publications are 'Innovation processes in large-scale public foodservice – Case findings from implementation of organic foods in a Danish county'. *Journal of Foodservice Business Research*. Volume 8, Issue 2 (with Mikkelsen, B.E., N.H. Kristensen, 2005) and 'Minimal and careful processing', in O. Schmid, A. Beck and U. Kretzschmar (eds.) *Underlying Principles in 'Low-Input' Food Processing. Switzerland* (2004).

Dimitris Papadopoulos, M.Sc., Ph.D., is a researcher in rural sociology at the National Agricultural Research Foundation, Greece. He graduated in agricultural economics and development, did his doctoral research at the Agricultural University of Athens, and studied the management of agro-ecological knowledge and social change at the Wageningen Agricultural University. He works in research on agricultural extension, the sociology of organizations and the management of agricultural knowledge systems. He is interested in the organizational aspects of developing sustainable food systems. Recent publications include 'An Analytical Approach for Law-Like Corruption' (2004) and 'Interaction Between State and Non-state Actors in the Implementation of the CAP Agri-environmental Measures' (with L. Louloudis and E. Arahoviti, 2001).

Margaritis Schinas has been Head of Cabinet to Commissioner Markos Kyprianou, Public Health, Consumer Protection and Food Safety, since 2004.

Liesbeth Schipper is junior researcher at the Applied Philosophy Group at Wageningen University. She studied philosophy at RijksUniversiteit Groningen, specializing in political philosophy and ethics. She finished her Masters in philosophy on the basis of the thesis *Liberalism and Animal Welfare Legislation; About the Concept of Neutrality*, in which the relationship between legislation concerning animal welfare and the concept of political liberal neutrality is the central issue.

Rosalind Sharpe is a writer and researcher. She has an MA in food policy and currently works at the Centre for Food Policy at City University, London, UK.

Agapi Vassiliou, M.Sc., Ph.D., is a research associate in agricultural economics and development at the National Agricultural Research Foundation, Greece. She graduated in agricultural economics at the Aristotle University of Thessaloniki, and studied economics and management at the Mediterranean Agronomic Institute of Chania and agricultural economics at the Institute of Rural Studies, University of Wales. She works in sustainable rural development, farm economics and co-operative theory and environmental policy. She is interested in developing organic markets and the economics of alternative farming systems. Recent publications includes 'Organic Farming in Greece. Trends and Perspectives'. In: M. Nikolaidis et al. (eds.) *The Market of Organic Products in the Mediterranean Region* (with Bitsaki, A. and E. Kabourakis, 2003).

Chapter 1
Ethical Traceability and Informed Food Choice

Christian Coff, Michiel Korthals, and David Barling

Introduction

The traceability of food and feed emerged as a focus for political attention and regulation at both national and international governmental levels at the turn of the millennium. The industrialization of food production and manufacture, and the complexities and anonymity of modern supply chains have been accompanied by a new wave of concerns around the safety and quality of the food supply. The emergent concept of keeping track of food products and their different ingredients through the various stages from field to plate offers a potential means of managing some of the recent safety and quality concerns around food. Food traceability covers a range of overlapping objectives, which are outlined below, and so has a wide potential appeal, to regulators, producers, processors, retailers and consumers alike.

In this chapter, we seek to establish the range of ethical concerns around food, drawing from an emerging canon of work on food ethics, and to look at the ways in which the concept of ethical traceability can enhance the public good of existing traceability systems. Traceability relates to where and how foods are produced. It follows that it has the potential to be developed as a tool for providing information to consumers that addresses their concerns about food production. As traceability retells the history of a food, it can address the ethical, as well as the practical and physical, aspects of that history, enabling more informed food choice. The importance of ethical traceability for consumers is essentially twofold: firstly, it can help them make informed food choices; and secondly, it can act as a (democratizing) means for enabling consumers to participate more fully as citizens in the shaping of the contemporary food supply. And ethical traceability has a third benefit, this time for food producers, who can use it as a tool for managing the ethical aspects of their own production practices and communicating ethical values about their products. In the following sections, the nature of food traceability and its differing but overlapping objectives are explained, and the role of ethical traceability is elaborated.

C. Coff et al. (eds.) *Ethical Traceability and Communicating Food*,

The Emergence of Traceability in the Food Chain

The idea of food traceability – i.e. the ability to track or trace food – has emerged in modern societies due to the professionalization of food production, whereby the production of food has been separated from its consumption. Food is very rarely produced and consumed by the same people, but is produced by persons in one (or indeed several) place(s) and consumed by others in other places. In more traditional societies, where production and consumption occur at the same place and are carried out by the same people, or where trade is dominated by face-to-face transactions in which buyer and seller can verify the qualities of the food, there is no need for conceptualizing or formalizing the idea of traceability; traceability is inherent in the transaction. This is because knowledge about production practices is part of such societies, or because the chain is very short and direct. The industrialization of food production and distribution has changed this.

During the last 200 years, major changes have taken place in food production practices. Mass food production accompanied growing urbanization and settlement. Agricultural production was increased through industrial upscaling and associated technological developments ranging from more rapid long-distance transportation to refrigeration and canning. The regular, face-to-face contact between buyer and seller declined, although this was not without its problems and consequent reactions. In the UK, adulteration of basic processed foodstuffs, such as sugar, led in 1844 to the creation, by the urban working class, of the first co-operative retail society, the Rochdale Society of Equitable Pioneers, in order to ensure supplies of unadulterated food. Such initiatives have since appeared in many countries all over the world, and today the direct marketing of farmers' produce is a growing phenomenon in many European countries, for instance in the form of box schemes, farm shops and farmers markets. The growth of these initiatives reflects the importance of traceability and the provenance of food for participating consumers.

One of the consequences of the industrial manufacture and long-distance transportation of food is that it can change profoundly during processing and transit. Fresh produce, such as vegetables and meat, is susceptible to deterioration. Products from different farmers can be mixed or mistaken. Hence, at the beginning of the 20th century, new record-keeping systems were developed in order to keep track of which grower delivered what, so that the grower could receive the proper price for his produce (USDA, 2004:12). These were early traceability systems, although the term was not used at that time.

Today, the specialization of food production practices means that food is increasingly processed outside the household, using industrial and scientific techniques that are not familiar to ordinary consumers. For example, few consumers are familiar with modern bread-processing techniques or the wide range of ingredients used in the industrial production of bread, although this is the main form of bread consumed in many European countries; nor are they aware of the extent to which olive oil from different sources is blended prior to retail; nor that a very large percentage of bacon sold in Denmark by Danish companies is not produced in Denmark, but in most cases comes from Germany or Poland. This industrialization and globalization of food production also mean that an increasing number of intermediaries, such as shippers, wholesalers, processors, repackers, brokers, importers and exporters, are involved in the process. All of these factors help to obscure how food is produced, how it is handled and from where it originates.

Industrialization has not only changed food products and production practices fundamentally, but has also generated new risks in the food production chain. The recent EC-enforced focus on traceability in the food sector occurred mainly as a response to food scandals, notably the outbreak of bovine spongiform encephalopathy (BSE or 'mad-cow' disease), in the 1990s in the UK and the discovery of dioxins in animal feed in Belgium in the late 1990s. (There are many other incidents, including for instance the contamination of Perrier water with benzene and the subsequent worldwide recall in 1990.) More generally, since the 1980s there has been growing attention to the presence of pathogenic micro-organisms, such as *Salmonella, Listeria, Clostridium* and *Escherichia coli* O157, and other contaminants in food. In the US alone food-borne pathogens are considered to cause 76 million illnesses per year (Hutter, 2004).

Fraudulent practices and adulteration are other problems of food supply chains that have recently attracted media attention. For instance, in 2005 it was discovered in Germany that waste from slaughterhouses, intended for pet food, had been used for human food products. In Germany and Denmark, the selling of old meat long after it was deemed unsafe for human consumption, with false and 'renewed' expiry dates, in the *alte Fleisch Skandal*, made headlines in the media and certainly contributed to a decrease in trust in the food chain among consumers in those countries. Fraud in the food chain is far from new, but with more extended and complex supply chains the implications and consequences have grown. In an era of mass consumption, serious faults and mistakes that occur during the production process may endanger the lives of (many) innocent consumers. In the longer term, such accidents also rebound on the producers, resulting in adverse media coverage, consumers deserting the product, and reaction from public authorities, which may impose regulatory sanctions or introduce reforms. Hence, it has become important in modern production systems to be able to trace faults rapidly when they occur during production.

Thus, traceability in its contemporary forms is intended to deal with the growing complexity of a food chain based on mass production and global distribution and consumption. It is used to keep records of the processing and transportation of food products through all production stages. It should make it possible to trace a specific product back through the chain at any time, and so isolate contaminated goods and expose frauds. On a practical level, the ideal is to set up record-keeping systems that make it possible to trace product flow through all production stages, enabling identification of the exact origin of food products and their ingredients, and logging the transformation processes that a product undergoes before reaching consumers.

Traceability in Contemporary Food Chains

Today, traceability has become common in the agri-food sector. Indeed, since 2005 EU law has required a certain level of traceability on the part of all food operators in the EU. Other traceability schemes are voluntarily implemented by actors in the food chain, as part of their business strategies or as part of quality assurance schemes, as we shall see in Part II. Many of the problems that are inherent in the modern food system, as described above, can be addressed by the introduction and implementation of traceability, which is thus used to meet a broad variety of commercial and regulatory objectives. Table 1.1 maps the key applications and objectives of traceability in contemporary food systems. The first four are widely used, while the fifth objective – consumer information and communication – is still in its initial phase. As we shall see, this fifth objective is, however, essential for developing traceability in the ethical direction of informed food choice.

Five objectives can be distinguished, even though there are some overlaps between them. The first category – risk management and food safety – has been a primary focus of regulatory attempts to introduce traceability. Food safety control has been built upon process-based auditing, such as Hazard Analysis of Critical Control Points (HACCP) standards and the International Organization for Standardization (ISO) 9000 (traceability is mentioned in ISO 9001:2000 as one aspect to be considered in quality management systems). The need to be able to recall contaminated products for public health reasons motivated food producers to incorporate traceability systems into supply chain management processes originally implemented to achieve efficiencies (Farm Foundation, 2004:8). The latter – supply chain management and efficiency – is the third category in Table 1.1, and its main concern is to allow food companies, notably the corporate retailers, to manage the flow of goods and information, link inventory to consumer purchasing, set product specifications for growers and processors under contract, and so on, in order to meet market demand and secure the efficient use of resources. Traceability is thus an instrument that can be deployed for a variety of purposes, often at the same time. Hence in practice there will usually be some overlap between the different categories depicted in Table 1.1, and traceability will rarely if ever be implemented for only one of the objectives mentioned. For example, the second category interweaves with the other two mentioned above. Keeping a record of the production history of a product can be used for surveillance and fraud prevention. It is interesting to observe that the two largest retail companies of the world, Wal-Mart and Carrefour, are increasingly asking for complete traceability from their suppliers (Bantham and Duval, 2004).

The fourth category is likewise linked to the second, as it concerns verification of quality claims and label schemes. Quality is a complex term, as the perception and dimensions of quality are continually shifting (an example would be the multiple uses and perceptions of the term 'fresh'). But the goals of traceability as set out in the fourth category can to some extent reflect ethical criteria for food production practices and also consumers' ethical concerns, and communicate them via labels. Four examples, out of many labelling schemes, are: the organic labels found in

Table 1.1 Key functions of traceability in the food sector

Objectives of traceability in food

1. **Risk management and food safety**
   - Risk assessment: mapping of foods and feed, food ingredients and processing technologies that have food safety implication (e.g. hygiene)
   - Food residue surveillance: food sampling at appropriate points testing for residues, e.g. pesticides
   - Public health recall systems: identification of breakdowns in food safety along the food supply chain, allowing recall of contaminated products for the purpose of protecting public health
2. **Control and verification**
   - Surveillance and auditing of producer and retailer activities
   - Avoidance of fraud and theft: control of products by chemical and molecular approaches (biological 'food-prints')
   - Identification of responsible actors (but also claims of innocence!)
   - Ingredients definition
   - Avoidance of negative claims (e.g. 'may contain GMO traces')
3. **Supply chain management and efficiency**
   - Cost-effective management of the supply chain
   - Computerized stock inventory and ordering systems linked to point of sale
   - Just-in-time delivery systems
   - Efficient use of resources (cost minimization)
4. **Provenance and quality assurance of products**
   - Marketing of health, ethical and other claims
   - Authenticity: identity of the product (food authentication) and the producer
   - Typicality: as with European schemes for Protected Designation of Origin (PDO) and Protected Geographical Indication (PGI)
   - Quality assurance of standards at different stages of production and/or processing (e.g. environmental protocols for production)
   - Final product quality assurance
5. **Information and communication to the consumer**
   - Transparency of the production history
   - Facilitation of informed food choice, through transparency and the ability to compare different products
   - Recognition of specific consumers concerns and information demands – where such concerns and demands are not static but may evolve
   - Public participation: consumer services, companies' 'care lines' and consultation to obtain consumer feedback

most European countries, the UK Red Tractor scheme, the UK Royal Society for the Protection of Animals' (RSPCA) Freedom Food label and the French Label Rouge. Increasingly, quality parameters of an ethical nature are integrated into supermarkets' own brands. However, we should note that this book does not directly address labelling. A motivating and decisive factor for taking up the idea of ethical traceability in this book was a growing awareness of the limits of labels: that they are symbolic representations of often rather huge quantities of information, which are rarely communicated to consumers and which are not accessible to ordinary consumers. Moreover, labels can in some cases create more confusion than enlightenment. This was shown in a European study called Welfare Quality on food labels in relation to animal welfare. The study showed that there were huge differences between labels in different EU countries and that there was disagreement as to definitions of animal welfare standards (Welfare Quality, 2005). In the light of the shortcomings of labels, ethical traceability was from the outset conceived as an alternative that could provide more complete information to consumers and thus respond more efficiently to consumers' ethical concerns (these are discussed in more detail later in this chapter).

The fifth category of objectives for traceability is far from fully developed. In some ways it is an aspiration that would facilitate consumers' understanding of food production practices and their ability to make informed choices about the foodstuffs they purchase and eat. It concerns the *communication* of production practices in the food chain. Here, the term communication is not restricted to information flows between producers, retailers and food authorities, but also includes making information available to consumers. In this sense, traceability is about visibility; it is about making the production history of food visible to the eyes of the consumers. It allows producers and retailers to establish a more advanced kind of communication with consumers about production practices. This more detailed communication could facilitate more informed choice by consumers.

The fourth category allows for some of these aspirations to be met, but the communication is shaped by the producers and/or the retailers (in some assurance schemes) and by the processors (in the EU's geographical origin schemes, the Protected Designations of Origin [PDO] and the Protected Geographical Indications [PGI]). The fifth category envisions more responsive and transparent systems, where traceability links to the ethics of food production practices.

This book concentrates on the fourth and fifth categories of traceability, as we seek to develop the idea of *ethical* traceability. It is, however, important to understand the different uses of traceability and to see it in its broad context, to get an idea of how the term has developed and how it is being used at present. Our focus on the ethical dimensions of traceability also means that this book does not include technical matters or enabling technologies for traceability (but see Annex to Chapter 13 for a discussion of some technological approaches to ethical traceability). The technical enabling of traceability is developing very rapidly in the current climate, seeking to deliver with greater and more rapid precision an expanding range of features. Traceability may involve keeping track of hundreds of inputs and processes, and the systems required to handle and transmit all this data need to be highly sophisticated. Most of the information in traceability systems would be irrelevant to consumers, as it concerns matters that are of interest only to actors in the supply chain. For instance, details about the moisture content of a consignment of wheat and the variety of grain used are essential to flour millers, who need the information to make decisions about how to process the grain and what type of flour to turn it into. Consumers will want to know what sort of flour they are buying, but in many cases are not interested in the technical details leading to the production of that type of flour.

The different uses to which traceability can be put have led several authors to speak of traceability as a *tool* (among others Clemens, 2003:3; EU Standing Committee on the Food chain and Animal Health, 2004:10; USDA, 2004:3; Farm

Foundation, 2004:22; CIES, 2005:6; GS1, 2006:6–8). So, looking at traceability from a different perspective from that presented in Table 1.1, three categories of traceability as a tool can be distinguished:

1. **Management tool**

Purpose: Supply chain management and internal management of resources in co-operations.

2. **Government tool**

Purpose: Political and administrative government of the food chain, anti-fraud measures and verification of product attributes and liability.

3. **Communication tool**

'Value-capture' of food qualities (such as animal welfare) for the purpose of informing consumers.

In this book we focus on traceability as a tool for informing and communicating with consumers on ethical concerns. This aspect of traceability is mentioned in several reports on traceability, but has so far never been treated in depth (see among others Food Strategy Division and Food Standards Agency, 2002:2; Farm Foundation, 2004:8; USDA, 2004:9).

To some extent, all traceability is ethical. Food safety is obviously an ethical issue since it aims at protecting consumers from food-borne diseases and pollution. Preventing fraud in the food chain is likewise inherently ethical, as is guaranteeing the accuracy of the information provided to consumers, and the verification of assurance and labelling schemes. However, it is at the communication level that specifically 'ethical' traceability gains a certain power. For actors in the food chain (be they processors, manufacturers or retailers) who wish to secure a minimum level of ethical behaviour among their suppliers, ethical traceability provides information on the ethics of a given product's production history, which is essential if the buyer is to be able to form an ethical judgement of the supplier (see Coff, 2006, for a description of the link between food ethics and food production histories). And the same goes for consumers: ethical traceability should provide the information necessary for consumers to exercise their ethical judgement about the production history of a given food, and thus allow consumers' informed choice. Such information is vital to ethical consumers who are concerned about the impact of food production on issues such as animal welfare, working conditions, the environment and sustainability.

The different uses of traceability make it a potential battlefield. There is widespread agreement that the need for fully documented traceability systems within the food chain has never been stronger (Morrison, 2003:459), but there is tremendous disagreement about the purpose of introducing traceability and about which aspects of production should be incorporated in traceability systems. These arguments about how to make use of traceability in supply chains expose disagreements about the role of food ethics in production practices. Many of these issues are discussed in ensuing chapters of this book.

Ethics, Traceability and Food

Most people are aware that in recent decades massive changes have taken place in agricultural and food production practices. This is clear not only from the radical changes that can be observed in the landscape and the bewildering array of goods available in contemporary supermarkets, but also from media headlines. Media coverage of food production practices tends to highlight negative aspects, such as food scandals, environmental and animal welfare problems, and so forth.

The physical, social and mental separation of production and consumption, which is characteristic of modern societies, means that in most cases producers and consumers do not know each other and that consumers do not know what happens during production processes. They are invisible to one another. In spite of this differentiation of the two spheres, and the obscurity of the food system, people, as citizens and consumers, may still seek to feel that they somehow are involved in agriculture and food production. Or, at least, that food production practices matter, in the sense that it makes a difference to consumer-citizens if food is produced in one way as opposed to another. But how can food production practices matter, even though production has been so clearly separated from consumption?

There is an old saying that ‘if you eat, you are involved in agriculture’. Shopping, preparing and eating are key notions for understanding the involvement of consumers in the agri-food sector. These three activities lead the thoughts in two different directions. We could say that our thoughts are led both backwards and forwards in time. Shopping, preparing and eating are, so to speak, specific points in a chain of events, from which it is possible to think both backwards and forwards. We think backwards when we consider *what* we are buying, preparing and eating. We cannot ascertain if something is edible, and for instance whether it is meat or a vegetable, without relating our sensuous perception to our knowledge of what is meat and what is a vegetable. We have an idea of what is meat and what is a vegetable, and from our experience we judge them to be either edible or non-edible. Now, this experience is often associated with many different stories. One very simple story is that meat comes from living animals. For some people, this knowledge is very important (vegetarians, for instance). This simple illustration shows how, in the act of eating, we consciously or unconsciously direct our thoughts towards the past.

We also direct our thoughts towards the future, when considering how a particular food will affect our bodies. Food is taken into the body. It is incorporated and incarnated. In this sense, food links our body with past events in the agri-food sector in a very physical sense. But we might also consider the pleasure of the food (taste, digestibility, effect on the mental state and so forth), the healthiness of the food, and the social and cultural contexts of shopping, cooking and eating. With the food we choose, we make a statement about our identity and connect ourselves with other people who make the same kind of food choices. Vegetarianism, or whatever kind of diet we choose, is as much about belonging to certain groups as about eating. This gives some idea of why production practices in the agri-food sector still matter for many people, despite the separation of production and consumption. Highlighting

Table 1.2 Research areas in food ethics

Research areas in food ethics
1. *Food security* deals with the just and fair supply of food to human beings. With more than 800 million starving or undernourished people in the world, this is probably the most pressing ethical question.
2. *Food safety* deals with the safety of the food: food should not endanger the health of consumers due to pathogens or pollution present in the food. There are ongoing discussions about what is safe enough and whose definition of safety should be followed.
3. New developments in *nutritional research* and technology, such as personalized nutrition, functional foods and health foods, challenge existing norms and values about food. This also includes food-related diseases such as obesity, cardiovascular diseases and cancer and their association with food culture, because they raise issues of responsibility and respect for 'non-healthy' lifestyles and production methods.
4. *Ethical questions raised by specific production practices and conditions in the food chain*: this concerns animal welfare, the environment, sustainability, working conditions, use of new (bio and nano) technology, research ethics and so forth. These ethics relate to the production history of the food, i.e. how and under what conditions it was produced.

the link between consumers' activities and the production history of foods takes us to the central theme of the book: i.e. *traceability*. To return to the meaning of the concept, it refers to the history of a product and the records kept of that history.[1] Thus, traceability seems to offer the possibility of making the link between production and consumption visible. As mentioned earlier, traceability is already a requirement of EU food law (regulation (EC) No. 178/2002). Thus, all European food businesses have a legal responsibility to implement traceability. The question raised in this book is: can this traceability be used to provide ethical information to consumers about the production history of foods, and thereby enable consumers to make informed choices on ethical issues?

A central concern, then, is how traceability can link to food ethics. That is why we have introduced the term 'ethical traceability'. Food ethics is a discipline that embraces many different ethical and philosophical studies on food. However, it is not just an academic discipline; it also describes more practical ways in which people think about food and act according to their values of right, injustice or good and bad in food production. At present there are roughly four main research areas within food ethics, presented in Table 1.2.

Traceability is about keeping track of the history of the food. *Ethical* traceability is about keeping track of the ethical aspects of food production practices and the conditions under which the food is produced. It is a means of capturing and mapping values and processes in the food production chain. It can be used as a verification

[1] The internationally most recognized definition of traceability belongs to ISO. ISO 9000/2000 refers to a set of quality management standards. To this set belongs ISO 8402. This standard defines traceability as 'the ability to trace the history, application or location of an entity by means of recorded identifications. More recently ISO has defined traceability in terms of management: 'A Traceability system is a useful tool to assist an organization operating within a feed and food chain to achieve defined objectives in a management system' (ISO, 2007:iv).

process of the methods and practices used, in response to consumers' ethical concerns. It can be defined in the following way:

Ethical traceability is the ability to trace and map ethical aspects of the food chain by means of recorded identifications.

Once the information on the ethical aspects of production practices has been captured and mapped, it can be used to communicate with interested stakeholders in the food chain, including producers, processors, retailers and consumers. It can be used as part of the 'value-capture' of products and also to enable stakeholders to make choices consistent with their own values. Tim Lang, co-author of Chapter 6, discerns a movement away from 'value for money', the idea that price is the fundamental determinant of choice, towards 'values for money', reflecting the notion that consumers are more than just wallets on legs, but are also citizens who will select and reward companies that behave in socially responsible ways (Lang, 2007).

Consumers' Ethical Concerns

Based on work by the philosopher Michiel Korthals (see among others Korthals, 2004) in the initial phase of the project, ten main ethical concerns relevant to food production were identified (see Table 1.3). These ten concerns were used to structure some of the philosophical work, and especially to structure the interviews used in the empirical research presented in Part II. The ten concerns can be divided into two categories. First, consumers have *substantive concerns* about the first seven ethical issues while shopping for food. These are issues that relate directly to the consequences of production practices or to the consequences or impacts of food consumption, for instance human health and food quality. They are substantive in that they are a matter of substance rather than a matter of procedure; we could also term them vertical or specific concerns. This leads to the second category, the *procedural concerns*, which includes the last three ethical issues listed. Procedural refers in this context to the communicative aspects of information sharing, feedback and listening procedures, participatory methods and co-production. They are procedural in the sense that they are not matters of substance, but are horizontal and cut across the various substantive or vertical concerns. They are about access to, and availability of, information, the reliability of information, and the opportunity for consumers to have a voice on the substantive concerns. The two categories are of a different nature and therefore they raise different problems and demand different solutions. Furthermore, each concern may embody more than ethics, and each concern may be interlinked with others from the list. For instance, is 'origin and place' a concern that works differently from 'terms of trade'? 'Origin and place' may not necessarily be an ethical parameter, but people make a lot of associations with origin and place that involve ethical judgements. Equally, 'origin and place' may be linked to concerns around 'working conditions', such as with food from developing countries.

Table 1.3 Ten ethical consumer concerns relevant to food production, used as a basis for the studies in this book

1.	Animal welfare
2.	Human health
3.	Methods of production and processing and their impact (e.g. environmental, landscape)
4.	Terms of trade (fair price etc.)
5.	Working conditions
6.	Quality (intrinsic qualities such as taste, composition, etc.)
7.	Origin and place
8.	Trust
9.	Voice (participation)
10.	Transparency

Also, trust is a complex concern that seems to be interlinked with the other procedural concerns of transparency, voice and participation.

Informed Food Choice

Ethical consumption mixes the role of consumer with that of citizen. The term 'consumer-citizen' refers to this duality (see for instance Scammell, 2003; Consumer Citizenship Network, 2006). Clive Barnett and his colleagues (2005b:4–5) consider consumer-oriented activism as a pathway to participation for ordinary people. Ethical consumption is a reconfiguration of the consumer's role, merging it with the citizen's role. The majority of consumers subscribe to at least one of the ten ethical concerns listed above. From a recent attitudinal survey we know that in Europe 60% of the population is worried about animal welfare when prompted (European Commission, 2006:28). However, in this study only a minority subscribed to two or more of the concerns.

Empirical sociology and psychology have taught us that there is often a gap between people's attitudes and their behaviour. Consumers' concerns as measured in surveys do not always translate into actual food-purchasing behaviour. This means that even if consumers express concerns about animal welfare (attitude) they will not necessarily purchase meat selectively (behaviour). Many factors contribute to this gap. For instance, it can be difficult to shop according to individual values due to a lack of reliable information or a lack of trust in the food system. Economic constraints or lack of easy access to food with the desired ethical attributes can also act as barriers. In Fig. 1.1 we have listed some of the major issues that influence consumers' food choice (in a simplified form, as there may be many more). From the figure it is clear that choosing food is not a simple matter and that many different and opposing interests must be weighed against each other. The priority given to different interests may vary over time, depending on the situation, the mood of the consumer or the social contexts at the moment of shopping.

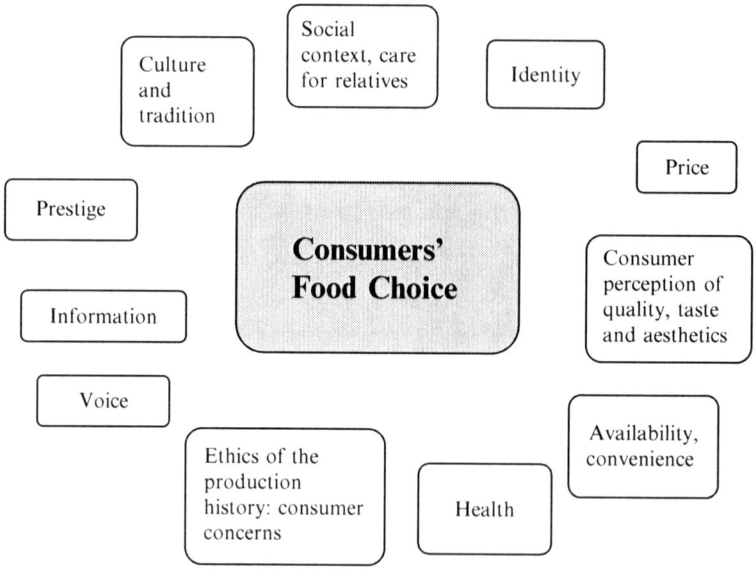

Fig. 1.1 Some factors affecting consumers' food choice

Some of the issues in Fig. 1.1 are related to self-interest. Price, quality, taste, prestige, health and convenience can all be part of self-interested considerations around the 'best buy'. Purely self-interested considerations can, in fact, be said to lack any ethical reflection and awareness, since they entail no concern for others. The ethical dimension of food purchase is opened when food is bought not only for one's own satisfaction, but also to take into consideration the needs of others.

Barnett and his colleagues (2005:99) speak of *caring at a distance*, because the others that we take into consideration or show care for in ethical consumption are not necessarily people we encounter face to face, but may be distant from us. The problem with such an approach to showing care is that it is often assumed that the consequences of our actions are unintelligible, as they are hidden by the 'space' in between. There are different approaches on how to make the long-range ethics required to care for these distant others functional. For instance, it has been proposed that long-range ethics could be based on short-range ethics (face-to-face interaction) by making the distant consequences visible (Coff, 2006:100). The intention behind such a strategy is to let food production practices appear in narrative forms, as stories, or production (hi)stories, so that consumer-citizens can take a stance on them and the consequences can be made explicit.

Whichever way ethical consumption is carried out and whatever strategy is used, it needs some degree of concerted action by organizations, institutions, consumers and so forth (Barnett et al., 2005a:8). As an individual consumer it can be impossible to gain access to the ethical information desired (see for instance Coff, 2006:175 for such an attempt). Cooperation among actors in the chain is necessary to ensure access to the necessary information. In a dynamic fashion, feedback from consumers to producers and processors may enhance the development of new and 'more ethical' food production practices. Mobilization of consumer support for ethical trading and consumption can also be promoted by organizational efforts, such as campaigns. The nature of the *agencies* involved and the collective organizations that serve as the *mediators* of engagement and participation are also important (Barnett et al., 2005:7). In Chapter 13 the organizational aspects of ethical traceability are developed in more detail.

Some issues in Fig. 1.1 are related to sociological and cultural aspects of food production and consumption. For instance, it is clear that food choice is linked to culture, social class and tradition. The selection of food – the matter or 'environment' to be incorporated in one's own body – confers identity not only through the social context in which it takes place but also, on a more individual level, through selection of particular foods, such as the avoidance of meat, preferences for organic food or animal-friendly meat, one's own particular preferences or dislikes, the avoidance of certain food ingredients because of their association with certain diseases and so forth.

Information plays a crucial role for most of the issues in Fig. 1.1, and for some of the issues it is paramount. It is well documented that many animals instinctively know which plants can cure diseases and also which plants/animals should be avoided. This is no longer the case for human beings: we need knowledge to help us distinguish what is edible. In fact, human food is embedded in a culture of knowledge. Food in its different social contexts relies heavily on knowledge, not so much about the food itself but on cultural traditions and habits.

However, consumers differ about which information they see as relevant; much of the information that is provided simply goes unnoticed because it is not relevant to the consumer's purposes. To be sure, in order to estimate the impact of food intake on health consumers need to be informed; but consumers have very different conceptions of health. Furthermore, information is essential for consumer decision-making as it allows for comparison between alternatives. The aim of making a comparison between different foods is to arrive at a judgement about which is the best food. 'Best' depends, of course, on what criteria one considers most important. Such a judgement cannot be made without information.

For the consumer who finds the ways and modes of production important, access to *relevant* information is a key concern. It has already been said that some consumers, called ethical consumers, have an interest in the ethical aspects of the production history of foods (see Harrison et al., 2005). If consumers want to choose food on the basis of ethical considerations, and to make *informed food choices*, it is necessary to make ethical information on the production history of foods accessible to consumers. The core question that we look into in this book is how ethical traceability can be linked to the idea and practice of informed food choice.

The Plan of the Book

Regulations on food traceability are gradually being implemented worldwide. Research on food traceability is increasing and so is the literature on traceability. Most food companies either have implemented traceability schemes or are in the process of implementing them. A great deal of attention is given to the development of traceability schemes. New tracing techniques that make use of computers, the Internet or molecular tests serve to make tracing more efficient and also to make it possible to include still more parameters, such as food quality or handling during processing, and increasingly to include ethical dimensions. There is no doubt that traceability has become a major issue in the creation of modern food policies and that it is an issue that involves all actors in the food chain.

However, little attention has so far been paid to how traceability could be used to 'trace' ethical dimensions in the agri-food sector and thus address the consumer concerns mentioned earlier in this chapter. This, therefore, is the key aim of this book: to address existing (and possible future) links between traceability on the one hand and ethics on the other. The authors of the book have explored in their research how traceability links to the ethical questions and concerns of the agri-food sector. No less important is the question of how traceability in the future *could be* related to ethical questions and concerns. Such reflections on ethical traceability – i.e. the tracing of ethical aspects of the food chain, and how this process could be used to facilitate informed food choice – have presented a common and major challenge for all the authors.

The book has four parts, which represent four aspects of the link between traceability and ethics.

Part I: Regulation, Governance and Narrative Strategies of Food Traceability

Part I presents the broader policy and social contexts within which the development and contemporary place of food traceability may be understood. It starts with the current status of the regulation of food traceability, and goes on to discuss how food traceability regulation and implementation have been developed through processes of governance. It ends with an analysis of how traceability in the form of narratives is used by food companies in advertising. In Chapter 2, Alessandro Arienzo, Christian Coff and David Barling offer an analysis of the status of traceability in EU food law and regulation. The authors argue that the democratizing potential of full agri-food traceability is missed. It is concluded that there is potential for a more comprehensive mode of (ethical) traceability, which could open the way for a more informed and participatory food system. David Barling in Chapter 3 examines the governing and wider governance of food and food traceability. Contemporary agri-food governance is portrayed as a process marked by both conflict and compromise, involving both public (state) and private actors (from the corporate sector and from civil society). The multilevel nature of agri-food governance frames the conflicts that have occurred over the development of food traceability standards at the international level, illustrated by the conflicts that have taken place in the Codex Alimentarius (Codex). In Chapter 4, Guido Nicolosi and Michiel Korthals examine advertising strategies in selected Italian and Spanish magazines and discuss how these make use of narratives that draw on the traceability of the food in addressing consumers. Tradition, nature, geographical and cultural origins are almost obsessively present in the advertisements examined. It is shown that the narrative strategies relate to some substantive ethical concerns but not to the procedural concerns of consumers.

Part II: Ethical Traceability in Three Food Supply Chains: Case Studies of Danish Bacon, UK Wheat-Bread and Greek Olive Oil

Part II presents three case studies on traceability and ethical traceability in different food supply chains. The case studies are: pigs into bacon in Denmark, wheat into bread in the UK and olives into olive oil in Greece. Each case study describes and analyses the current status of traceability in the chain and looks at the extent to which ethical traceability is being addressed and how it is being handled. The case studies present empirical data collected from interviews with stakeholders and consumers from the three chains. The interviews included questions about the ten ethical concerns outlined earlier in this chapter, and about information flows in the chains studied. The research found that some of these concerns are already addressed by existing traceability or assurance schemes. In all three case studies it was found that producers in the chain felt well supplied with information whereas consumers, by contrast, felt that information was withheld and unreliable.

In Chapter 5, Thorkild Nielsen and Niels Heine Kristensen describe how Danish consumers feel a need for more information about bacon production practices, especially about some of the invisible attributes, such as origin, use of medicine and animal welfare, even though there is a long tradition for highly developed traceability systems. Traceability is *reactive* in this chain and is not intended to transmit information on the safety, production practices or quality of the final product proactively downstream to firms or end consumers. Rosalind Sharpe, David Barling and Tim Lang show, in Chapter 6, how traceability in the UK wheat-bread chain is limited by the routine practice of blending wheat for convenient handling and to manipulate quality and cost. However, some examples of traceability back to farm were found. This chain, which is highly industrialized, is subject to many regulatory and quasi-regulatory controls (such as the regulations governing the development of wheat varieties, or the assurance schemes which impose quality standards on all wheat destined for human consumption, from farm to mill) which incorporate traceability and which include ethical dimensions. In Chapter 7, Agapi Vassiliou, Emmanouil Kabourakis and Dimitris Papadopoulos explore the olive oil chain and describe how traceability and potentially ethical traceability were widely said to be limited by the practice of blending oil by olive mills and packing houses, in order to manipulate quality and cost or for convenience. The dominant ethical concerns for stakeholders and consumers in the olive oil supply chain are trust and transparency.

Part III: Ethical Traceability and Its Philosophical Implications for Civil Society, Market, State and Democracy

The philosophical studies all deal with the challenges and philosophical problems that the notion of ethical traceability raises. Of course, not all these challenges and problems can be solved here and disagreements are likely to persist. However, the chapters expose most of the important challenges raised by ethical traceability, which deserve further attention and reflection in the future. The studies do not stop at mapping and exposing the challenges; they go on to present some of the ways in which ethical traceability could assist in solving problems that face actors in all parts of the food chain, including consumers.

In Chapter 8, Christian Coff opens with a discussion of the challenges that ethical traceability presents to common perceptions of the structural organization of society. Ethical traceability breaks with many mainstream ideas about what should be considered as private and public concerns. Implementing ethical traceability and informed food choice entails creating new kinds of public spheres and a new kind of civil society. The two subsequent chapters also address the tasks of the market (private) and the government (public). In Chapter 9, Liesbeth Schipper asks whether the issue of animal welfare should be dealt with by the market or by the government, and whether traceability should be used as a communication tool or a government tool for improving animal welfare. In Chapter 10, Volkert Beekman also examines what the roles of the market and government should be and links this to an analysis of which kinds of ethical traceability could be justified from the perspective of liberal political philosophy. In Chapter 11, Michiel Korthals's focus is on how consumer concerns can democratically and practically be incorporated in the market and in food chains by participatory methods. The concept of 'Ethical Room for Manoeuvre' is constructed to specify the ethically desirable conditions under which the identification and weighing of values and their dilemmas can be processed. Marco Castagna emphasizes in Chapter 12 that tracing always involves interpretation. Interpretation by consumers, it is argued, opens up new ways of consumer participation and involvement.

Part IV: Conclusions and Outlook

Part IV opens with Chapter 13, where Volkert Beekman, Christain Coff, Michiel Korthals and Liesbeth Schipper map different ways of providing information to consumers on ethical traceability and establishing communication with consumers.

Participative strategies are discussed in the light of ethical traceability, and a three-step process is recommended which involves (1) providing sound information to consumers; (2) facilitation of everyday dialogue between consumers and producers; and (3) deeper engagement between dedicated consumer-citizens and producers. In Chapter 14, Christian Coff, David Barling and Michiel Korthals summarize the main conclusions and results of the book. The authors start with a presentation of the main findings of the sociological investigations and the main conclusions of the philosophical reflections. Implementing ethical traceability also entails problems, so the major risks associated with ethical traceability are listed. The chapter ends with a set of recommendations on how to develop the idea of ethical traceability in practice as well as in future research.

Finally, two political speeches on ethical traceability are presented in an Annex to Part IV. Both speeches were presented at a conference entitled Ethical Traceability in the Food Chain held in Brussels on 20 September 2006. The first speech is by Margaritis Schinas, Head of Cabinet to Markus Kyprianou, Commissioner for Health and Consumer Protection. He presents the work and views of Directorate General Health and Consumer Protection (DG SANCO) on consumers' informed choice. The second speech, by Mariann Fischer Boel, European Commissioner for Agriculture and Rural Development, addresses the role of traceability, ethics and food quality in the EU's Common Agricultural Policy.

References

Bantham, A. and J-L. Duval (2004) *Connecting Food Chain Information for Food Safety/Security and New Value*. John Deere, Food Origins.

Barnett, C., N. Clarke, P. Cloke and A. Malpass (2005a) 'Articulating Ethics and Consumption', pp. 99–112, in M. Böstrom, F. Andreas et al. (eds.) *Political Consumerism: Its Motivations, Power and Conditions in the Nordic Countries and Elsewhere*. Proceedings from the 2nd International Seminar on Political Consumerism, Oslo, August 26–29, 2004, TemaNord.

Barnett, C., N. Clarke, P. Cloke and A. Malpass (2005b) 'Citizenship between individualization and participation: relocating agency in the growth of ethical consumerism in the United Kingdom'. http://www.open.ac.uk/socialsciences/staff/cbarnett/ethicalconsumption.htm. Accessed June 2007.

Bingen, J. and L. Busch (eds.) (2006) *Agricultural Standards. The Shape of the Global Food and Fiber System*. Dordrecht, The Netherlands: Springer. Series: The International Library of Environmental, Agricultural and Food Ethics, Vol. 6.

CIES (2005) *Implementing Traceability in the Food Supply Chain*. Paris: CIES – The Food Business Forum.

Clemens, R. (2003) *Meat Traceability in Japan*. Review Paper (IAR 9:4:4–5). Centre for Agricultural and Rural Development, Iowa State University, Iowa.

Coff, C. (2006) *The Taste for Ethics. An Ethic of Food Consumption*. Dordrecht, The Netherlands: Springer. Series: The International Library of Environmental, Agricultural and Food Ethics, Vol. 7.

Consumer Citizenship Network (2006) http://www.hihm.no/eway/default.aspx?pid=252. Accessed October 2006.

EU Standing Committee on the Food chain and Animal Health (2004) Guidance on the Implementation of Articles 11, 12, 16, 17, 18, 19 and 20 of Regulation (EC) No. 178/2002 on General Food Law.

European Commission (2006) *Special Eurobarometer: Risk Issues*. 238/Wave 64.1 – TNS Opinion and Social, Brussels.
Farm Foundation (2004) *Food Traceability and Assurance in the Global Food System*. Farm Foundation's traceability and Assurance Panel Report, July 2004.
Food Strategy Division and Food Standards Agency (2002) *Traceability in the Food Chain. A Preliminary Study*. March 2002.
GS1 (2006) *The Global Traceability Standard*. GS1. www.ciesnet.com.
Harrison, R., D. Shaw and T. Newsholm (2005) *The Ethical Consumer*. London: Sage.
Hutter, L. (2004) *Assuring Food Safety in an Ever More Complex, Price Sensitive World*. PPP at the FoodTrace conference (EU concerted action programme).
ISO (2007) *Traceability in Feed and Food Chains – General Principles and Basic Requirements for System Design and Implementation. ISO 22005*, Geneva: International Standards Organisation.
Korthals, M. (2004) *Before Dinner. Philosophy and Ethics of Food*. Dordrecht, The Netherlands: Springer. Series: The International Library of Environmental, Agricultural and Food Ethics, Vol. 5.
Lang, T. (2007) 'The new order is values-for-money', *The Grocer*, 230, 27 January: 29.
Morrison, C. (2003) 'Traceability in food processing: an introduction', pp. 458–470 in M. Lees (ed.) *Food anthenticity and traceability*. Cambridge: Woodhead.
USDA (2004) *Traceability in the U.S. Food Supply: Economic Theory and Industry Studies*. United States Department of Agriculture, Agricultural Economic Report Number 830, Washington, DC.
Welfare Quality (2005) *Science and Society Improving Animal Welfare in the Food Quality Chain*. EU funded project FOOD-CT-2004-506508. www.welfarequality.net

Part I
Regulation, Governance and Narrative Strategies of Food Traceability

During the past 10 years, the notion of traceability has become increasingly important within food legislation and food policy. First of all, the concept of food traceability has become an integrated part of the European Union (EU) food law and regulation, as well as the subject of definition by the Codex Alimentarius and the International Standards Organization (ISO). The process leading to these regulations included not only governments but also, to a large extent, private and civil actors (such as non-governmental organizations [NGOs]). Thus the process of formulating these regulations can be depicted as a governance process, meaning less top-down control and more emphasis on collaborative outcomes involving many actors. Private actors' interest in traceability has also led to private and voluntary traceability schemes, which are implemented at different points in the food chain (see Part II for examples). Private actors' interest in traceability is also illustrated by the way traceability is used as a narrative element in advertising and the marketing of food.

The European setting for agri-food traceability is a key focus of this book. The empirical research in Part II investigates the traceability systems that exist, and the dimensions of ethical traceability present, in three different agri-food chains at the national level in Europe: pigs to pork and bacon in Denmark, wheat to flour to bread in the UK; and olives to olive oil in Greece. To provide the context for this research, Part I starts with an account of the EU's regulation of food traceability. The emergence and forms of regulation of food traceability, as developed by the EU institutions, are critically assessed. This assessment juxtaposes the EU's reforms of the regulation and the institutional forms for governing food safety against the proposed reform of the EU's own wider governance.

Both streams of reform are portrayed as interrelated attempts to buttress the legitimacy of the European integration project during a period of more general discontent amongst European publics. This discontent was voiced over the pace of integration, notably the formalization of a European constitution, in different member states. Also, the working and efficiencies of the single European market were shaken by a succession of food safety crises and the subsequent administrative weaknesses that emerged from the handling of these crises, from bovine spongiform encephalopathy (BSE) through to the dioxin contamination of chicken, and the subsequent erosion of public trust in the EU's management capabilities. Out of these crises, the need for better and more comprehensive forms of traceability emerged as a concern in the management of food and feed chains, to enable effective product recall and facilitate the public health response to food safety crises. Overall, traceability regulation has covered a range of purposes in European regulation, from food safety recall to food provenance and the conservation of natural resources. However, Chapter 2 argues that the bounded nature of traceability in the general principles of EU food law reflects a top-down, managerialistic approach, with the result that the democratizing potential of full agri-food traceability is missed. This potential would allow a more comprehensive mode of (ethical) traceability, which could open the way for a more informed and participatory food system, where consumers could identify a range of information at their own request rather than be restricted to information provided for them by vendors.

Chapter 3 explores the wider governance of food traceability. Agri-food governance is understood as crossing both public and private sectors, in the latter case through the setting of standards which can act as private regulatory forms, based upon differing criteria of safety and quality. The setting of standards in buyer-driven supply chains effectively regulates the suppliers in that chain. In the case of private corporate interests the drive for commercial advantage, both through profit and through brand esteem or corporate social responsibility, acts as a regulative mechanism that can drive standards upwards. Civil society organizations have entered this process by advocating or initiating standards that benchmark more equitable terms of trade or improved standards of animal welfare, resource conservation or sustainability – introducing new ethical dimensions into food standards. These private forms of regulation exist alongside public regulations. 'Governance' has emerged as an umbrella term to depict this merging of public modes of governing with private forms of governance. In the case of traceability, private forms outstripped the public regulatory response in the 1990s. This was most graphically illustrated through the systems set up to segregate genetically modified (GM) crops in large-scale commodity supply chains and their derivates, now widely deployed across food processing and manufacturing in Europe.

The public regulation of traceability and the forms of traceability that emerged from the EU have also entered into regulatory debates in the global institutions enlisted to govern biodiversity and aspects of agri-food trade under the World Trade Organization (WTO). These include the Joint United Nations Food and Agriculture Organization (FAO) and World Health Organization (WHO) Food Standards Programme, known as the Codex Alimentarius (Codex). Chapter 3 examines the multilevel nature of agri-food governance that is conveyed through the debates and conflict around the regulation of traceability that have taken place in these global regulatory institutions.

Chapter 4 explores the narrative or 'historical' character of traceability at the interface between food production and food consumption by examining the marketing and advertising strategies of selected Italian and Spanish magazines *Marketing and Advertising Strategies* conceptualize products, consumers, their concerns, preferences and behaviour according to cultural images that differ according to national and cultural contexts. Purity, danger, integrity, order and naturalness are considered central categories, on the basis of which consumers and producers tackle their daily discourse on food preparation and consumption. The hypothesis put forward is that these categories can help to illuminate the way consumer concerns are addressed in advertising strategies. Tradition, nature and geographical and cultural origins are almost obsessively present in the advertisements examined. Everything seems to revolve around the 'quasi-mythical' celebration of these elements, presented as actual and fixed absolute entities. However, these advertisements do not 'trace' a genuine story, since they do not show a historical development from a point of departure through a process to a final goal. As for ethical concerns, some of the substantive concerns (such as methods of production and processing and their impact, terms of trade, working conditions, quality and origin and place) are addressed. But procedural concerns are not addressed at all: the issues of the reliability of information, trust and voice (participation) are left out. Transparency is only implicitly addressed by mentioning phone numbers or web sites.

Chapter 2
The European Union and the Regulation of Food Traceability: From Risk Management to Informed Choice?

Alessandro Arienzo, Christian Coff, and David Barling

The European Union (EU) embarked on a major reform of its regulation of food safety and its food law from the mid to late 1990s, in an effort to ensure a safer and more trustworthy food supply after the political fallout from the spread of bovine spongiform encephalopathy (BSE) and infected products across the continent. The EU's regulation of food has been driven to varying degrees by the need to integrate the European market, from agricultural production subsidy and production management controls, to the harmonization of food standards around the principle of mutual recognition, to food safety and hygiene standards. In the case of BSE, the management of the single market was seen to have failed. Furthermore, the rise of public concerns over food safety resulted in a period of 'contested governance', signalling 'a pervasive sense of distrust that challenges the legitimacy of existing institutional arrangements' (Ansell and Vogel, 2006:10). This distrust went beyond policy disagreement to embrace deeper concerns about the ability of the prevailing institutions and processes to manage risk in the food supply. This contested governance over food safety coincided with a more general review by the EU of its governance arrangements, and the reform efforts around food safety became tied up in the EU's political efforts to renew its legitimacy in the eyes of the European publics. The food safety focus led to a reform of the EU's risk analysis institutions for food safety and the European Commission's responsibilities around food law, with a revision of the general principles of food law. In short, the reforms for food safety were part of a wider political management effort to rebuild both consumer and citizen trust in the European institutions and processes for the longer term. Within these reforms food traceability appeared as an important element for the operation of risk management.

The European market is subject to the pressures of economic globalization, including the challenges of international economic competition, globalizing trends in food sourcing and in food consumption through the mass fast food industry, all underpinned by the neo-liberal reform of international trading rules. Yet, there is a strong pressure to valorize and protect regional varieties of food cultures, habits and customs, both from European producers and processors, and from consumers. Furthermore, the regulators of the European market are pressed to respect the growing ethical concerns of citizens for the preservation of the environment, animal welfare, social justice and solidarity and

C. Coff et al. (eds.) *Ethical Traceability and Communicating Food*,

fair trade. From the political and institutional standpoint, the themes of governance, democratic citizenship, political participation and sustainable development are confronted with the compelling need to establish institutions that are capable of delivering efficacy and maintaining legitimacy both in society and in the market. The European effort may seem a Sisyphean enterprise, for it aims to virtuously link different (sometimes diverging) social, economic and political concerns. The reforms around food safety and food law provide some response to these demands. Food traceability has been deployed, in other EU regulations, with the aim of conveying the specificity or typicality of locally produced foods (and the specific processes of production) in order to enable consumer choice in the marketplace. However, the application of traceability within the revision of food law was more circumscribed and limiting for consumers wishing to exercise their ethical concerns in their food purchasing and consumption.

In this chapter, we review the EU's approaches to the regulation of food traceability, which embrace a range of priorities covering risk management, provenance of food by its place and nature of production, and the enabling of consumer choice. In particular, the focus is on the reforms to food safety regulation and food law since 1997, and the way traceability has emerged as a risk management tool from this reform period. The reforms to food safety regulation are placed in the wider context of the EU's review of its own governance arrangements and procedures. The role of traceability as a risk management tool for food safety and public health recall is critically assessed. Two main consequences are elaborated as a result of this appraisal. Firstly, it is argued that the nature of EU governance has not changed to any notable extent in the case of the food safety regulatory reforms. Indeed, despite the reforms to food safety governance, the essentially technocratic approach to governance by the EU and the European Commission remains in place with regard to food. Within the food safety regulatory reforms, the introduction and definition of traceability as a general principle of EU food law (Regulation 178/2002) is essentially a precautionary and procedural instrument for food safety and risk management that is based on a model of liberal governance whose main purpose is the regulation and unification of the European market. Secondly, it is argued that food traceability in a European context needs to be considered in a different way. That is, traceability should also be employed as a means to facilitate and promote informed food choice. Informed food choice allows consumers to take a more active role, and a more central place, in determining the nature and type of information provided by traceability about our food. In this way trust may be re-embedded in the European food system in a more substantial fashion.

EU Governance and Its Review

The EU has extended its regulatory scope over the past few decades in its effort to achieve a common and then a single and internal market. Regulation of the agri-food sector has been an important element of this drive to a single internal market.

There have been three main elements in the Europeanization of the regulation of the agri-food sector: firstly, the development of the Common Agricultural Policy (CAP), which remains an ongoing political project; secondly, the harmonization of national food standards for the common market, and most recently, the reform of food safety regulation as part of the EU's growing public health agenda. The CAP was born from the experiences of the immediate post-Second World War years of food shortage and insecurity, and remains the single largest element of EU expenditure and the site for ongoing political reform and contest. The establishment of a common market and a unified trade space also necessitated market harmonization of food standards and definitions, and the mutual recognition of national standards was established as a guiding principle. The growing complexity of the arrangements, combined with the fears and concerns raised by numerous food scandals, placed the governance of food safety among the highest priorities on the EU's political agenda as the 1990s progressed. All of these elements of European regulation are driven by the aspiration to establish and maintain an internal European market to underpin the EU as a political project. The focus on food safety occurred at a time when the legitimacy of the European political project was under increasing stress.

The use of the term governance is very widespread, so much so that there is danger of a loss of clarity and utility. In this book, governance in the food sector is described as 'a dynamic process both occurring across and involving actors from not just the public sector but also the corporate sector and civil society' (Chapter 3). This process may vary significantly according to a number of different variables, such as the actors involved, their goals and the political, legal and social environment. The agri-food sector encompasses different forms of governance that relate public and private actors in a variety of overlapping ways, creating a somewhat novel sphere that is at the core of the relations between the public sector, the corporate sector and civil society. In addition, food governance crosses a range of policy dimensions, from agricultural production to environmental impact, or within public health from food safety to nutritional composition. The term governance may also be used to describe a certain structure of institutions and actors and to analyze the complexities of their interactions, such as in the conceptualization of European multilevel governance (Sharpf, 1997; Pernice, 1999; Hooghe and Marks, 2000). During its political drive towards an internal market, the EU has also sought to Europeanize its governance arrangements. For example, the European Commission promotes an inclusive approach to policy formulation, inviting key stakeholders to round table discussions of policy, particularly those who can speak for member organizations across the EU, the so-called European peak associations (Greenwood and Aspinwall, 1998). On the one hand, this is a technocratic approach to policy-making, seeking to defuse potential or existing political disagreements at an early stage (Radaelli, 1999). On the other hand, it is a deliberative approach, involving stakeholders at the early stages of the formulation of new regulations (Laffan et al., 2000). This governance style has evolved as a preferred operating mode for EU policy-making, steered by the Commission. At an intergovernmental level this approach is reinforced by the EU's comitology, in particular the regulatory management committees, chaired by the Commission. This results in a relatively elite stakeholder form of participation, essentially technocratic and functional, removed from the public gaze and lacking in wider public participation (Héritier, 1997).

The EU itself has sought to redefine and overhaul what it terms its modes of governance. Yet, despite the long and detailed discussion raised by the *White Paper on European Governance* (CEC, 2001), there is little evidence as to the significance of any changes in its governance approach (see Héritier, 2000; De Búrca and Scott, 2001; Fukuda, 2003). At the outset the ambitions were for a more democratic and open approach. In a preparatory document to the *White Paper on European Governance*, entitled *Enhancing Democracy in the European Union*, the Commission had expressed the intention to set up rules capable of producing a wider consensus. In this sense, European governance had 'to be placed in a wider context. It should underline the necessity and opportunity for the Union to promote better world governance in harmony with its own internal governance' (SEC, 2000:13). By emphasizing a multilevel system of government based on the principles of transparency, responsibility and efficiency the Commission wanted to start a new democratic process, for 'the reform of European modes of governance is all about improving democracy in Europe' (SEC, 2000:14). This reform had to make European policies and their normative architecture more efficient and effective through:

- Simplification and diversification of juridical instruments
- Definition of *standards* and good practices of consultation
- Recourse to experts and committees
- Establishment of a framework for regulatory agencies
- Promotion of the participation of European citizenship in the processes of *policy-making* and *policy-formulation* by supporting consultations, agreements and partnerships with non-institutional actors
- Reform of the role of European institutions – especially in their executive function – in order to offer a greater coherence and legitimacy to their policies, to give clear responsibilities to institutional actors, and to define long-term convincing strategies of development and economic growth

The much debated *White Paper on European Governance* represented an attempt to transform and make more efficient and responsive the executive and administrative functions of the Union. In this document, governance was defined as: 'rules, processes and behaviour that affect the way in which powers are exercised at European level, particularly as regards openness, participation, accountability, effectiveness and coherence' (CEC, 2001:8n), such as to offer to the Union 'a less top-down approach and complement its policy tools more effectively with non-legislative instruments' (CEC, 2001:4). The reforms that arose around food safety and the market, as part of the attempts to restore public trust and so buttress the legitimacy of the European project, are explained in further detail below. However, the argument is made that these reforms fell within the existing modes of EU governance and while their breadth is significant, they offered little in the way of greater public participation for the individual citizen in the operation of food traceability.

EU Regulation of Traceability

As with governance, there is not a single definition of traceability or a single system for its implementation. There is widespread agreement that the need for fully documented traceability systems within the food chain has never been stronger (Morrison, 2003), but there is divergence concerning the purpose of introducing traceability in the food sector. Disagreements about the role of food ethics in production practices are reflected and exposed in the opinions of how to make use of traceability and which of its models should be implemented. In general, the following concepts are used to distinguish theoretically between different kinds of traceability systems:

- *Internal traceability* is traceability *within one link* (usually a company) within the chain. This concerns the internal management or record-keeping of flow and use of materials within a company.
- The '*one step back, one step forward*' approach says that each link in the chain is only obliged to know where the products come from and where they are delivered.
- *Chain traceability* is traceability between links in the chain through all stages of production, processing and distribution. It makes it possible to trace the history, application or location of an entity along the food chain by way of recorded identification

The three kinds of traceability can be illuminated by an example. For a producer of frozen pizzas, *internal traceability* means keeping a record of the flow and use of all ingredients within the company. For instance, tomatoes from different batches should not be mixed during production as this would make tracing to a single batch impossible. Conversely, a record should be kept of which pizzas a given batch of tomatoes is used for. Internal traceability creates a link from raw material/ingredients to products within a company.

The '*one step back, one step forward*' approach links the pizza producer to suppliers and buyers. For every frozen pizza, the company must be able to trace all suppliers of ingredients (one step back), and must also know to whom the pizzas were sold (one step forward), although the forward step does not apply at the very end of the chain, to those selling pizzas to consumers. Through 'one step back, one step forward' traceability, frozen pizzas can in this case be linked to a specific supplier of tomatoes (and all the other ingredients, of course) and to buyers of finished pizzas (e.g. retailers). The 'one step back, one step forward' approach should require internal traceability to be at work.

Chain traceability builds on 'one step back, one step forward'. It links the whole change from farm to end retailer. In the case of the frozen pizza, traceability is present from the tomato grower (and maybe even tomato breeder) to the final retailer selling the frozen pizza.

At present, the three main international organizations engaged in defining traceability are the EU, the International Standards Organization (ISO) and the Codex Alimentarius (see Table 2.1). The ISO definition (ISO 9000:2000) refers to a set of

Table 2.1 Definitions of traceability

ISO 8402: The ability to trace the history, application or location of an entity by means of recorded identifications.	
ISO 9000: 2000: The organization [e.g. a food company] should take steps to identify the status of the product/service insofar as concerns the required measurement and verification activities and should, where necessary, identify the product and/or service using the appropriate means throughout the process. This should apply to all parties involved in the product and/or service where their interaction has a bearing on the conformity to requirements. When traceability is a requirement, the organization should control and record the unique identity of the product and/or service.	
Codex Alimentarius 2004: Traceability, product tracing: the ability to follow the movement of a food through specified stage(s) of production, processing and distribution.	
Regulation (EC) No 178/2002: the ability to trace and follow a food, feed, food-producing animal or substance intended to be, or expected to be, incorporated into food or feed through all stages of production, processing and distribution.	

quality management standards and is the most widely recognized, although it is generic in its application across all sectors (it is derived from instrumentation standards). The ISO has also agreed to a set of guidelines for the design of traceability systems for food quality management (see Chapter 3). In June 2004, Codex Alimentarius reached internal agreement on defining traceability in relation to food as the ability to trace food products in every stage of the chain (CCGP, 2004a, b). However, the debates in Codex have reflected different approaches to the role and aims of traceability (see Chapter 3 for further discussion).

The EU has forms of traceability in place for foods produced by certain processes and to allow for the provenance of certain foods to be traced and verified. In the early 1990s, the EU recognized the national-level regulation of foods according to their place and method of production through the system of Geographical Indications (GIs). In addition, organic standards were regulated by the EU from 1991, providing systems for verifying the authenticity of the means of production (CEC, 1991). In other words, the EU's concerns for traceability have gone beyond food safety and recall needs.

During the years between 2000 and 2005 the EU passed traceability legislation for specific types of food and feed products (see Table 2.2). These regulations covered bovine animals and beef products, fisheries and genetically modified organisms (GMOs). Risk management was a key motivation for these regulations. The traceability of beef products stemmed from the BSE crisis, and traceability for fish stemmed from conservation management needs to reduce the risk of further destruction of fisheries. However, in the case of GMOs the European Commission made it clear that traceability and associated labelling was also to allow for consumer choice, as only those GM products that had been deemed safe through the EU's risk assessment processes would be allowed on to the market (see Chapter 3 for a more detailed discussion).

Table 2.2 EU regulations and directives including food traceability

1. **Food provenance, place of origin and production methods:**
PGI – Protected Geographical Indications. This labelling scheme is for individual products that have a specific *characteristic* or *reputation* associating them with a given geographical area. At least one stage in the production, processing and preparation process is carried out in that area.
PDO – Protected Designations of Origin. The product has proven *characteristics* which can only result from the *natural environment* and *abilities of producers* in the region of production it is associated with.
TSG – Traditional Speciality Guaranteed. TSG implies that the product has distinctive features, which either have traditional ingredients or are made by traditional methods.
Organic food. In the traceability system of organic foods, each link in the food production chain (farm to fork) must be documented to show compliance with approved organic methods.
2. **Cattle and beef products**
Since September 2000 an extensive labelling and registration scheme has been obligatory for beef in the EU. Beef and veal on the market in the EU must be labelled with information on: country of origin, country of slaughter, slaughter company, country(-ies) of further processing, company(-ies) of further processing. The aim was to establish traceability between a carcass, quarter or pieces of meat to an individual animal or a group of animals, for food safety reasons. The regulation bids each member state to establish a system for identification and registration of bovine animals, comprising ear tags for the individual animal, databases, animal passports, and individual registers kept on each holding (farm records). The legislation is also connected to a more specific regulation (Commission Regulation (EC) No 1825/2000 of 25 August 2000) laying down detailed rules for the application of Regulation (EC) No 1760/2000 as regards the labelling of beef and beef products. In this regulation the sizes of batches, the demands for labelling for minced meat, control and sanctions are specified.
3. **Fisheries and fish products**
From 1 January 2002 a new labelling scheme was put into effect for a wide range of fish products. The aim was to supply consumers with information on catch area, species and production method (caught in freshwater, or farmed). The regulation requires the information on species and catch area to be made available to the consumer through either labelling or trade documents. The regulation is Commission Regulation (EC) No 2065/2001 of 22 October 2001, laying down detailed rules for the application of Council Regulation (EC) No 104/2000 as regards informing consumers about fishery and aquaculture products.
4. **GMO traceability**
On 22 September 2003 the Council adopted new rules for improved traceability and labelling of GMOs (Regulation 1830/2003). According to these rules, all products (food and feed) consisting of or containing GMOs shall be labelled. This also applies to GMO products with no protein or DNA residue (e.g. GM soy oil). Traces of GMOs are allowed in unlabelled food, provided they are adventitious or technically unavoidable and in a proportion no higher than 0.9%.
5. **Packaging materials**
Since 27 October 2006, processors have been required to have a traceability system in place for packaging materials. This new requirement is a provision of EC Regulation 1935/2004, which deals with materials and articles that may come into contact with foods. It covers materials such as rubbers, ceramics, plastics, paper, glass, metals, inks, textiles, waxes, cork and wood. It applies to all food, animal feed, food-producing animals and all types of food chain operators.

In 2002 the EU established a legislative framework for traceability through the food production chain. Regulation (EC) No 178/2002 (referred to hereafter as the General Principles), laid down the general principles and requirements of EU food law. It also established the European Food Safety Authority (EFSA) and laid down procedures in matters of food safety. The regulation came into force on 21 February 2002, with its provisions on traceability to be applied from 1 January 2005, although official implementation was delayed in some member states. The General Principles regulation defines the concept of traceability as the ability to trace and follow a food through all stages of production, processing and distribution (see Table 2.1). According to this definition, traceability is to be an end-to-end and integrated supply chain process. But by specifying production, processing and distribution, the definition excludes consumers, who are not normally part of this process (see Fig. 2.1). Article 18, 1 of the EU food law, on Applications, contains some general provisions for traceability through the food production and distribution chain. Traceability covers all food and feed, and all food and feed business operators (without prejudice to existing legislation on specific sectors such as beef, fish, GMOs, etc.). Importers are similarly affected, as they are required to identify the exporter in the country of origin. The regulations have had several later amendments, in particular Regulation (EC) No 1642/2003, dealing with: (1) Food safety and consumer protection against 'misguidance' – to be able to supply consumers and control authorities with relevant information. (2) A simpler, faster and more targeted recall of products in emergencies.

Unless specific provisions for further traceability exist, the requirement for traceability is limited to the 'one step back, one step forward' approach outlined above, ensuring that businesses are at least able to identify the immediate supplier of the product in question and the immediate recipient, with the exemption of retailers to final consumers (Art. 18, 2.-3.). To satisfy these requirements for traceability, food or feed must be adequately labelled or identified through relevant documentation or information in accordance with the relevant requirements of more specific

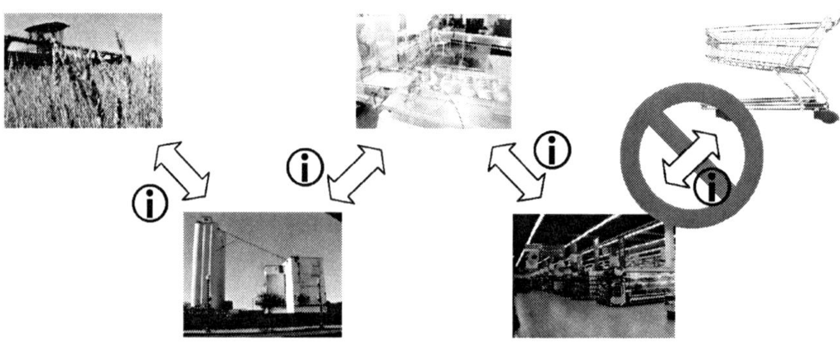

Fig. 2.1 Illustration of the flow if information, ***i***, in the wheat-bread supply chain. None or only limited information is made accessible to the end users, the consumers. Also the possibilities for consumer feedback are almost non-existent

provisions (Art. 18, 4). The information must be made available to the competent authorities on demand. Minimum documentation requirements are:

- Who is the supplier?
- What is supplied?
- When was it received?
- What is sold?
- To whom is it delivered?
- When is it delivered?

The chain traceability introduced by the General Principles offered a potential vehicle for recording and passing on the sort of information that could enhance the scope for choice by the final consumer. This would have reflected the existing food provenance and consumer choice elements found in private quality assurance schemes, GIs and GMO traceability legislation. Two key principles were that the food producer bore primary responsibility for food safety and that a whole chain approach, from 'farm to fork' should be adopted. Traceability provided a crucial mechanism for ensuring food identification along the supply chain. However, the step procedure specified for implementation of Regulation 178/2002 stressed the food contamination and public health recall imperatives. The clear impulse for the introduction of traceability into the General Principles of Food Law lay in the risk management of food safety, rather than informing consumers about the history of their food.

The Reform of Food Safety Regulation and Food Law in the EU: Risk Management and Traceability as Control

The European Commission's reform of food safety regulation and food law was precipitated by its mishandling of the BSE crisis. The European Parliament found the Commission guilty of serious maladministration and threatened to censure the Commission should it fail to act, which would in effect have dismissed the European Commissioners from office. Responding in 1997 the European Commissioner Jacques Santer acknowledged shortcomings in the protection of consumer health and promised radical reform of the Commission's machinery. He called for 'nothing short of a revolution in our way of looking at food and agriculture' (Santer, 1997). As a result, in 1997 the Commission's scientific committees were put under the supervision of the Consumer Protection Directorate General, and two Green Papers were issued laying out plans for food safety reform and for a revision of EU food law (CEC, 1997a, b).

Key aims laid out at the beginning of the reform process were excellence, independence and transparency. The *Green Paper on Consumer Health and Food Safety* placed 'food safety and consumer health at the core of a new political departure' (CEC, 1997a:3). The reform was based on the adoption of the risk analysis model, where there was a separation of risk assessment (in the form of scientific advice) from risk management (i.e. the political management of risk). Risk assessment became the province of the scientific committees, based on excellence of expertise and independence, which in turn became part of the EFSA (see below). Risk communication, the third element of risk analysis, was dealt with less clearly, with the primary role of conveying the communication going to the Commission (as the political actor and risk manager), but at the direction of the scientific risk assessors. A deeper problem with the model of dividing risk management and assessment was the lack of acknowledgement that risk assessment is framed by the assessors' political considerations – notably regarding how the regulation(s) concerned already frame the risk – and social assumptions. That is, any scientific assessment is socially constructed, a reality ignored by the Commission in framing these food safety reforms (van Zwanenberg and Millstone, 2005). Independence of inspection was to result in the revamping of the veterinary and inspection service into an independent agency named the Food and Veterinary Office, based in Ireland. In this Green Paper the proper function and management of the internal market was a declared objective: 'food safety is not only of concern to the consumer, but is also at the very root of a proper functioning of the market' (CEC, 1997:6).

After the collapse of the Santer Commission in 1999, due to another political scandal, the new Prodi Commission kept food safety as a priority and furthered the reform process, including a reorganization of the Commission's Directorates General (DGs). The Consumer Protection DG was renamed the Health and Consumer Protection DG (SANCO), taking over food safety and food law policy-making responsibilities, which had been previously housed in the DGs for Industry and Agriculture respectively. In other words, the DGs responsible for promoting the agricultural and food industries lost their regulatory responsibilities in these areas; these responsibilities were moved to a new DG oriented towards consumer safety and public health. This was a potentially important departure in policy-making focus.

The *White Paper on Food Safety* released early in 2000 spelled out more clearly a wide-ranging consolidation and revision of the European food law (CEC, 2000). It also proposed the creation of a new European Food Authority which became the EFSA, established by a further regulation in January 2002, heralding the birth of the Authority in 2003. The importance given by the Prodi Commission to food safety reform was underpinned by continuing food scandals and controversies (e.g. dioxin contamination and GM foods). EFSA's mission, as laid down in the General Principles, is to provide 'scientific advice and scientific and technical support for the Community's legislation and policy in all fields which have a direct or indirect impact on food and feed safety. It shall provide independent information … and communicate on risks' (OJL, 2002:31/12). The remit reaches along the whole food and feed supply chain, but the scientific opinions are limited to food safety only. The remit does include scientific advice on human nutrition in relation to Community legislation, and assistance on communication on nutritional issues within the Community's health programme, but only at the request of the Commission. The scientific committees from SANCO were transferred to EFSA, reconstituted and newly appointed as eight committees – a process completed in

May 2003. This institutional reorganization and renewed legislative agenda can be seen more broadly as part of a strategy by the Commission to restore citizens' confidence in the safety of the food supply in the EU, and to retain the legitimacy of the EU and the single market through the provision of safe food. As the European Commissioners Fischler (Agriculture) and Byrne (Health and Consumer Protection) stressed in a joint statement: 'The real issue here is one of consumer confidence in the ability of the whole food chain, including public regulators, to satisfy public demand for safe quality food' (European Commission, 2002).

With its *Green Paper on General Principles of Food Law* (CEC, 1997b) the Commission hoped to give greater coherence to its laws and policies. The Commission proposed a series of guidelines to ensure a high level of public health and food safety, to improve the protection of consumer's rights, and to enhance the process of market unification while leaving a door open for competition, expertise, public debate and stakeholders' involvement. Emphasis was placed on the use of policy instruments such as labelling, standardization of procedures and the use of voluntary self-regulation (CEC, 1997b). Traceability was not mentioned. However, in a section on the application of the principle of product liability in the foodstuffs sector there is mention of 'tracing the origin of the foodstuff from the point of sale to the consumer back to the point of production' (CEC, 1997b:48). While noting that in the case of bovine products some measures had already been taken, it is only proposed to evaluate the possibility of applying this policy to other animal products. In this regard, information was intended to be conveyed to the consumer primarily through labelling, and it only related to basic ingredients and more general information on the product and about the company.

The *White Paper on Food Safety* introduced traceability as a new principle, establishing 'the obligation for feed and food businesses to ensure that adequate procedures are in place to withdraw feed and food from the market where a risk to the health of the consumer is posed' (CEC, 2000:8). Moreover, the role played by all the actors in the food chain had to be clearly defined, and the development of this approach needed 'to be transparent, involving all the stakeholders, and allowing them to make effective contributions to new development' (CEC, 2000:8). Traceability is proposed as a tool to reform the European food market and its legislation in order to provide citizens and consumers with safe food, to offer greater transparency in the processes of production and distribution of agri-food and feed products, and to develop a substantive customer/consumer trust in the European economic and political system. From its very beginning, in this regulatory form, traceability was related to the categories of risk, safety, transparency and involvement, which operated as a conceptual grid within which to frame the future application of the policy. Nonetheless, at that time no action was taken to establish traceability along the food chain and the document presents only a few very general proposals in the annex about 'Actions' (see Table 2.3). Thus, in this document traceability was not an autonomous policy but only a principle yet to be implemented. Due to the high sensibility of European citizens on these matters, traceability came to be expressly related to the need for clear labelling, clear scientific assessment of risk and enforceable control of processes. Subsequently, the regulation

Table 2.3 White Paper on Food Safety Annex: 'Action Plan on Food Safety'

Action 3 (*Proposals for a General Food Law Directive*): To lay down the common principles underlying food legislation (in particular: scientific basis, responsibility of producers and suppliers, traceability along the food chain
Action 5 (*Regulation on feed*) proposing to lay down the principle of traceability working within the EU feed legislation
Action 25 (*Proposal for amending Directive 95/69/EEC laying down the conditions and arrangements for approving and registration of certain establishments and intermediaries operating in the feeding stuffs sector*) proposing to improve traceability of feed materials and identification of 'critical points' (CEC 2000a)

of the General Principles of Food Law (178/2002) aimed to promote the safe movement of food and feeds within a clear, coherent and well-regulated framework of risk analysis, management and communication. The regulation's preamble provides its rationale, stating that the 'free movement of safe and wholesome food is an essential aspect of the internal market and contributes significantly to the health and well-being of citizens, and to their social and economic interests' while ensuring 'a high level of protection of human life and health' (OJL, 2002:L 31/1). Within these objectives, traceability becomes a tool to support the free movement of goods as well as food safety and public health recall.

Didier Torny (2003:78) has argued that traceability is not so much a means of identifying subjects of juridical imputation as a device for the attribution of responsibilities and reduction of risks:

> The establishing of traceability does not yet guarantee the certainty of an imputation of a juridical kind. But it offers a visibility *ex ante* of the complexity of the processes of production or distribution, and tends to extend responsibility rather than dissolution. A requirement for traceability hides a new definition of the limits of responsibility that are removed from the physical borders of the entities concerned.[1]

Paolo Napoli (2003) points out that traceability may be usefully ascribed to the long-lasting history of the administrative power ('*administratio*') rather than to the concept of imputation that came from the modern history of jurisdiction. This effort toward '*administratio*' is in itself an attempt to regulate and put 'at work' the uncertainty that exists in any social environment through a politics of risk management. Thus European food politics implements a system of agri-food governance centred around political and managerial presuppositions that give birth to a continuous and circular motion between: *more freedom > more uncertainty/risks > more governance/government > more intervention/control /[self]regulation > less freedom*. The more freedom given to the system, the higher the level of uncertainty and risks we will probably face. The need to govern and manage a higher level of risk and

[1] '*La mise en place d'une traçabilité ne garantit pas pour autant la certitude d'une imputation de type juridique. Mais elle donne à voir ex ante la complexité de processus de fabrication ou de distribution et tend à étendre la responsabilité plutôt qu'à la dissoudre. Une exigence de traçabilité recouvre une nouvelle définition des limites de la responsabilité, qui s'éloignent des frontières physiques des entités concernées.*'

uncertainty carries with it the need for more control as well as the permanent possibility of limiting that freedom we had sought to offer the system. The instruments of control being applied are unavoidable managerial devices whose purpose is the guarantee of an adequate level of security. As a paradox, the necessity to establish a wider, open and free market constitutes the reason to support broader and rigid controls of the food chain and to implement regulatory policies such as traceability.

Traceability represents a significant part of European food governance as it works:

- By recourse to authorities and expertise in data gathering, evaluation processes and consulting phases and in the phase of risk assessment
- Under the Commission's direct control of risk management and communication and legislation and within its strategic supervision
- By the attribution to national authorities of direct control and inspection of the processes and juridical imputation
- By the mandatory implementation of traceability systems by companies: while the choice of the system of traceability is left to them, the data to be collected and stored is defined by the law

When compared with the contents of the *White Paper on European Governance*, the issues of participation and involvement have disappeared and the overall framework is much more the expression of a top-down governmental approach. The extent to which the food safety reforms have embraced wider participation remains firmly within the more elitist stakeholder consultation model. For example, a new advisory body set up in August 2004 to provide a whole food chain stakeholder forum called the *Food Chain and Animal and Plant Health* provided a further example of deliberation among representatives from the main European peak associations and civil society organizations from the food chain.

While documents such as the *Green Paper on European Food Law* and the *White Paper on Food Safety* intended to promote consumers and business involvement, in the regulative frame ultimately established by Regulation 178/2002, traceability is not intended as a tool for consumer information. In other words, traceability is only expected to give the European institutions a system to control, monitor and intervene in the food market in cases of public health need. Companies are offered the opportunity to fit their business into a uniform, competitive and regulated market. Consumers are provided with minimum standards of food safety and a risk management device to be operated in the interests of their health in the case of contamination.

Contemporary EU policy on food safety is also based on political and strategic aims. Strategically, through the definition of minimal common standards and the adoption of a coherent legislative framework, it provides a minimum standardization of practice for the European internal market while affirming openness. Accordingly, the establishment of an articulated system of traceability aims to provide safe food through fast and efficient procedures for the withdrawal of products in cases of emergency that, in the last resort, are decided by the Commission (on the advice of EFSA). The establishment of a system to track and monitor potentially harmful products is linked to a set of shared managerial and organizational practices framed within a process of market-building.

Politically, the procedures adopted by the Commission aim at the acquisition of a functional legitimacy on the grounds of efficient policies supported by epistemic communities, expert groups and open consultations. The Commission's relationship with corporations and consumers follows the consultative and consensual model, where there is no place for an open confrontation on strategic programmes and political agendas. Moreover, the individual consumer is a passive subject of food safety policies. Hence, we can see that European food policies are intended to offer a high standard of food safety to the consumers. At the same time, because they have to guarantee the concurrent rights of producers and distributors and the needs of market development, they provide the consumer with only limited information. Furthermore, this implies a restrictive interpretation of the subject-consumer. The consumer is deemed to act as a rational actor who chooses what and how to buy within a given market, and according to a market analysis. Even where a consumer might have strong ethical motivations, the consumer's logic is postulated as a cost/benefits analysis, in which ethics is a variable among others (price, quality, availability). Within this approach, ethical or political beliefs do not belong to the sphere of rights but to that of market availability; therefore it is up to the market, framed by the regulatory action of public authorities, to answer the ethical demands of consumers.

Food Traceability as a Relational Tool for Information, Communication and Participation

At present, traceability in EU legislation appears to be part of an ongoing reform of the European legislative framework, and may be understood as a governance tool for risk management and emergency prevention that works within a regulative political system whose legitimacy is provided by the efficacy of its intervention. This policy, by putting so much emphasis on political risk management and expertise, prevents the EU from adopting a more participative and communicative approach toward traceability and informed food choice. A different approach is to be found, however, in some European documents such as the *Green Paper on Food Law*, the *White Paper on Food Safety* and the designations of *PDO* and *PGI*, which relate traceability not only to the issues of safety and health but also to consumer information. The two dominant positions, which we shall call for convenience the *risk management position* and the *informed choice position*, have their own specific objectives and ways of implementing traceability. The former position has already been explained in the preceding sections of this chapter, but the informed choice position needs further elaboration.

A better balancing of the relations among actors in the food chain cannot leave out a consideration of their informational competencies. Actors can only choose safely and ethically if they have access to accurate and reliable information. Moreover, despite the pronunciations of White and Green Papers, there is still a need to offer European citizens (both as individuals and as collective actors) a suitable space for participation in the shaping and enacting of European polices. Today citizens' participation is limited to the consensual implementation of policies already determined, or in wide and too often ineffectual consultations on very broad and vague issues, as in the case of the reform of European governance. Therefore, at present, in the political arena, the room for ethical confrontation is limited and this is particularly true in a highly contested sector such as food.

So what are the possibilities for the informed choice position? If traceability is to be used as 'valorization' of food products, by quality surveillance and assurance schemes, and as a tool for informing consumers, then it becomes essential to communicate traceability information to consumers. This also means that the demands on traceability systems increase, because more detailed and complex information must follow the products. In order to let the consumer claim an active role in the communicative process on food and ethics, informed choice plays a prominent role. This is expressed by several official reports and debates on traceability in Europe. A paper from the UK's Food Standards Agency (FSA) stated that 'traceability systems are of interest to consumers, as part of systems which... enable real choice to be exercised between food produced in different ways' (FSA, 2002:14). Similarly, the Danish Ministry of Foods, Agriculture and Fisheries observed that 'knowledge about the product history gives the consumer the possibility of taking responsibility in the shopping situation' (Danish Ministry of Foods, Agriculture and Fisheries, 2004:11). Here the phrase 'taking responsibility in the shopping situation' refers to ethical consumption. The commodification of food has disguised immanent social relations brought about by food consumption; however, in this Danish report knowledge about the production history (i.e. traceability) is considered a possible means to establish more explicit relations through food consumption. The Danish report continues that such traceability is likely to be developed in the future, but does not discuss how this could be done in practice. It notes only that such information is difficult to handle (Danish Ministry of Foods, Agriculture and Fisheries, 2004:27). In 2006 the Danish Consumer Ministry and the Danish government went even further in their support for integrating consumers as the last link in traceability schemes of the food chain:

> Good consumer information and transparency mean that the consumers gain access to information on, for instance, the background of the product, production practices and terms of trade. For consumers, such information can have the same influence on choice as, for instance, price. Traceability shall be used to assure complete consumer information and avoid misinformation – especially when the production chain is long and complex. The government supports the highest possible transparency in the food chain, and in the EU the government will work for new technology, like for instance RFID, that can support access to additional information on goods and thereby secure transparency and ethical traceability (Danish Ministry of Family and Consumer Affairs, 2006).

In the food sector, political participation could be taken into greater account as it potentially offers a way of strengthening the processes of legislation and policy-making on food in relation to health, social accountability, the environment and animal welfare. In this sense, the growing importance of 'political consumerism' is the expression of a new sensibility that should be politically and institutionally recognized. New social movements transform the act of consuming into a highly political *and ethical* act, using 'the market as an arena for politics and consumer choice as political choice' (Boström et al., 2005:9). Political consumerism: 'draws on the observation that consumer choice and the rising politics of products is an increasingly important form of political participation that exists parallel to conventional party-centred and national level politics' (Boström et al., 2005:9). It is within this complex frame that traceability and labelling could play a crucial role, supported by the developments of information and communication technology (ICT). Information, communication and participation are political tools that must be used for the shaping of European agri-food politics, and traceability can play a significant role in making this politics available to citizens. In order to propose traceability as a tool to improve information, communication and participation, we consider each of these aspects in turn.

Information

At present, norms on labelling define the level of mandatory information provided to consumers – typically, they require information on the ingredients, the place of production/processing, the producer, the distributor or the retailer. Traceability and labelling are two important instruments for providing information on products and their circulation, as well as two crucial tools to control the circulation of products and to regulate the market. They put a minimum amount of important information at the disposal of companies and consumers. Nonetheless, these instruments constitute a one-directional flow of data from one actor to another and can only be represented as a vertical line of communication. Furthermore, at present the data accessible to consumers is relatively scarce, as the UK FSA (2002:1) affirmed: 'consumers gain mostly hidden benefits [from traceability schemes]'. The consumer, while benefiting from provisions for mandatory information in European law on labelling, is not provided with data about the production, processing and distribution of products – data that is already registered and stored through traceability schemes. This is true in spite of the fact that: 'traceability has also a role to play in the promotion of informed consumer choice because it offers the potential to verify label information on product and ingredient history' (FSA, 2002:1).

The additional information for consumers should not be limited to ingredients and nutritional data, as this information meets only a small part of consumers' concerns. Moreover, 'the provision of knowledge of origin direct to consumer does not make them [i.e., the consumers] safer' (FSA, 2002:15). Such a system will never allow any of us to make informed food choices, since the food system is kept hidden: 'the maintenance of a verifiable and robust traceability system means that consumer safety is likely to be increased in the event of an emergency. However, such a system does not require the knowledge to be passed on to the consumer. The provision of knowledge through traceability is more clearly linked to an increase in consumer choice based on product identity and origin' (FSA, 2002:15). Therefore, a unique system of labelling/traceability, which was accessible to the final consumer and which provided a range of information, from mandatory data to data that companies as well as consumers' organizations (or any other stakeholder group) might provide about their products and/or their ethical standards or procedures, would be a great improvement. Continuous technological development makes it possible to unify traceability and labelling.[2] A thorough consideration is needed of the technical and informational aspects of such a system: which information is relevant and should be more visible; how data and information can be collected, organized and managed; how to avoid an overload of data (which would make the system useless) or discriminatory practices. Traceability and labelling could then represent a complex system designed for consumers' informed choice as well as a powerful instrument for citizens' knowledge building. In other words, traceability may become an instrument for informed (ethical) choice.

Communication

Traceability/labelling may represent a useful instrument for communication between different actors if:

(a) The provision of information is widened in order to increase the number of actors interested in the process
(b) Feedback tools are promoted to make the interactive circulation of information and data possible
(c) Procedures are adopted for public and open dialogue and negotiation on the information and data provided

The definition of instruments for a continuous, horizontal and autonomous dialogue between interested actors should be part of a more inclusive European system of agri-food governance. In that case, communication and dialogue between actors on the process of traceability/labelling may become an important part of a thorough economic, ethical and political debate.

Participation

Contemporary European systems for management of the food chain take for granted a liberal model of development, mainly organized around big producers and distributors/retailers that operate on a global scale. The opportunity for European citizens to participate and decide should not be confined to a debate about existing markets. It is necessary to empower them by offering them the opportunity to develop alternative and different markets. In this sense, the creation of spheres of ethical and political participation should imply the development of institutions

[2] For more on this subject, see the studies by Morrison (2000, 2003). For different techniques of traceability see Lees (2003); for a proposal of bio-markers' traceability, Raspor (2005). For informatics models see: Bechini et al., 2005; Cimino et al., 2005; Lo Bello et al., 2004.

framing the participation of European citizens in food strategies and political agendas, and the establishment of instruments to develop different and perhaps alternative systems of production, processing and distribution.

Conclusion

The two-faced European discourse on food safety, one expressing the need for consumer information that emerges in some of the Commission's policy documents and the other represented by the much more political, risk-management-oriented General Principles of Food Law reflects the difficulty in bridging the need for safety and the exigencies of market and political stability. Keeping the information within the food chain hidden and not allowing consumers any access runs the risk of not fulfilling important expectations associated with the concept of traceability. To avoid such a risk, traceability should not be used as a purely administrative tool or as a mere safety system, because it is related to highly contested and sensitive issues such as animal welfare, fair trade, traditions and beliefs, environmental protection and sustainability. Rather; it should represent an instrument for establishing effective and responsive policies and institutions based on involvement via informed food choices by citizens/consumers.

The reform of food safety regulation and the overhaul of the EU's food law have engendered some substantial regulatory and institutional changes. A key principle of the revision of the General Principles of EU food law was to extend responsibility for safety and due diligence right along the food supply chain back to the farm. Within this reform, food traceability emerged as an important instrument for identifying points of contamination, and for ascertaining responsibility. It allowed for surveillance along the whole length of the chain, permitting more effective risk management by the European Commission and the other political actors in Europe. This risk management approach was informed by the need to underpin both consumer and citizen confidence in the workings of the European food system and so the internal market. The successful functioning of the internal market is a condition for the continued legitimacy of the European integration project. Within the body of EU regulations, there are examples of food traceability being used as a means to inform the public about the origins and production processes of food that go beyond risk management goals. The regulations allow for informed choice about some elements of the nature of the food and its journey to the point of purchase. However, such regulations are few and cover only certain dimensions where traceability is necessary to ensure authenticity of origin and production process. Also, the economic interests of the producers (often producer processors) are safeguarded, as with the GIs.

Conversely, the informed choice position on food traceability can only be made possible by establishing a different system of European governance around food. Such a system would not only be aimed at safety, risk management, political legitimacy and market efficiency, but would also serve to promote debates and discussions among actors in and around the food chain. Providing information is only the first step to achieving these goals. It is also necessary to make the information more responsive and transparent, and to build processes that allow actors not only to choose in accordance with the information provided, but also to choose the information they wish to be provided with. This will be a dynamic process, as the demands from the actors alter over time and as changing priorities come to the fore. Food traceability as a governance instrument should be a communicative process among interests and beliefs and be underpinned by regulatory mandate, and not be just a regulative system oriented at the market and the management of risk and safety. By establishing open and transparent channels for the movement of information along the food chain, flowing from one actor to another, through different processes, traceability may become a powerful instrument. For that to happen, it should not be used merely as a safety tool for risk management, but also as a communicative and relational device.

References

Ansell, C. and D. Vogel (eds.) (2006) *What's the Beef? The Contested Governance of European Food Safety*. Cambridge, MA: MIT Press.

Bechini, A., M.G.C.A Cimino, B. Lazzerini, F. Marcelloni, A. Tomasi (2005) 'A General Framework for Food Traceability', *Proceedings of the 2005 Symposium on Applications and the Internet Workshops*, Washington, DC: IEEE Computer Society, pp. 366–369.

Böstrom, M., A. Føllesdal, M. Klintman, M. Micheletti, and M.P. Sørensen (2005) 'Studying Political Consumerism', pp. 9–23 in Böstrom, M., A. Føllesdal, M. Klintman, M. Micheletti, and M.P. Sørensen (eds.) *Political Consumerism: Its Motivations, Power and Conditions in the Nordic Countries and Elsewhere. Proceedings from the 2nd International Seminar on Political Consumerism, August 26–29, 2004*. Oslo: TemaNord.

CCGP (2004a) 'Definition of Traceability/Product Tracing of Foodstuffs (Prepared by France)', Codex Committee on General Principles Agenda Item 6, *CX/GP 04/20/6*. Rome: Codex Alimentarius Commission.

CCGP (2004b) 'Report of the Twentieth Session of the Codex Committee on General Principles', *Alinorm 04/27/33A*. Rome: Codex Alimentarius Commission.

CEC (1991) Council Regulation (EEC) No 2092/91 organic production of agricultural products and indications referring thereto on agricultural products and foodstuffs. *Official Journal L 198*, 22 July.

CEC (1997a) *Communication on Consumer Health and Food Safety*, COM (1997) 183 final. Brussels: Commission of the European Communities.

CEC (1997b) *Communication on the General Principles of Food Law in the European Union*, COM (1997) 179 final. Brussels: Commission of the European Communities.

CEC (2000a) *White Paper on Food Safety*, COM (1999) 719 final, 12 January. Brussels: Commission of the European Communities.

CEC (2000b) *A White Paper on European Governance, 'Enhancing Democracy in the European Union'*, Work Programme, SEC (2000) 1547/7 final 11 October. Brussels: Commission of the European Communities.

CEC (2001) *European Governance, A White Paper*, COM (2001) 428 final. Brussels: Commission of the European Communities.

Cimino, M.G.C.A., B. Lazzerini, F. Marcelloni and A. Tomasi (2005) '*Cerere*: an information system supporting traceability in the food supply chain', *Proceedings of the 2005 Symposium on Applications and the Internet Workshops*. Washington, DC: IEEE Computer Society, pp. 90–98.

Danish Ministry of Family and Consumer Affairs (2006) *Fødevarepolitisk redegørelse – i et forbrugerperspektiv [Report on Food Politics – from a Consumer Perspective]*. Copenhagen: Danish Ministry of Family and Consumer Affairs.
Danish Ministry of Foods, Agriculture and Fisheries (2004) *Sporbarhed i fødevarekæden. Traceability in the food production chain*). Copenhagen: Danish Ministry of Foods, Agriculture and Fisheries.
De Búrca, G. and J. Scott (eds.) (2001) *Constitutional Change in the EU: From Uniformity to Flexibility?* Oxford: Hart.
European Commission (2002) *Fischler and Byrne Final Round Table on Agriculture and Food*. EU Institutions Press Release IP/02/700, Brussels, 13 May.
FSA (2002) *Traceability in the Food Chain. A Preliminary Study*. London: Food Chain Strategy Division, Food Standards Agency.
Fukuda, K (2003) *European Governance after Nice*. London: Routledge.
Greenwood, J. and M. Aspinwall (eds.) (1998) *Collective Action in the European Union*. London: Routledge.
Héritier, A. (1997) 'Policymaking by Subterfuge: Interest Accommodation, Innovation and Substitute Democratic Legitimation in Europe. Perspective from Distinctive Policy Areas'. *Journal of European Public Policy*, 4: 170–185.
Héritier, A. (2000) 'New modes of governance in Europe: policy making without legislating?' Political Science Series (81). Vienna: Institut für Höhere Studien (IHS).
Hooghe, L. and G. Marks (2000) *Multilevel Governance and European Integration*. Boulder: Rowmann & Littlefield.
Laffan, B., R. O'Donnell and M. Smith (2000) *Europe's Experimental Union*. London: Routledge.
Lees, M. (ed.) (2003) *Food Authenticity and Traceability*. Cambridge: Woodhead.
Lo Bello, L., O. Mirabella, and N. Torrisi (2004) 'Modelling traccability systems in food manufacturing chains'. *IEEE International Workshops on Enabling Technologies*, Vol. 13. Washington, DC: IEEE Computer Society, pp. 173–179.
Morrison, C. (2000) 'The role of traceability in food labelling', pp. 267–280 in R.J. Blanchfield (ed.) *Food Labelling*. London: Woodhead.
Morrison, C. (2003) 'Traceability in food processing: an introduction', pp. 458–470 in M. Lees (ed.) *Food Authenticity and Traceability*. Cambridge: Woodhead.
Napoli, P. (2003) 'Administrare *et* curare. *Les origines gestionnaires de la traçabilité*', pp. 45–70 in P. Pedrot (ed.), *Traçabilité et responsabilité*. Paris: Economica.
OJL (2002) 'Regulation (EC) No 178/2002 of the European Parliament and of the Council of 28 January 2002 laying down the general principles and requirements of food law, establishing the European Food Safety Authority and laying down procedures in matters of food safety'. *Official Journal of the European Communities* 1.2.2002, L31/1–24.
Pernice, I. (1999) 'Multilevel Constitutionalism and the treaty of Amsterdam: European Constitution-making Revised?' *Common Market Law Review*, 36: 703–750.
Raspor, P. (2005) 'Bio-markers: traceability in food safety issues', *Acta Biochimica Polonica*, 52(3):659–664.
Radaelli, C.M. (1999) *Technocracy in the European Union*. London: Routledge.
Santer, J. (1997) Speech by Jacques Santer President of the European commission. Debate on the report by the Committee of Inquiry into BSE, European Parliament, Strasbourg, 18 February. *Speech 97/39*. Brussels: European Commission.
Sharpf, F.W. (1997), 'Introduction: the problem solving capacity of multi-level governance', *Journal of European Public Policy*, 4(4): 520–538.
Torny, D. (2003) '*Une mémoire ou le futur. La traçabilité comme allocateur de responsabilité*', pp. 72–86 in P. Pedrot (ed.), *Traçabilité et responsabilité*. Paris: Economica.
Van Zwanenberg, P. and E. Millstone (2005) *BSE: Risk, Science and Governance*. Oxford: Oxford University Press.

Chapter 3
Governing and Governance in the Agri-Food Sector and Traceability

David Barling

Introduction

In this chapter food governance is depicted as a dynamic process, both occurring across and involving actors from the public sector, the corporate sector and civil society. Governing has given way to governance in the sense that private governance forms are increasingly enrolled and recognized by the state. Indeed public governance can follow in the wake of initiatives pioneered in the private sector. Governance according to this interpretation thus hybridizes public forms of governing with private schemes of governance. This approach to understanding the nature of governance provides a context for further evaluation of the differing but overlapping forms of agri-food traceability that were introduced in Chapters 1 and 2. The political and institutional policy contexts for traceability are located within a multi-level governance framework for food reaching up from national (and sub-national, or local and regional) to European and global levels, and involving a multiplicity of attentive actors from across the public, corporate and non-governmental sectors. The core of this dynamic is the interaction of public and private governance schemes in the agri-food sector which can result in new governance forms for agri-food standards.

The interpretations and conceptualizations of governance and the forms that governance can take are examined more closely in the next section, followed by some illustrations of the dynamics of governance in the agri-food sector, drawing particularly upon some selected recent experiences in the UK and the EU and at the global-international levels around standards setting. A model of food governance operating across public, corporate and civil society sectors and at multilevels is presented (see also Barling, 2004; Lang, 2006). The selected examples for agri-food standards-setting and the model of contemporary food governance provide contexts within which the emergence of traceability systems in the agri-food sector can be placed. Conceptualizing ethical traceability and its potential for communicating about food in contemporary markets and societies requires consideration of the dynamics of the governance for this sector. The translation of traceability for food and feed into national standards, as well as European standards, also has to fall in line with international trade rules. The internationalization of traceability

C. Coff et al. (eds.) *Ethical Traceability and Communicating Food*,

standards has been debated within the UN's joint Food and Agriculture Organization (FAO) and World Health Organization (WHO) Food Standards Programme the Codex Alimentarius (Codex). The debates in Codex illustrate some of the dynamics of international governance, including the political contest around agri-food traceability and its potential objectives which have also arisen in other international regimes.

From Governing to Governance

The concept of governing depicts a command and control approach to law-making and enforcement from the nation state which in modern times has been underpinned by the state's successful claim to political legitimacy. In the past three decades the state has lessened its control and command over economic sectors and society on the one hand, while on the other it has sought to extend its regulatory and strategic reach, partly through new governance forms (Pierre, 2000; Pierre and Peters, 2000). Governance implies more indirect and softer forms of direction from the state than command and control, and reflects collaborative outcomes, involving a wide range of actors often from the private sector as well as from government bureaucracy, as much as deliberate interventions by the state. The relationship of the state to such governance (as opposed to only governing) forms provides an underpinning for this analysis and for a focus on the emerging forms of governing and governance in the contemporary agri-food sector.

A dominant theme of the research into how governing has given way to looser governance arrangements has focused on the role of networks. Central government authority is more dispersed and dependent upon a multiplicity of actors located in a variety of arenas to reach policy solutions. The actors provide necessary resources for these solutions and engage in bargaining and compromise within institutional norms and rules that, in turn, can shape the outcomes (Kooiman, 1993; Scharpf, 1997). Resource dependency is a feature of policy networks identified in UK studies of modern governance arrangements, leading to governance through networks, some of which are seemingly self-sustaining and separated from the government in implementing and administering public policy (Rhodes, 1996; Rhodes, 2000).

Governance depicts a less clear distinction between public and private realms (see Chapter 8 for a further discussion of the relationship between public, private and civil society). This may reflect the spread of the new public management credo of government and public administration focusing more effort on steering rather than rowing the ship of state and public policy, diversifying the range of service providers and the criteria for efficient public services. The state relies on a range of mechanisms, professions and actors to shape order in society. Nikolas Rose has portrayed the maintenance of 'liberal rule' as being 'inextricably bound to the activities and calculations of a proliferation of independent authorities – philanthropists, doctors, hygienists, managers, planners, parents and social workers' (Rose, 1999:49). In addition, the state may be taking the opportunity to further its regulatory reach, relying more on regulation rather than distributive or redistributive policies. Regulatory policies, unlike the other types, potentially transfer the bulk of the economic costs to the regulated, while extending and enhancing the reach and influence of the state and its officials (Majone, 1996; Moran, 2003). These trends involve embracing a range of non-state actors in service or policy delivery. The regulated can become the deliverers. This is not an entirely new phenomenon, of course. Traditionally, the state has had to look outside for expertise in framing its laws and in ensuring their effective implementation, making for functional representation. Hence, representative bodies for specific economic interests and economic or social groups can become incorporated by the state (or pan-national bodies such as the EC) into the policy process. The corporatist state was the most formal and extreme model of incorporation, involving key peak interest groups from the economy and society. In more pluralistic polities and times, interest groups may gain insider status – as for example was the case for the National Farmers Union in UK agricultural policy from the late 1930s until the 1980s and 1990s (Smith, 1990). Today, the agri-food sector exhibits many of the governance features identified above in its operation, such as: policy networks, private-public-civil society resource dependencies, multilevel governance and calculative practices.

Agri-Food Governance: The Interaction of Public and Private Forms

Governance in the agri-food sector can occur in the absence of direct state involvement as private and societal interests seek to exert forms of control within the market. Examples include: standards-setting and grading of produce, process- and product-based food assurance schemes, contractual specifications from food manufacturers and retailers to growers, or from retailers to manufacturers through own-brand labelled foods (see Busch, 2000; Reardon and Farina, 2001; Barling and Lang, 2005; Bingen and Busch, 2005; Henson and Reardon, 2005). These governance forms are underpinned by audit measures and practices. Richard Le Heron has illustrated how these calculative practices, such as auditing and benchmarking, are used to reformulate and to judge what a good farming practice is in a globalizing economy (Le Heron, 1999; Larner and Le Heron, 2004). Retailers entered the Ethical Trading Initiative (ETI) to monitor adherence to national labour laws by suppliers along international (agri-food) supply chains and these codes are applied within the UK wheat to bread chain (see Chapter 6). Traceability provides a further example of calculative practice, presenting new modes of ordering for food. The key concerns for this chapter are: what modes of ordering and forms of governance are being advocated for traceability, what are they trying to achieve, and how are they being politically contested, by whom and for what ends?

The meaning and scope of traceability for food and feed are contested by both private and public sector actors. The operation and interaction of private and public governance forms, and the articulation of different economic and social interests through the more formal mechanisms of government, provide a fuller picture of the political contest around traceability. This is illuminated both by the development of private governance schemes around the entry of genetically modified (GM) foods and feed into European markets, and by the subsequent regulatory debates over traceability in international regulatory institutions and regimes, such as Codex (as will be explained below).

Indeed, new and alternative modes of ordering may emerge from civil society as 'alternative food networks' arise to challenge the dominant supply chains, setting new criteria, as in the case of fair trade foods (Whatmore and Thorne, 1997). The growth of the fair trade movement provides a good example of how the governance of agri-food standards can originate from civil society-based non-governmental organizations (NGOs), or what we term here as civil society organizations (CSOs), as has been the case with animal welfare standards, such as the Freedom Foods classification developed by a UK animal welfare NGO. CSOs and corporations can combine to produce their own criteria, traceability and labelling systems for food produce, as for example with the Marine Stewardship Council, set up as a result of collaboration between the World Wildlife Fund (WWF) and Unilever in 1997 to certify produce from sustainable fisheries.

Private governance forms throw up new power relationships along supply chains, particularly through the extraction of value. Academic concepts such as global commodity chain analysis and global value chains have focused on the governance strategies and forms that are deployed along international supply chains (Gereffi, 2005; Ponte and Gibbon, 2005). In the UK, retailers are passing their quality specifications and demands on to their overseas suppliers, giving a few specialist importers the management role of ensuring standards are met. These importers are in turn replacing traditional fresh food wholesale markets as the domestic entry point for fresh food imports (Dolan and Humphrey, 2000). These supply chain trends are being witnessed across the EU to differing degrees, notwithstanding historical national regulations protecting regional wholesale markets in some member states. In turn, the replacement of wholesalers by fewer specialist importers is favouring contracts with larger estate producers over small-scale growers in Africa who cannot deliver the same economies of scale and the same clear traceability (Barrett et al., 1999; Dolan and Humphrey, 2000). In domestic supply chains in the UK, policy conflict has emerged over the 'arm lock' that the corporate retailers hold over the marketplace and over the farmers and growers who supply them. This allows the extraction of value from those at the production end of the chain, as reflected in the comparatively low farm-gate prices for domestic products such as milk, compared to the retail price (Lang and Barling, 2007).

Corporate retailers are at the forefront of what are now characterized as buyer-driven food supply chains. The positions of control gained by supermarkets – as the buyers with a dominant market position and as gatekeepers to the consumer – have altered relationships and changed who adds value and appropriates profits along the supply chain. Product and process specifications are set out by individual retailers or consortia of different retailers. At the national level, UK supermarkets set tight specifications for suppliers (Competition Commission, 2000: ch. 11). It is clear that the setting of standards entails forms of private sector governance of supply chains. This governance can impact upon both the social and environmental standards of contracted suppliers and their livelihoods. For example, at the cross-national level, the Euro Retailer Produce Working Group (Eurep) was set up in 1997 by 13 large European retailers to set minimum standards for Integrated Crop Management production (Van der Grijp, 2003). Eurep's Good Agricultural Practice (EurepGAP) protocol for fruit and vegetables has evolved from its initial defensive role in trying to set environment-friendly pesticide standards into setting standards for many more characteristics and systems (such as traceability). EurepGAP's reach is increasingly global, the scheme was renamed GLOBALGAP in 2007, providing another example of the significance of such calculative practices (Campbell, 2005). Individual supermarkets may offer further enhanced standards for ranges of fresh produce as competitive bait for a market niche, as with Tesco's Nature's Choice range in the UK.

There has been a rapid expansion of buying consortia and alliances among European (including UK) retailers, a phenomenon nearly two decades old but now increasing in range and scope across national boundaries (Dobson et al., 2003). In 1999, it was estimated that the joint turnover of the members of seven main cross-border buyer alliances accounted for about 40% (or €340 billion) of total EU supermarket turnover (Dobson et al., 2003:116). One business overview of retailer dominance of the supply chain in Europe identified 600 supermarket formats and 110 buying desks acting as mediators for 90 million shoppers purchasing for a further 160 million consumers (Grievenik, 2003). This analysis depicts the concentration in the form of a funnel (or hourglass) that narrows considerably in the middle as the buyer desks of the retail formats are reached. Here the picture is of retailer dominance, but studies of individual commodities might produce a differing geometry for each case. For example, Bill Vorley (2003) has adapted the hourglass diagram to fit the differing concentration points for a range of commodities and their supply chains, from soybean to coffee, where processors and manufacturers are the dominant concentration points in the funnel.

The development of private sector forms of governance is being redirected by the state and intergovernmental organizations and polities (such as the EU) to address specific public policy goals around food safety and public health recall. For example, Hazard Analysis Critical Control Points (HACCP) standards were developed from the mid-1980s in the agri-food sector, although their origins lay with the NASA space programme in earlier decades. HACCP is a process-based system aimed at good hygiene practice and inspection in food businesses, which focuses on the points where contamination of food is most likely to occur at each stage of the food supply chain. Codex developed HACCP-based standards for countries to use. Governments quickly adopted HACCP in the form of statutory legislation in the 1990s. The EU adopted this approach under its Food Hygiene directive in 1993. However, HACCP is a regulatory process that relies upon voluntary compliance by industry, and government (approved) inspection is the only form of state enforcement (Flynn et al., 1999). The widespread adoption of these essentially voluntary regulatory instruments by governments fitted the deregulatory turn adopted by many states during this era, where a more liberal approach to government regulation of business was prevalent. For food businesses, this was reinforced by the due diligence requirements of the UK's 1990 Food Safety Act, a measure that was incorporated into the general principles of EU food law in 2002.

One of the goals ascribed to traceability is to provide details of the provenance of food. Examples are the growing number of food assurance schemes. In the UK, there was a rapid growth in farm assurance schemes aimed at relaying their information to the final consumer in the 1990s. The growth was also stimulated by the requirements for due diligence of the 1990 UK Food Safety Act, as was the case with the Assured Combinable Crops Scheme (ACCS), which included HACCP procedures within the protocols (IGD, 2003). The ACCS is one example of the traceability schemes that can be found linking discrete stages of the wheat to flour to bread supply chain in the UK (see Chapter 6 for a more detailed survey of these schemes). Hence, assurance schemes can also serve to assure businesses at different stages of the food supply chain that the product they are receiving meets their required standards and that problems can be traced back. In other words, traceability as represented by these assurance schemes may be reserved to specific stages of the chain or for informing the whole length of the chain from producer to consumer. The setting up of assurance schemes, which include elements of traceability systems, provides a good example of the private governance forms that emerged increasingly in the 1990s, stimulated by state legislation in the case of the UK.

The retailers' embrace of higher standards and the subsequent instruction to their farmers and growers comes under the watchful eye of environmental and other NGOs. The watchdog role of civil society-based NGOs and the potential for (negative) publicity can act as a pressure on corporate actions with attendant marketplace effect, from ensuring dolphin-free tuna to preventing the use of child labour by contract suppliers. The rise of corporate social responsibility (CSR) has gone hand in hand with both civil society scrutiny and state regulation. In the UK, there has been a rush of CSR initiatives by the corporate food retailers (Lang and Barling, 2007). This follows growing civil society scrutiny of aspects of their product choice and supply chain instructions to farmers and growers; including exercises that score the supermarkets according to environmental and other criteria, such as the health-supporting nature of their products (Friends of the Earth, 2004; Dibb, 2005). In the case of pesticide residues, the UK regulators had made public the results of governmental inspections of residue levels found in store on fresh produce from 1998. Friends of the Earth then tabulated the results to compare the environmental and health performance of the retailers (Friends of the Earth, 2004). Two of the smaller high-street supermarkets which specialize in own-brands responded to these pressures by raising their standards. Marks and Spencer, which has a strong high-value-added food profile, phased out 79 pesticides, some of which remained state approved (Buffin, 2001a). Its stated goal of selling residue-free produce would have a significant impact on its 47 fresh produce suppliers, who in turn work with 1,000 farmers worldwide. The Co-operative Group, a retailer with a lower socio-economic customer base but a strong ethical tradition and with around 4% UK market share, unilaterally banned 24 pesticides

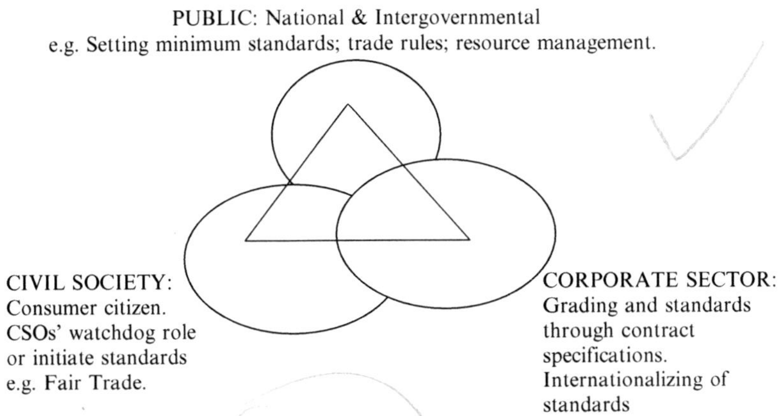

Fig. 3.1 Food governance – overlapping forms: public, corporate and civil society sectors (Barling, 2008)

for which there are alternative growing options, six of them approved by the UK regulatory system (Buffin, 2001b). The picture here reflects the overlap area in Fig. 3.1 where the interaction of governance forms between corporate retailers, civil society organizations and the state produces a raising of standards for consumers.

The reaction of the European marketplace to GM food and feed provides a good example of the interaction between different actors in the food supply chain and illustrates some of the dynamics of contemporary food governance. Also, the entry on a large scale of GM soybean and maize commodities and the use of their processed derivatives of oils and flours in manufactured foods in the late 1990s brought the realities of traceability for cross-continental supply chains onto the policy agenda. The goals of the big life science and agrichemical input industries (such as Monsanto), and grain-trading and first-stage processing corporations (such as Cargill) clashed with the reaction of the large corporate retailers and their customers in Northern Europe. In the UK, the large membership environmental organizations also played a key role in protest and challenge to the entry of GM crops and their derivatives. The public authorities, through the EU, followed with hastily drafted regulations for the labelling of GM soybean and maize imports and, in turn, their traceability.

The first large-scale shipment from the US of GM soybean mixed in with non-modified grain reached Europe via Antwerp in 1996, where Greenp eace organized an unsuccessful attempt to prevent the unloading of the cargo. The protests did succeed in raising public awareness, notably amongst northern continental consumers, of the entry of GM commodities into the European food chain. The disquiet amongst the German public led some of the main food processors and distributors, including Unilever and Nestlé, to seek to remove soy oil derived from these modified soybeans from their ingredients at the end of 1996 (Nottingham, 1998:133). However they refused at this stage to remove the oil from their products sold in other European countries.

The publicity attached to the imports of the soybean and maize was reinforced by a renewed focus on the potential impacts on biodiversity and farming practice in the European agri-environment from herbicide- and pesticide-modified GM crops. Monsanto had launched a high-profile public relations campaign in the UK in the summer of 1998 to persuade the public of the benefits of food biotechnology. For example in a full-page advertisement placed in the press, Monsanto extolled the belief that 'biotechnology is one way to cut down on the amount of pesticides used in agriculture'.[1] However, NGOs, which were keen to expand the debate beyond the narrow confines of environmental policy networks and to involve the wider public, swiftly countered such assertions. Apart from providing this role of counter-expertise, some NGOs used symbolic protest events, such as the destruction of plants in GM-crop field trials, to amplify the message, hopeful that latent public concerns would turn to more vocal support. The public in turn voiced their concerns as food consumers. The sluggish response of the regulatory process was quickly outpaced by the response of the key commercial players in the food chain as they picked up the cues from their customers.

Thus, by the end of 1998, public discontent at the entry of these GM products into the food chain had spread to other major European nations, notably France and the UK. The public's unease in the UK reached fever pitch in the early months of 1999, as was reflected by campaigns on the issue of GM foods by the middle-range tabloid newspapers, the *Daily Express* and the *Daily Mail*. The sensitivity of the UK market was illustrated by the case of a GM tomato paste that had been launched in both the Sainsbury and Safeway supermarket chains in the UK in 1996. The launch of this product was seen as an example of good practice for a new technology product entry, with full labelling and explanatory leaflets. Initially, it outsold its non-modified alternative, which it undercut in price. However, by the end of the decade the supermarkets had withdrawn the product due to falling sales. Sainsbury responded to the increased media reports about GM food in early 1999 by opening a dedicated customer call line on the subject. They received 300 calls in the first 4 h, and reacted accordingly (Sainsbury, 1999).

Initial pleas from European food retailers and processors for the segregation of modified from non-modified soybean for the European market were met with resistance in the US by the large grain companies and the American Soybean Association (WFFR, 1997). The UK frozen food retailer Iceland responded by leading a search for non-modified sources of soybean for its own-brand foods. It sourced non-modified soybean derivatives from Canada and Brazil. The company sought to verify their supplies using the detection methodology for GM DNA of Polymerase Chain Reaction (PCR) technology, developed by a US company, which they then brought over to labs in the UK. However, the reliability of these detection methods was not complete. In response, Iceland supported them with a clear audit trail of testing through the different stages of the food chain, from the field through the processing and manufacturing phases. In March 1998 they were able to announce

[1] *The Guardian* Weekend (1998) June 6: 47

their own-brand foods as being non-GM. To quote Bill Wadsworth, technical director of Iceland, who helped to pioneer the search for a non-GM soybean supply for Europe: 'The only way that any European consumer was given a choice was because we fundamentally broke the supply chain. We set up a totally unique supply chain' (House of Commons Select Committee on Science and Technology, 1999).

The major supermarket retailers in the UK and northern Europe soon followed Iceland's lead. In March 1999, a consortium of Sainsburys, Marks & Spencer, Carrefour (France), SuperQuinn (Ireland), Effelunga (Italy), Migros (Switzerland) and Delhaize (Belgium) was announced. Their intention was to create a market presence with enough buying power to ensure the maintenance of a non-GM supply chain, fully tested and audited, for their own-brand products (ENDS, 1999). The major processors followed suit, under pressure from the retailers, as did some of the major restaurant and catering groups. A survey by Friends of the Earth found that in March 2000 most of the world's top 26 food manufacturers which sold in the European market had adopted non-GM policies for that market (Friends of the Earth, 2000).

Private governance systems came to the fore in this period of rejection of GM commodities in Europe. The importance of traceability systems was highlighted. Real challenges remain in maintaining effective segregation of GM crops and their derivatives. The problem of non-GM contamination by GM equivalents exists from the production of seed varieties through to the farm stages of planting, harvesting and storage and movement off farm to the grain merchants. The storage, handling and transportation of the grain, from elevator to silo to port to container and into the processing chain overseas, provide further challenges to effective segregation. The contamination of seeds of oilseed rape by Advanta's Hyola GM variety and sold through Europe (traced back to crop production in Canada) and of human food in the US by Aventis's GM Starlink maize (approved only for animal feed use in the US) in 2000 further illustrated the difficulties of segregation.

The EU continued to frame a regulatory response as traceability and labelling regulation around GM food and feed was passed in the form of Regulation EC 1830/2003 (OJL, 2003), among other revised regulations and guidance, and as traceability became part of the reframing of the principles of EU food law (see Chapter 2). The conflict between large-scale commodity exporters and exporting nation states, on the one hand, and EU member states and their attentive publics, on the other, over traceability, labelling and segregation of imported grain and derivatives continued. The Codex debates over traceability have been marked by these political divisions; as were the World Trade Organization's (WTO) discussions of Geographical Indications (GIs) of food provenance, which led to a complaint to the disputes settlement board (see following section on the international governance of traceability). Also, the commodity exporting nation states co-sponsored another WTO dispute against the EU based on the so-called moratorium on GM crop approvals that had occurred in the late 1990s and early 2000s. The multilevel nature of agri-food governance was reflected in the relocation of these debates from Europe to these WTO-related fora, adding to the ongoing negotiations in the International Biosafety Protocol under the Convention on Biodiversity.

Multilevel Governance of Food and Agriculture

The concept of multilevel governance was developed first in the context of the EU and its member states and sub-national governments (Marks et al., 1996). Some recent academic work on food governance has extended the concept for the agri-food sector to include international-global level agreements and rules (Lang et al., 2001; Barling, 2004; Morgan et al., 2006). Notably, these include the impact upon national and local levels of decision making of the WTO's Agreement on Agriculture (AoA), the Sanitary and Phytosanitary (SPS) Agreement (based in part on Codex standards), the Technical Barriers to Trade (TBT) Agreement and the Trade Related Intellectual Property Rights Agreement (TRIPS).

The determination of some key decisions by the EU and its member states has been explained within a framework of multilevel governance, in the form of two-level or multilevel strategic bargaining and decision-making (Moravcsik, 1993; Marks et al., 1996; Scharpf, 1997). Domestic considerations impact on international decisions, but international decisions also catalyze domestic considerations (Putnam, 1988). There may be a range of decisions taking place simultaneously according to different institutional rules and in slightly different domestic policy contexts, which are shaping each other to some extent.

In agriculture, the negotiation of the General Agreement on Tariffs and Trade (GATT) Uruguay Round agreements and the so-called McSharry reforms of the CAP (achieved in 1992) took place simultaneously. The negotiations involved international-EU-national level bargaining games (including the US, countries in the Cairns group of commodity exporting countries, developing countries, and so on) (Paarlberg, 1997; Moyer and Josling, 2002). The Agenda 2000 reforms of the CAP were shaped, at least in part, by the terms of the trade rules laid out in the AoA (Moyer and Josling, 2002). The agreement allowed for government supports for agriculture that have 'no, or at most minimal, trade distorting effects or effects on production' nor have 'the effect of providing price support to producers'. Such supports were seen as truly decoupled from production and so were put into the 'green box'. Direct payments to farmers, that seek to reduce production under Agenda 2000, such as arable area and livestock headage payments, were also allowed under the AoA. However, such supports are supposed to be phased out over a period of time under the 'blue box' classification.

There were several ambiguities in the wording of the AoA, reflecting the fraught diplomatic negotiations and compromises that produced it, and a review was built in from 2000 originally due to be completed in 2003. The AoA review was subsumed within the Doha or development round of negotiations over revision of the WTO agreements agreed in 2002. By early 2008 the review negotiations remained unresolved and becalmed. In part, the completion of the AoA review had been seen as being contingent upon the outcomes of the mid-term review of the Agenda 2000 reform of the CAP subsidies which was completed in 2003 with the agreement for single farm payments. Hence, the EU claimed reform and a common support regime that was in compliance with the existing terms of the AoA. Progress in further liberalizing developed countries' agricultural subsidies is seen as essential for the completion of the Doha round of trade liberalization negotiations, but ground to a halt in July 2006.

This process of mutually contingent regime reviews is illustrative of the context within which multilevel governance leads to strategic policy choices, and multilevel bargaining and game playing, by states and other participants such as international organizations. In the case of the CAP Agenda 2000 midterm review, the EC led a further shift from production subsidies, a process termed decoupling, to more qualitative supports under the single farm payment schemes. The EU has sought to frame these supports as green box compliant and non- or minimally trade distorting under the AoA. The UK Government has supported the Commission's policy direction on CAP reform. The introduction of the Department for the Environment, Food and Rural Affairs (DEFRA) as a government department merging environment with agriculture and rural affairs marked an institutional affirmation of this policy approach of wider agri-environment and rural development supports for British agriculture. Hence, an important factor in the UK's domestic approach to food policy reform was the calculation of the likely direction of trade liberalization agreements and CAP reform. Strategic decision-making, of course, is just that. It is highly contingent upon calculations of policy at other levels playing out in certain directions. Exactly what type of supports are indeed *green box* compliant is still open to interpretation, and has been challenged by the 'G20' developing countries in WTO negotiations. At the same time, then, multilevel governance may not only lead to a process of gradual harmonization through compromise but also witness political contest, conflict and stalemate.

Corporate, industrial and civil society organizations monitor and seek to influence the different policy arenas at these multilevels. European and international levels of governance have seen a multiplication of new NGO alliances and trade associations and their sponsored organizations engaging as transnational networks entering a transnational policy space (Coleman et al., 2004). In addition, professional bodies and experts, notably in the form of epistemic communities, that is bodies of experts who claim authoritative knowledge and are recognized within international regimes, can be instrumental in international policy formulation. International organizations themselves, such as the EC and the secretariat of the WTO, play important roles in prompting policy solutions by helping to frame the terms of the debate. Traceability has become an agenda item at global levels of agri-food policy-making in recent years, engendering a degree of political conflict that is explained below.

The International Governance of Agri-Food Traceability

Traceability has been and still is the subject of regulation and definition within the multilevels of agri-food governance. Industry-led standards increasingly are being adopted within state-based systems, and at the international level private governance agreements of the International Standards Organization (ISO) are informing intergovernmental decisions at the EU and at Codex. This hybridization of public-private standards incorporates internationally audited systems by organizations such as the European Committee for Standardization (CEN) and the ISO. The ISO has a general definition for traceability that covers industrial products, drawing in particular from existing definitions pertinent to comparable national standards for measurement instrumentation: 'Traceability: ability to trace the history, application or location of that which is under consideration'. In terms of applying traceability to food, this is covered under the quality management systems standards for food (see Chapter 2), and the international standard agreed in 2007 for: *Traceability in feed and food chain – General principles and basic requirements for system design and implementation (ISO 22005:2007)*. The text explained that: 'A Traceability system is a useful tool to assist an organization operating within a feed and food chain to achieve defined objectives in a management system' (ISO, 2007:iv). At the international political levels the disagreement has centred on what the objectives of traceability systems that needed to be regulated should be.

Within the EU, traceability was defined in Regulation 178/2002 on the General Principles and requirements of Food Law regulation that also established the European Food Safety Authority. This was part of a much wider wave of EU food policy reform that arose from the 1990s onwards (for a more detailed discussion see Barling, 2004; Barling and Lang, 2005; Ansell and Vogel, 2006). The EU definition of traceability as laid out in the General Principles is that traceability 'means the ability to trace and follow a food, feed, food producing animal or substance intended to be or expected to be incorporated into a food or feed, through all stages of production, processing and distribution' (Article 3 (15)) (OJL, 2002). There is a full food and animal feed chain approach, as the stages are defined as originating with primary production, and the regulation includes imports, and extends 'up to the final consumer' (Article 3 (16)). However, in terms of its operation and implementation it is a one-step-back and one-step-forward approach to record keeping (Article 18). Nonetheless, this effectively transfers responsibility to each stage of the food chain back as far as the farm. This regulation added to a corpus of EU law encompassing agri-food traceability, including the previously mentioned traceability of GM food and feed (see Chapter 2 for more details on this body of EU legislation).

The EU also has in place a series of GIs designed to protect the origin and authenticity of approved European food and drink products. The most notable of these are: the Protected Designation of Origin (PDO) for a product originating in a specific region where the quality of that product is due exclusively to a particular geographical environment with its inherent human and natural factors; and the Protected Geographical Indication (PGI), which relates to a product originating in a particular region which possesses a specific quality, reputation or other characteristics attributable to that geographical origin (but not necessarily due to its natural environment). PDOs were derived from the producer group designations of *Appellation d'origine contrôlée* (AOC) in France, and from those in Italy (Ilbery et al., 2000). The validity of these regulations under world trade rules was challenged under the WTO's dispute resolution process in 2003, a challenge that sheds light on the international policy conflict over traceability, and to which we return below.

The establishment of the Codex in 1963, a joint initiative of the UN's FAO and WHO's Food Standards Programme, marked an attempt to raise the minimum food standards of the member states. Codex's role as laid out in its articles is effectively a dual mandate of 'protecting the health of the consumers and ensuring fair practices in the food trade' (Article 1a). The balancing of public health, on the one hand, with trade facilitation, on the other, is a source of potential tension in its workings. The international convergence of agri-food standards accelerated from the mid-1990s with the introduction of a dispute resolution process on the legality of national standards under the WTO agreements. In the post-WTO global regulatory environment, signatory countries are required to base their domestic standards or technical regulations on those developed by international organizations. These organizations include: Codex, the Office International des Epizooites (OIE) for animal health, and the International Plant Protection Convention (IPPC) for plant health.

Prior to the completion of the GATT Uruguay Round and the SPS and TBT agreements, adoption of Codex standards at the national level was voluntary. Now there are legal obligations on countries to observe Codex standards. A member state can adopt higher levels of standards than those set by Codex in order to protect consumer health, but such actions must not be judged as discriminatory or as technical barriers to trade. Member states of the WTO can challenge a fellow member state's standards as discriminatory through the WTO dispute process. The standards, agreements and guidelines of Codex are used as a reference point in such dispute rulings. This means that the WTO does not initiate disputes itself but responds to member nations' complaints. The threat of a dispute being invoked no doubt acts as a diplomatic lever. In addition, SPS details are discussed and communicated through the WTO's SPS committee, sitting in Geneva, which is another important forum in the harmonization of food standards driven by trade concerns.

The main workings of Codex take place in some 24 active subsidiary bodies: commodity or cross-cutting issues committees, joint expert scientific advisory bodies or *ad hoc* task forces. The process is both highly technical and slow moving. A senior European member of Codex noted the problem 'with interplay between the different committees with issues moving back and forth' between them (Barling and Lang, 2005:47). Decisions are highly negotiated. They go through a number of steps and are supposed to be based on consensus at each stage. To some extent, these negotiations have always been implicitly political but with the new authority given to Codex in informing standards for the high politics of international trade disputes, the politics of decision-making are becoming more explicit.

In 2001, the Codex Executive Committee put forward a proposal on how to approach traceability within the Codex framework. Traceability had emerged in various elements of previous Codex work, including the Intergovernmental Task Force on Foods Derived from Biotechnology. Codex defined its role within the neo-liberal trade rules by identifying that traceability fell under Codex's purview as a food safety objective or SPS measure and as a 'legitimate objective' for a TBT measure. The latter included the 'use of traceability for product integrity, authenticity and identification' (CCGP, 2004a). Priority was given to traceability as a food safety objective, and as a risk management instrument. Various committees were deemed as having relevant work around traceability, covering: risk assessment, food hygiene, labelling, import and export certification and inspection, and general principles. Between 2001 and 2006, however, the Codex's main work on traceability centred on the Codex Committees on General Principles (CCGP) to provide a definition, and on the Codex Committee on Food Import and Export Inspection and Certification Systems (CCFICS) for setting of requirements for cross-border trade.

Within Codex the very use of the term traceability was challenged. The US led a preference for 'product tracing' as the more appropriate term. Product tracing had been adopted in the discussions of Codex's Biotechnology task force. Product tracing reflected the US emphasis on tracing a product back along the food chain, principally for reasons of product recall in the case of a food safety concern. That is, as a risk management procedure by governing authorities. The notion of product tracing also emphasizes a step by step process. Conversely, the term traceability implies a whole food chain perspective, which implicitly would allow a food's history to be told in full, and is the common term in Europe. The Codex Executive Committee sanctioned the use of a combined term of traceability/product tracing. The debate over terminology mirrored a deeper political conflict in which the main protagonists were the US and the Cairns Group of agricultural commodity exporting nations, on the one hand, and the EU member states, on the other. Similarly, the general principles committee, CCGP, was chaired by France, and the import/export committee, CCFICS, chaired by Australia, a leading member of the Cairns Group.[2]

The US identified CCFICS as a more favourable arena to manage the negotiations over the regulatory reach of traceability as a Codex standard and its implementation. The US, in the consultations around the definition, re-emphasized its belief that the term should be product tracing and that the alternative traceability should be dispensed with. The US also stated that it 'does not believe that information on raw materials used, on how a product was changed, or on controls which the product has been subject to are elements of all product tracing systems. Requirements of this type of information would be determined on a case-by-case basis' (CCGP, 2004b:8).

Conversely, the European regional co-ordinating committee for Codex stressed the importance of traceability to ensure the authenticity of the product to be of equal importance to food safety concerns (CCGP, 2003: paras 30–32). The European Community emphasized that the definition should take into account the need to identify specific characteristics of the product such as 'organic', 'halal' or 'kosher' (CCGP, 2004b:2). An additional dimension that came from representatives of developing countries was the costs and difficulties of implementing systems of traceability, and so the prescription of particular forms of technology was avoided by Codex.

[2] The Cairns Group figures predominantly in the WTO negotiations, particularly around the Agreement on Agriculture further liberalization of agricultural trade. The Cairns Group members are: Argentina, Australia, Bolivia, Brazil, Canada, Chile, Colombia, Costa Rica, Guatemala, Indonesia, Malaysia, New Zealand, Paraguay, Philippines, South Africa, Thailand and Uruguay.

The political divisions were a feature of the committee negotiations of the definition of the combined term. In 2004, the plenary meeting of Codex approved a negotiated definition for traceability/product tracing put forward by the CCGP as 'the ability to follow the movement of a food through specified stage(s) of production, processing and distribution' (CCGP, 2004c: Appendix IV). Disagreement was recorded over the specification of stages, the use of the terms 'trace' or 'track', and the use of the term 'identify' (CCGP, 2004c: paras 85–96). The elaboration of a single stage as well as plural stages served to emphasize the step approach to product tracing, and that tracing may be confined to specific stages of the food chain, not extended along the whole chain. The possible restriction of tracing to a certain stage coincided with ISO's elaboration on the principle of traceability for food. Also, animal feed was omitted, reflecting Codex's food remit, and deferred to the deliberations in the Intergovernmental Task Force on Animal Feed (CCGP, 2004c: paras 85–96). This definition stands in contrast to the EU definition which includes the term 'trace', specifies all stages and covers animal feed (see above).

The draft principles for inspection and certification were agreed at the CCFICS meeting in 2006 and forwarded to the 29th session of Codex for final approval in July 2006 (CCFICS, 2006). The emphasis of the key principles was to offer some protection to exporting countries in regard to the traceability requirements that may be made by importing countries in their inspection and certification requirements. For example: 'Exporting countries should not have to replicate the traceability/product tracing tool of the importing country' (CCFICS, 2006: para 51).

The disputes in Codex over traceability reflected political divisions between the large-scale commodity grain exporters that had emerged in the Convention on Biodiversity negotiations over the Biosafety Protocol in the 1990s. Although the US was not a formal member of this convention it played a steering role in negotiations, notably through the Miami Group of commodity exporters (namely, Canada, Argentina, Australia, Uruguay and Chile) (see Barling, 2006). Australia and the US also launched a dispute under the WTO process against the legitimacy of the PDO and PGI designations in 2003. Part of the challenge made was that the GIs contradicted the TRIPS agreement by permitting the coexistence of GIs with pre-existing trademarks with similar-sounding names. This complaint was rejected in the ruling in 2005. However, the ruling did uphold another complaint that certain aspects of the GIs discriminated against other countries' products. Namely, that the EU regulation should not require other WTO members to offer levels of protection for GIs similar to the those required by the EU in order to receive protection for their trademarks (e.g. Florida Oranges) in Europe. The success of this particular part of the complaint did result in the rewriting of the EU regulations to comply with the panel's findings (Council of the European Union, 2006).

The international governance and negotiations over agri-food traceability have witnessed conflict and interest aggregation between states representing the large-scale (mainly grain) commodity exporters against member states of the EU, in particular. The use of traceability and GIs for the purposes of communicating a food's provenance and its history has also been challenged. The debates over definitions and application of the different forms of traceability reflect a deeper conflict of economic interests within the agri-food system. This conflict in turn signals deeper potential barriers that exist at the global level to the realization of more communicative forms of traceability to the consumer.

Conclusion: The Governance Contexts for Realizing Ethical Traceability

Contemporary rule in liberal democracies has seen a shift from state-directed governing to more diffuse governance forms. The agri-food sector has a tradition of state intervention around agricultural support and subsidy. The state, in its various forms, has sought to extend its regulatory reach in areas relating to food safety and standards along the supply chain in the past 15 years or more. These regulatory ambitions have involved an increasing array of stakeholders and a need to engage supply chain actors in the delivery of regulatory processes, as with HACCP systems, watched and audited by state bodies. In the shifting marketplace, the large-scale production of commodities and their fractionation into ingredients for complex manufactured foods (as with soybean oils and wheat flour, for example) coexist with demands for quality food, often with clear provenance. Civil society organizations campaigning on fair trade(terms of trade), animal welfare or conservation concerns have set up alternative supply networks within existing supply chains to bring forward foods with key provenance features to the differentiating consumer. There is increasing corporate concentration along supply chains, from agri-chemical inputs and seed variety ownership, to large-scale grain commodity and meat trading and processing, to supermarket retailing. The retailers in particular are now a dominant influence along these chains as they act as gatekeepers to the consuming publics. State regulators, and indeed civil society-based networks, have to negotiate with these corporate and private sector governance forms.

Traceability systems that have emerged have done so against this interaction of state, corporate and civil society governance. The regulation of standards within Europe has moved to the EU political level. Since the early 1990s, the WTO and the rules of its trade agreements have become institutional parameters within which any standards have to fit. This multilevel dimension of agri-food governance has encompassed the regulatory boundaries and demands upon traceability systems. Behind the market trends and demands, more specific clusters of economic and political interests exist around specific traceability systems, which are seeking to validate contrasting goals for their type of system. In the Codex debates over traceability there was a clash of national priorities representing different combinations of economic and civil society interests and concerns.

In the current governance arrangements there are ethical dimensions apparent in traceability systems applied to specific foods, from animal welfare to labour practice to terms of trade (fair trade). Within the EU regulations traceability has a strong food safety and public health recall focus, but there exists regulation designed to promote food choice around process of production (GM food and traceability, organic standards) and origin and typicality (GIs). The conflicts at the global level in the international regimes mapped out reflect a resistance, based on economic interest, to formally recognizing such broader approaches to traceability. The broader tone of contemporary governance is of different forms of public–private sector interaction. This sees public authorities moving in to approve food traceability schemes, such as assurance schemes and organic standards. These schemes remain relatively closed and closely managed, however. That is, they represent managed responses by sectors along the supply chains to the presentation of information to meet the demands of the regulators and to appease and try to direct the consumer.

Ethical traceability is put forward as a potential goal for traceability systems to allow for, and to enable, a more open and democratic approach for consumers to act as citizens in the marketplace through their purchasing decisions, by asking for and obtaining the information they desire about food production practices (see Chapters 1 and 2). The realization of ethical traceability will need to negotiate both these modern supply chain complexities and their governance, and the existing private sector and public sector endorsed traceability forms in the food system. Pressures emanating from civil society will continue to play an important role. Realization of ethical traceability will not just be a morally approved step supported by appropriate technology and communication strategies, but is likely to entail politically negotiated processes. This political negotiation will need to be international as well as national or European in its setting and scope.

References

Ansell, C. and D. Vogel (2006) *What's the Beef? The Contested Governance of European Food Safety*. Cambridge, MA: MIT Press.

Barling, D. (2004) 'Food Agencies as an institutional response to policy failure by the UK and the EU', pp. 107–128 in M. Harvey, A. McMeekin and A. Warde (eds.) *Qualities of Food*. Manchester: Manchester University Press.

Barling, D. (2006) 'The international regulation of agri-food traceability: commodities, provenance and trade,' paper presented to the *Bar Coding the Food of the Future: The Social and Political Implications of Agri-food Supply Chain Re-governance* panel of RC-40: Sociology of Agriculture and Food at the *XVI ISA World Congress of Sociology*, Durban, July 25.

Barling, D. and T. Lang (2005) 'Trading on health: cross-continental production-consumption tensions and the governance of international food standards' pp. 39–51 in N. Fold and B. Pritchard (eds.) *Cross-Continental Food Chains*, London: Routledge.

Barrett, H., B. Ilbery, A. Browne and T. Binns (1999) 'Globalization and the changing networks of food supply: the importation of fresh horticultural produce from Kenya into the UK'. *Transactions of the Institute of British Geographers*, 24: 159–174.

Buffin, D. (2001a) 'Food retailer aims to restrict pesticide use'. *Pesticides News*, 54: 3.

Buffin, D. (2001b) 'Retailer bans suspect pesticides'. *Pesticides News*, 53: 3.

Busch, L. (2000) 'The moral economy of grades and standards'. *Journal of Rural Studies*, 16: 273–283.

Bingen, J. and L. Busch (eds.) (2005) *Agricultural Standards: The Shape of the Global Food and Fiber System*. Dordrecht, The Netherlands: Springer.

Campbell, H. (2005) 'The Rise and Rise of EurepGAP: Europe's (re)invention of colonial food relations'. *International Journal of the Sociology of Food and Agriculture*, 13(2): 1–19.
CCGP (2003) 'Consideration of Traceability/Product Tracing' Codex Committee on General Principles *CX/GP 03/07*. Rome: Codex Alimentarius Commission. *Alinorm* 03/19: paras 30–32.
CCGP (2004a) 'Definition of Traceability/Product Tracing of Foodstuffs (Prepared by France)', Codex Committee on General Principles Agenda Item 6, *CX/GP 04/20/6*. Rome: Codex Alimentarius Commission.
CCGP (2004b) 'Definition of Traceability/Product Tracing of Foodstuffs Government Comments (Prepared by France)', Codex Committee on General Principles, Agenda Item 6, *CX/GP 04/20/6 – Add.1*. Rome: Codex Alimentarius Commission.
CCGP (2004c) 'Report of the Twentieth Session of the Codex Committee on General Principles', *Alinorm 04/27/33A*. Rome: Codex Alimentarius Commission.
CCFICS (2006) 'Report of the Twentieth Session of the Codex Committee on Food Import and Export Inspection and Certification Systems,' *Alinorm 06/29/30*. Rome: Codex Alimentarius Commission.
Coleman, W., W. Grant and T. Josling (2004) *Agriculture in the New Global Economy*. Cheltenham: Edward Elgar.
Competition Commission (2000) *Supermarkets: A Report on the Supply of Groceries from Multiple Stores in the United Kingdom*. London: Competition Commission.
Council of the European Union (2006) 2720th Council Meeting Agriculture and Fisheries, Brussels, 20 March. *Press Release 7049/06*.
Dibb, S.E. (2005) *Healthy Competition: How Supermarkets can Affect Your Chances of a Healthy Diet*. London: National Consumer Council.
Dobson, P.W., M. Waterson and S.W. Davies (2003) 'The Patterns and Implications of Increasing Concentration in European Food Retailing'. *Journal of Agricultural Economics*, 54(1): 111–126.
Dolan, C. and J. Humphrey (2000) 'Governance and trade in fresh vegetables: The impact of UK supermarkets on the African horticulture market'. *The Journal of Development Studies*, 37: 145–176.
ENDS (1999) 'Sainsbury's, M&S in "GM-free" retailer consortium'. *ENDS Report*, 290: 33–4.
Flynn, A., T. Marsden and M. Harrison (1999) 'The regulation of food in Britain in the 1990s'. *Policy and Politics*, 27(4): 435–446.
Friends of the Earth (2000) *European Food Manufacturers Shun GMOs but Consumers Urged to Keep Up the Pressure- Press Release*, March 7, London: Friends of the Earth.
Friends of the Earth (2004) *Pesticides in Supermarket Food*. London: Friends of the Earth.
Grievenik, J.W. (2003) 'The changing face of the global food industry', presentation at the *OECD conference on Changing Dimensions of the Food Economy: Exploring the Policy Issues*, The Hague, 6 February.
Henson, S. and T. Reardon (2005) 'Private agri-food standards: Implications for food policy and agri-food system'. *Food Policy*, 30: 241–253.
House of Commons Select Committee on Science and Technology (1999) Minutes of evidence: Examination of Witnesses (Questions 1–19) 3 March 1999. http://www.parliament.the-stationary-office.co.uk/pa/cm199899/cmselect/c.../9030303.ht.
IGD (2003) *Working Group Report- Farm Assurance Schemes in UK: Summary Review and Issues for Development*. Letchworth: Institute of Grocery Distribution.
Ilbery, B., M. Kneafsey and M. Bamford (2000) 'Protecting and promoting regional speciality food and drink products in the European Union'. *Outlook on Agriculture*, 29(1): 31–37.
ISO (2007) *Traceability in feed and food chain – General principles and basic requirements for system design and implementation. ISO 22005*, Geneva: International Standards Organisation.
Lang, T. (2006) 'Food, the law and public health: Three models of the relationship'. *Public Health*, 120, 30–41 October.
Lang, T., D. Barling and M. Caraher (2001) 'Food, Social Policy and the Environment: Towards a New Model'. *Social Policy and Administration*, 35(5): 538–559.

Lang, T. and D. Barling (2007) 'The Environmental Impact of Supermarkets: Mapping the Terrain and the Policy Problems in the UK', pp. 192–215 in D. Burch and G. Lawrence (eds.) *Supermarkets and Agri-Food Supply Chains: Transformations in the Production and Consumption of Foods*. Cheltenham: Edward Elgar.

Larner, W. and R. Le Heron (2004) 'Global Benchmarking: Participating at a Distance in the Global Economy', pp. 212–232 in W. Larner and W. Walters (eds.) *Global Governmentality: New Perspectives on International Rule*. London: Routledge.

Le Heron, R. (1999) 'Creating food futures: reflections on food governance issues in New Zealand's agri-food sector'. *Journal of Rural Studies*, 19(1): 111–125.

Majone, G. (1996) *Regulating Europe*. London: Routledge.

Marks, G., L. Hooghe and K. Blank (1996) 'European integration since the 1980s: state-centric verses multi-level governance'. *Journal of Common Market Studies*, 34(3): 341–78.

Moran, M. (2003) *The British Regulatory State: High Modernism and Hyper Innovation*. Oxford: Oxford University Press.

Moravcsik, A. (1993) 'Preferences and power in the European Community'. *Journal of Common Market Studies*, 31: 473–524.

Morgan, K., T. Marsden and J. Murdoch (2006) *Worlds of Food: Place, Power and Provenance in the Food Chain*. Oxford: Oxford University Press.

OJL (2002) 'Regulation (EC) No 178/2002 of the European Parliament and of the Council of 28 January 2002 laying down the general principles and requirements of food law, establishing the European Food Safety Authority and laying down procedures in matters of food safety'. *Official Journal of the European Communities*, 1.2.2002, L31/1–24.

OJL (2003) 'Regulation (EC) No 1830/2003 of the European Parliament and of the Council of 22 September 2003 concerning the traceability and labelling of genetically modified organisms and the traceability of food and feed products produced from genetically modified organisms and amending Directive 2001/18/EC'. *Official Journal of the European Communities*, 1.2.2002, L268/24–28.

Moyer, W. and T. Josling (2002) *Agricultural Policy Reform: Politics and Process in the EU and US in the 1990s*. Aldershot: Ashgate.

Nottingham, S. (1998) *Eat Your Genes: How Genetically Modified Food is Entering our Diet*. London: Zed Books.

Paarlberg, R. (1997) 'Agricultural policy reform and the Uruguay Round: synergistic linkage in a two-level game?' *International Organisation*, 51(3): 413–444.

Pierre, J. (2000) *Governance, Politics and the State*. Basingstoke: Palgrave.

Pierre, J. and B.G. Peters (eds.) (2000) *Debating Governance*. Oxford: Oxford University Press.

Ponte, S. and P. Gibbon (2005) 'Quality standards, conventions and governance of global value chains'. *Economy and Society*, 34(1): 1–31.

Putnam, R. (1988) 'Diplomacy and domestic politics: the logic of two-level games'. *International Organisation*, 42(3): 427–460.

Reardon, T. and E. Farina (2001) 'The rise of private food quality and safety standards: illustrations from Brazil'. *International Food and Agribusiness Management Review*, 4(4): 413–421.

Rhodes, R. (1996) 'The new governance: governing without government'. *Political Studies*, 44: 652–667.

Rhodes, R. (2000) 'The governance narrative: key findings and lessons from the ESRC's Whitehall Programme'. *Public Administration*, 78(2): 345–363.

Rose, N. (1999) *Powers of Freedom: Reframing Political Thought*. Cambridge: Cambridge University Press.

Sainsbury (1999) Memorandum from J. Sainsbury plc to the House of Commons Select Committee on Agriculture, 15 November 1999; see http//www.publications.parliament.uk/pa/cm199900/cmselect/cmagric/71/7111.htm

Scharpf, F. (1997) 'Introduction: the problem-solving capacity of multi-level governance'. *Journal of European Public Policy*, 4(4): 520–538.

Smith, M.J. (1990) *The Politics of Agricultural Support in Britain*. Aldershot: Dartmouth.

Vorley, B. (2003) *Food Inc. Corporate Concentration from Farm to Consumer*. London: UK Food Group.
WFFR (1997) 'Boycott begins against products containing US genetically engineered soybeans'. *World Food Regulation Review*, 6(6): 22–23.
Whatmore, S. and L. Thorne (1997) 'Nourishing Networks: Alternative Geographies of Food', pp. 287–304 in D. Goodman and M.J. Watts (eds.) *Globalising Food. Agrarian Questions and Global Restructuring*. London: Routledge.

Chapter 4
Narrative Strategies in Food Advertising

Guido Nicolosi and Michiel Korthals

Introduction

Communication between food production and food consumption consists, from the producers' side, mainly of marketing and advertising strategies. Traceability, and in particular ethical traceability, covers this communication, as the 'historical' or narrative character of traceability encompasses themes that connect producers and consumers in certain ways (see Chapter 1 for definitions of traceability and ethical traceability). In this chapter we will show the results of our empirical research (supporting our claim that both philosophical and sociological research can elucidate the intricacies of ethical traceability) in marketing and advertising strategies. These strategies conceptualize products, consumers, their concerns, preferences and behaviour according to cultural images that differ according to national and cultural contexts. One can call these strategies and the 'images' of consumers they incorporate 'narrative strategies', because they tell stories within which groups of producers and consumers recognize their main activities and interests. These stories of food intermingle with personal life stories and by doing so they contribute to the formation of personal identities. The narrative strategies in food advertising oscillate between evocative and mythical images, such as nature, naturalness and tradition, as a representation of a 'healthy' and ethically 'right' order, and rational and scientific images in which the main role is played by information. They are never only about information, but also about 'lifestyles'. This form of symbolism in advertising is often independent of the material qualities of the product and lacks a strong relation with what, in traditional economics, is termed exchange value (Gorz, 2003:41). We will show that at least two narrative strategies of food advertising can be discerned in the Mediterranean area of Italy and Spain, and that their relationship is dynamic and ambivalent.

Four Possible Narrative Strategies of Food Advertising

The main issue for food advertising is how to conceptualize food and food consumption successfully, i.e., what kind of images, stories, symbols and types of information are attractive to consumers. These images say something not only

C. Coff et al. (eds.) *Ethical Traceability and Communicating Food*,

about the food product and its origin, but also about the act of consumption and its connection with the human body (does it give pleasure or health etc.), and about consumers; in other words, they frame and illustrate the way the relationship between food, body, society and nature is seen by different groups of producers and consumers.

In the work of Claude Fischler (1988), Mary Douglas (1970, 1996), Pasi Falk (1994) and Guido Ferraro (1999) one can find interesting ideas about the relationship between food, nature, society and body. Purity, danger, integrity, order and naturalness are, according to these authors, central categories on the basis of which consumers and producers tackle their daily activities in relation to food, which include discourse as well as preparation and consumption. With their categories they take into account that human beings have different views of their relationship with food and nature. Our hypothesis is that these categories can help to illuminate the way consumer concerns are addressed in various differing advertising strategies. We organized these ideas with the help of the four quadrants (Fig. 4.1) from the work of Mary Douglas (1970, 1996) and Guido Ferraro (1999), which are the result of two intersecting axes. The horizontal and vertical axes cover values and issues respectively, from relative to absolute and from subjective to objective. On the horizontal axis, 'relative' means that reality is seen as socially constructed and 'absolute' means that reality is seen as nature – i.e. as something independent of human intervention. On the vertical axis, 'subjective' means that individuals are seen as most important while 'objective' means that society (or the group) is seen as most important. The horizontal and vertical axes intersect, resulting in four quadrants. The four quadrants define (clockwise) four different narrative strategies or 'regimes': *causal, positional, perspectival* and *multi-perspectival*. Together, the four quadrants classify orientations in societies and cultures, in a fairly general and universal manner. According to Douglas and Ferraro, these four orientations or strategies mutually exclude each other, which means, for example, that it is impossible to use the causal and the positional perspectives simultaneously. The orientations cannot be compromised because they are incompatible organizing principles, as Douglas (1996:100) states: 'Any choice which is made in favour of one is at the same time a choice against the others'.

The *causal* regime is defined by the objective and relative axes: this strategy is structured in a series of logical steps based on cause-and-effect relations. The approach tends towards objectivity, observability and rationality. The central values of this model are *information, effectiveness, force, energy* and *power*. What counts are the concrete facts, and not values associated with images. Advertising in this field tends to promote the product as a convenient and effective solution capable of solving even complex problems. In texts belonging to this quadrant, a central role is played by compromises, that is, the attractiveness of 'taking into account' different and apparently irreconcilable needs like taste and physical fitness, quality and savings. As Ferraro writes, 'It is not a perfect world that is discussed, but a somewhat improved world, a more rational, significant, comfortable world, in which we have more resources and more energy to face the problems that

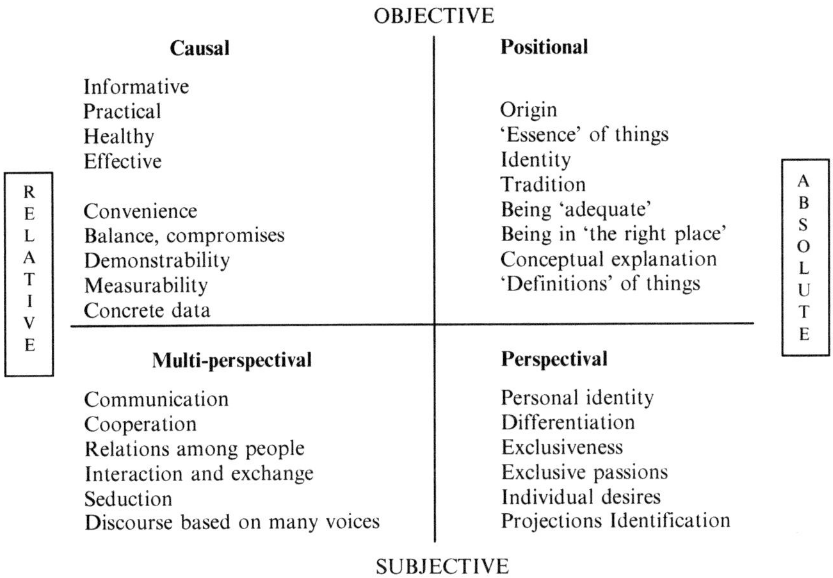

Fig. 4.1 Four quadrants of narrative strategies and regimes based on Mary Douglas (1970, 1996) and Guido Ferraro (1999), showing the narrative forms and values involved

constantly arise' (Ferraro, 1999:68). He calls this type of advertising 'informative'. The advertising texts that belong to this quadrant emphasize the *practical, measurable* and *objective* qualities of food (such as the nutritional value and calories). *Rules* and *measurements* are central aspects and *diet* (in the broad sense of organizing food consumption) is the texts' central message. Nature does not dictate what humans should eat; scientific (human) knowledge tells us what is healthy and good.

The *positional* regime is located in the objective/absolute quadrant; where the fundamental value is the intrinsic and 'objective' qualities of the product. The narrative formula used in texts belonging to this quadrant emphasizes the value of the *essence* of things, their true nature. According to this perspective, food products have static qualities, referring to their traditional origin. *Authenticity* holds a supreme, *absolute* value (Ferraro, 1999:65–66). Food advertising belonging to this regime constantly evokes *tradition, nature* and the value of traditional things. Ferraro calls the actual advertising associated with this regime 'identity advertising'; the message is based on 'the expression of the full adequacy of a product to its essence' (Ferraro, 1999:72), an essence that is usually original, but can also be the final result of a process.

The *perspectival* regime is defined by the subjective/absolute axes. This regime is presented as coming from a specific subject or from a 'system of values' expressing a specific (and usually exclusive) 'world vision' (Ferraro, 1999:69). The central aspects of this quadrant are *exclusiveness* and *uniqueness*. Products are associated with prestige and luxury. They are presented as powerfully captivating, irresistible and as the fulfilment of irrepressible desire. Frequent use is made of irony and simile. Typical objects of this narrative strategy are clothing, jewels and perfume. The *exclusiveness* and *uniqueness* that define this quadrant are expressed in the food sector by the centrality of 'gourmet food', with the consumer perceived as a 'gourmet'. The characteristics of food are not defined as objective elements in the real world, but presented as dependent on the taste of an exclusive and unique subject, the gourmet, i.e. the expert and connoisseur of good food.

Finally, the *multi-perspectival* regime of the subjective/relative quadrant is characterized by a strategy that presents the interaction and integration of different perspectives as various systems of values playfully interacting. The diversity of food in the world is seen as one continuous conversation between cultures. 'Food adventurers' enjoy the rich variety of cuisines. According to this perspective, food advertising centres on the deregulation of traditions and associates itself with creativity, with the 'de-ritualizing' of meals, and with cooking innovations such as new dishes, new origins sources and new combinations.

Empirical Research into Narrative Strategies used in Italian and Spanish Journals

In our empirical research we analyzed food advertisements that appeared in 2004 in three weekly periodicals (one Italian and two Spanish). The research was based on two different counting systems. First, we counted how many ads belonged to each narrative regime. In this case, each ad was counted only once and therefore repeating appearances of an ad are not counted. The result showed how much the advertising world used the four regimes with a given text (we called this text 'models' in Table 4.1). Secondly, we also sought to evaluate the 'diachronic impact' on the readers of the journals, by counting the number of times the same ad, expressing a given narrative regime, was published (the 'frequency' in the table). Results were analyzed separately for each publication. Thirdly, we analyzed some representative advertisements in more depth, to highlight some interesting features of the dominant regime.

Table 4.1 Models and Frequency *D di Repubblica*

Regime	Models	Frequency
Causal	8	13
Positional	33	96
Perspectival	24	86
Multi-perspectival	–	–
Totals	65	195

Data from Italy: D di Repubblica

We analyzed a weekly magazine called *D di Repubblica: La Repubblica delle Donne*,[1] distributed on Saturdays along with the newspaper *La Repubblica*, and targeted at a mainly female middle-upper class public (the 'D' in the title stands for *donna*, woman). We examined the issues published between March 14, 2004 (no. 392) and April 30, 2005 (no. 447). The research therefore included a detailed analysis of 54 issues[2] and 195 ads. The results are presented in Table 4.1 and Fig. 4.2.

The majority of food advertisements we observed belonged to the positional quadrant. More specifically, 33 out of 65 ads with a given text (sometimes the same brand and/or product is presented through different texts) belong to this quadrant. Out of the 195 texts surveyed, a total of 96 (49%) belong to this quadrant.[3]

Quantitative analysis, however, revealed that a large number of the ads belonging to the positional quadrant also used the typical tools of *causal* advertising to suggest the health qualities of the products. There are references to the ingredients of the product, often presented in terms of percentages, references to the process of production or information on nutritional values (such as minerals and fibres). Specifically, 22 out of 33 advertisements belonging to the positional narrative regime adopt rational and informative elements in support of their 'argument'. If we consider the 'diachronic impact' (frequency, including repetitions), this

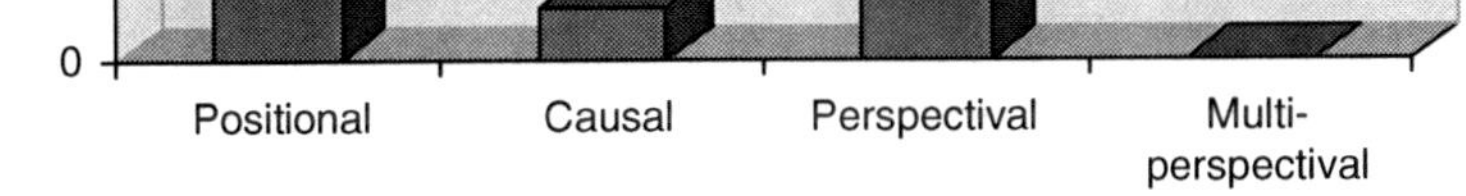

Fig. 4.2 Graph of the four narrative strategies in *'D'* di Repubblica

[1] This study was done in collaboration with Dr. Venera Trepiccione, University of Catania, Italy.

[2] The only gap is issue 421 which we were unable to obtain.

[3] In reality, we are dealing with a rather uncommon discursive 'polarization' between the positional regime (48% of the cases) and the perspectival regime (43% of the cases). The high proportion of ads belonging to the perspectival regime can probably be explained in terms of the social characteristics of the audience of the magazine: high social status and high level of education. This audience looks for highly distinctive, even exclusive, consumer products. This is confirmed by the analysis of the textual content of the magazine's columns and articles (fashion, design, architecture, etc.). It is worth remembering also that many ads belonging to the perspectival quadrant advertise ice cream and coffee.

Table 4.2 Models and frequency in *El Semanal* and *Mujer Hoy*

El Semanal		
Regime	Models	Frequency
Causal	16	28
Positional	13	21
Perspectival	1	5
Multi-perspectival	2	4
Total	32	58
Mujer Hoy		
Regime	Models	Frequency
Causal	17	26
Positional	4	6
Perspectival	1	2
Multi-perspectival	2	3
Total	24	37

approach is found in 49 out of 96 appearances (51%). It seems as if the narrative strategies are not so exclusive vis-à-vis each other as theoreticians want us to believe!

Another interesting aspect is the frequent use of brands and symbols (such as logos of consortia for the protection of consumers) that guarantee that the product has been constantly monitored and its quality and safety have been certified. This aspect is present in 36 appearances out of 96 belonging to the positional strategy (38%). Only one case from the causal regime (out of 13) appeals to brands and symbols for the protection of consumers.

Discussion of Two Advertisements

The first advertisement that we consider particularly significant is that of 'Grana Padano', Fig. 4.3.

A central theme of the advertisement is the relation of the product to milk – but not to cows. By talking only of milk, and not alluding to cows, the ad distracts consumers' attention from recent anxieties such as 'mad cow' disease. This metaphorical 'slippage' is evident in the visual headline, where the rational and linear argument is supported and reinforced by the photo-collage (cheese as a cup of milk). Besides, milk always functions very well as a synecdoche. What could be more reassuring than milk? Milk is our first meal as newborn and it makes up the only diet in the first months of our lives. So, it represents nature and purity and evokes, through another considerable metaphorical slippage, the idea of maternal love.

The advertisement reassures, stressing that the milk is particularly safe insofar as it is Italian (instead of English, for example). Furthermore, the authenticity of the product is confirmed by the origin (space, in the sense of place) and by tradition

Fig. 4.3 *D di Repubblica* no. 397, April 17, 2004

(time, in the sense of history); the latter is guaranteed with of '50 years on the side of quality' ('*50 anni dalla parte della qualità*').

There are other devices that refer to the authenticity and quality of the product, such as the reference to the 'Grana Padano Protection Consortium' that supervises cheese production through 'continuous and rigorous verifications'. Reference is also made to its Protected Denomination of Origin (PDO) certification, which is an additional guarantee of quality. This certification and the existence of a consortium serve to reassure consumers that this is a 'safe cheese', but also a cheese that respects regional traditions and therefore contributes to the continuation of local identity – and even to the life stories of the consumers.

A second highly significant example of a food's territorial and cultural associations is found in the ad for 'Parmigiano Reggiano' (Fig. 4.4). In the foreground there is a whole Parmesan cheese with the brand name clearly visible on the cheese itself. The cheese is surmounted by photographs of five important monuments from

Fig. 4.4 *D di Repubblica* no 422, October 16, 2004

five well-known cities of the Emilia region.[4] This collage of photos underlines the 'monumental' status of the product. Like the depicted monuments, the cheese also functions as a symbol of a geographical area famous for its food products. For this reason, the product is presented as unique as its origin' (see sub-headline). The first part of the sub-headline in the upper half suggests the long tradition behind the product ('70 years') stressing the fact that 'this extraordinary example of nature and wisdom' is a 'gift' from 'generations of the place of origin'. Tradition, territory and nature are merged in a single message.

The idea of the product as traditional and natural is reinforced by another element: the phrase in the payoff: 'Parmigiano Reggiano. You don't build it, you make it'.

---

[4] Note, incidentally, that Parma is currently the seat of the European Food Safety Authority. Parma is well known in Italy and abroad for the quality of its food products though its reputation recently received a blow because of a financial scandal involving Parmalat, the city's largest food company. The Italian government lobbied heavily for Parma to be the seat of the European agency.

The product is presented as the result of a tradition, which stays clear of industrial locations and procedures. It is 'prepared solely with milk from cows of the area of origin', as specified in the body copy. The latter also states that the product is 'without additives', reassuring customers about possible chemical contamination. The Consortium vouches for all these qualities, ensuring that the product is 'controlled unit after unit, day after day'.

Data from Two Spanish Journals

We performed an analysis[5] on *El Semanal* and *Mujer Hoy*, two of Spain's most popular magazines. Both are weekly supplements of *ABC*, a national newspaper, as well as of 25 important regional newspapers. Both magazines sell many copies, but they differ significantly in terms of content and target audience. In this case, too, research focused on food ads published throughout 2004 and, specifically, in issues 844 to 895 for *El Semanal* and 247 to 298 for *Mujer Hoy*, a total of 52 issues per magazine. Ad frequency was again measured in two different ways (see above).

In both cases, we found various types of advertisement occurring repeatedly in one or both magazines, with variations. More specifically, we found 32 ad models in *El Semanal* and 24 in *Mujer Hoy* (Fig. 4.5), which translated respectively into 58 and 37 ads (95 total). Ads were analyzed and classified on the basis of the four narrative strategies or regimes. Results for both magazines show a clear prevalence of causal and positional regimes.

Combining the data for the two magazines shows that out of a total of 95 ads, more than half (54) adopt the causal regime as the dominant discursive form, while 27 adopt a positional regime. Only 14 use perspectival or multi-perspectival regimes (7 times each).

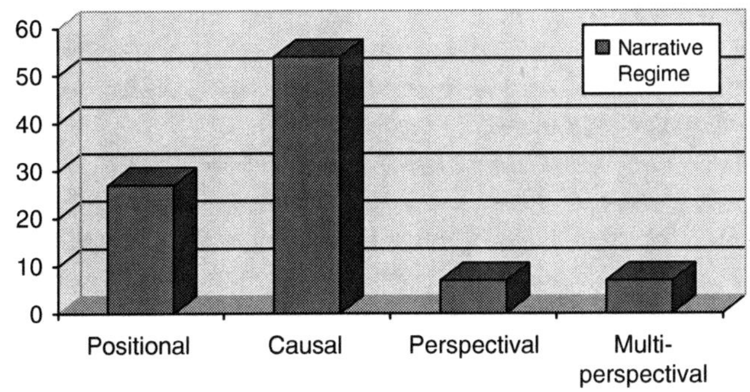

Fig. 4.5 Graph of the four narrative strategies in *El Semanal* and *Mujer Hoy*

[5] Dr. Alessandra Fatuzzo (University of Catania, Italy) collaborated on this research and thanks to an Erasmus scholarship was able to stay in Valencia to gather the material.

The two dominant narrative regimes are causal and positional, i.e. two advertising strategies belonging to the objective side of our model (with the two variables relative/absolute). Further confirmation of the emphasis on information is provided by the fact that the ads were often accompanied by symbols attesting to the approval or patronage of prestigious national or international agencies (ministries, towns, EU bodies, etc.). Specifically, such symbols occurred in seven different models, some present in both magazines, totalling 25 ads.

The belief that consumers need to be reassured on the safety of food is evidenced by the tendency of producers to offer web site addresses and phone numbers for information on advertised products. This was done in 10 models, some found in both magazines, totalling 41 ads. Besides offering more information on products, phone numbers and web sites, additional services were sometimes offered, as in the case of Danone's Vitalinea line, through which consumers can obtain a personalized diet.

While in both magazines the causal regime is dominant (as can be seen from the long informative descriptions of the product and its effects along with other practical information), there is also frequent reference to health and fitness, two primary and unquestionable values. Specifically, of 16 causal ads found in *El Semanal* (28 appearances), 12 (24 appearances) were based on these values. *Mujer Hoy* evidenced the same tendency: out of 17 ad models (26 appearances) classified as belonging to the causal regime, 14 (21 appearances) focus on health and fitness. However, compared to *El Semanal*, these advertisements referred more to fitness from an aesthetic perspective.

Discussion of Two Spanish Advertisements

As with the Italian research, we will discuss a few emblematic examples of Spanish the advertisements we analyzed in more detail. The first example (Fig. 4.6) falls into the category of *identity* advertising (positional quadrant). The text advertises a brand of mineral water popular in Spain: Lanjarón, which is also the name of the town where the water is collected.

The text is based on values generally associated with well-being, such as longevity, tranquillity, the absence of stress, nature and clean air. To vouch for the value of Lanjarón water, the ad rallies the quintessence of wisdom, merging the wisdom of old people and the wisdom of parents (tradition), specifically of a mother. Who could be wiser than the mother of a 76-year-old gentleman? The ad shows an elderly man who, giving the thumbs-up sign in a juvenile gesture, declares '*Estoy hecho un chaval, dice mi madre*' ('I look like a kid, says my mother'). The headline immediately stimulates the curiosity of the reader. The body copy provides the testimony of the man, called Pepe. This communicative strategy establishes a colloquial and reassuring context: consumers know they are dealing with a simple and transparent person, an old-timer, leading a serene life in some little rural town. Pepe in Spanish is a typically juvenile nickname. The man's language, too, is somewhat

Fig. 4.6 *Semanal* 855–867 – *Mujer Hoy* 268

childlike: 'my mother says', 'TV says'. These aspects all hint at the longevity of the inhabitants of Lanjarón, an oasis of peace and tranquillity in the midst of nature. It is said by Pepe's mother that the full quality of this water stems from the water of Lanjarón, springing from the highest mountain of Spain, the purest water or, as the village's old people say, 'the most water of water'.

The thermal baths of Lanjarón are also well known (they are the most important ones in Andalusia) and so is the purity of their water. Therefore, the ad needs only to remind consumers of these qualities, through the words of Pepe's mother, who functions as both witness and living proof: an exceptional witness since, as a mother, 'she knows what she's talking about'; and an exceptional living proof of the longevity of the inhabitants of Lanjarón, as the mother of a 76-year-old man. The payoff again takes up the three elements responsible for the vitality of the protagonist: the location, the thermal baths and finally and obviously, the water.

The second ad (Fig. 4.7) is part of an advertising campaign organized by the '*Fondo de Regulación y Organización del Mercado de los Productos de la Pesca*

Fig. 4.7 *El Semanal* no. 882, 883, 884, 885 – *Mujer Hoy* no. 285

y Cultivos Marinos' (FROM), an independent body whose goal is to promote fish consumption by offering subsidies to businesses that guarantee the quality of fish and respect for the marine environment. This ad is one of the most frequent in the two magazines. Between September 19 and October 19, it occurred five times, with small variations in order to avoid repetitiveness. The dominant discursive regime is undoubtedly the causal one. Main topics are fitness and health. In the ad, the reader is immediately confronted with the picture of an appetizing tuna-based dish, which can be easily prepared following the recipe contained in the ad itself. What immediately captures the attention of the reader is the main ingredient of the dish: in the headline, in a much larger font, we find the word '*ATÚN*' ('tuna fish'). The reader is entertained with a pun on *el tomo* ('the back', the best part of the fish) and '*tome*' to eat': '*Lo tomes como lo tomes*': 'in whatever way you eat it'). Tuna is presented as the perfect food: 'good for everything'.

The body copy outlines the qualities and advantages of the product. It refers to its nutritional characteristics, to the abundance of vitamins, the presence of omega 3, and closes by restating the characteristics that make tuna an excellent food: nutritional value, convenience, taste and price.

The logos of three important national and European bodies are displayed: the above-mentioned FROM, INTERATUN (Interprofessional Tuna Organization)[6] and Financial Instrument for the Orientation of Fishing (IFOP).[7] These logos, along with that of the Ministry of Agriculture, Fishing and Nutrition, enhance the credibility of the ad, offering greater security to customers.

Narratives and Advertising Strategies

Tradition, nature and geographical and cultural origins are almost obsessively present in the ads examined. Everything seems to revolve around the 'quasi-mythical' celebration of these elements, presented as actual and fixed absolute entities. In a sense these advertisements do not express a genuine story, since they do not show a historical development from a point of departure through a process to a final goal. The readers are immersed in a realm of immutable elements and are drawn in a movement of eternal return to the same elements again and again. The idea of authentic purity acquires an almost divine quality and the authenticity of food is its most marked expression. Eating is presented as a sacral act in which the essential things in life are intimately preserved and maintained. However, apparently in contradiction to these absolute items, the texts of the advertisements do refer to human and scientific items, such as information on the safety and health of the product.

The narratives used in the advertisements studied rarely show one strategy in its pure sense. In particular, ads belonging to the 'objective' (or upper) side of the four quadrants (causal and positional) tend to influence each other. In the Italian case it is evident that positional advertising, which is the most common, often shows elements typical of causal advertising (objective characteristic encounters relative characteristic). This fact is interesting insofar as it is unexpected according to the theories applied from the outset and hence in need of further interpretation.

These two features give these texts a certain ambivalent or hybrid character. The hybridizing consists in the frequent use of a technique which could be defined as 'information support'. It results in a significant type of advertising halfway between identity and informative advertising, aimed at providing rational reassurance. Measurability, reference to concrete data and the need to (scientifically) *demonstrate* the positional strategy are the most significant aspects we have detected. In a sense science is needed to support the cultural (positional) elements: science and cultural myths of origin and authenticity fuse in one new way of advertising.

Seen from this perspective of hybridization, the food advertisements examined often seem to break, somewhat unpredictably, the traditional distinction between

[6] An organization officially acknowledged by the Ministry of Agriculture, whose role is to represent the interests of the tuna industry.

[7] IFOP is an organization of the EU whose main objectives are the valorization of fish and agricultural products and the improvement of fishing methods for the purpose of ensuring quality and respect for the environment.

mythical and scientific thought. Lévi-Strauss (1978) has studied the differences and similarities between science and cultural myths and indicates that both have the function of ordering elements of reality. Science and cultural myths both refer to concepts of 'order' and 'rule' (code). However, they differ in their method of ordering reality. Scientific thought and reductionism initially fragment problems into as many parts as necessary, with the aim of ordering them and solving them afterwards; they reduce complex questions to single problems that can be analyzed and explained. Mythical thought, on the other hand, is a *totalizing* form of thought, which gives the illusion of understanding the world while incapable of providing much control over it (Lévi-Strauss, 1978). The ads we have analyzed seem to try to overcome this distinction through a persuasive act based on both forms of thought. They sometimes 'incorporate' both thoughts in attempting to persuade as many consumers as possible.

In the Spanish case, the most common narrative form is the causal one (informative advertising). Consumers are confronted with a text that refers to 'demonstrable' data and provides measurements with the purpose of presenting and legitimating the *truth* of the product. Here, however it is necessary to make a qualification similar to the one made in relation to the Italian research. In a significant number of cases (see Table 4.3), informative advertising takes on a hybrid character that is reminiscent of the typical form of identity advertising. But more than tending towards a *relative* configuration (like comparison and improvement) the texts appear to present some values as *absolute* and *ethical*, which is typical of the positional quadrant, such as respect for the body. The advertisements do not address the improvement of imperfect bodies but rather represent the achievement of a perfect and healthy body through a complex process of giving both analytical information and an authentic value. The healthy and perfect body is considered as something of sacral value, and health is sanctified in *sanctity* (compare Douglas, 1996: 94–96; Fischler 1996 refers to the same phenomenon with his concepts of *sanitas* and *sanctitas*). The moralistic aspect of these advertisements consists not so much in the perfection of the product as in the perfection of a lifestyle that one can achieve through this product. It expresses a late-modern ideal of a scientifically and rationally achievable 'sanctity'. Perfection is not given, it can be acquired: this could be the meta-headline of all these advertising texts. Perfection does not belong to an a-historical and a-temporal *non-place* independent of human actions and volition (as in the case of the positional model). The texts suggest that we are not born gods, but that each and any of us can become one, so long as he or she follows the

Table 4.3 Variation within the causal regime in *El Semanal* and *Mujer Hoy*

El Semanal	Models	Frequency
Health	10	22
Looks	2	2
Total	12	24
Mujer Hoy	Models	Frequency
Health	10	15
Looks	4	6
Total	14	21

indications and respects the rules (the *diet*), the principle of a 'correct' and healthy life, a life founded on correct and healthy food, i.e. food interpreted according to the doctrine of natural law: the divine law expressed as a law of *nature*.

Implications of Advertising Strategies for Ethical Traceability

As discussed in Chapter 1 there are at least two groups of consumer concerns, of which the first group consists of substantive ethical concerns, such as animal welfare, human health, methods of production and processing and their impact (e.g. environmental, landscape), terms of trade (fair price), working conditions, quality (taste and composition) and origin and place. Some of these concerns are addressed in the advertising texts we analyzed, but not all.

The second group of concerns, the procedural concerns, is not addressed at all. The issues of the reliability of information, trust and voice (participation) are left out. Transparency is implicitly addressed by mentioning phone numbers or web sites. Mostly, the texts do not consider the ambivalent and plural meaning of these concerns: rather, their style is persuasive and certain.

The advertisements analyzed show that consumers are addressed with cultural, social and personal images. These images exercise great influence on consumers, something which is widely known and often discussed in journals of marketing and advertising in terms of these images (Mason, 2005). With respect to consumer concerns, it is necessary to think about the symbolic and cultural forms of their presentation. Some consumer concerns, such as those connected to animal welfare, the conservation of nature and the environment, probably appeal more to positional narrative forms and values than to causal forms. Consumer concerns such as transparency and voice seem to appeal more to causal and multi-perspectival values. Suppose that food chains incorporated these values in systems of ethical traceability, and that consumers, at the end of the chains, could trace back, e.g. animal welfare systems to their successive operational locations in the chains. In this situation, ethical traceability systems should take into account that these appeals to values exist. In other words, ethical traceability systems, whatever they may look like, cannot be implemented without appealing to these deep-seated differential narrative forms and value systems. For successful systems of ethical traceability, it is necessary to construct narrative strategies for consumer concerns that best represent them and that are recognized by consumers.

Very often ethical consumer concerns are perceived as 'pure', 'holy', 'natural' and 'authentic'. In that case, the positional strategy is used. However, consumer concerns also have informative and quantitative aspects (it could even be information about the square centimetre of free space a chicken has) and in that case the causal strategy can be helpful. For example, the Dutch Food Agency, which in particular assesses food for its health and environmental effects, stresses these quantitative aspects by urging consumers to count, weigh and calculate calories, proteins, their own weight, etc. At first sight their assessments seem to be done from an impartial and neutral point of view, but it is clear from our analysis that this point of view is not neutral but based on a preference for the informative narrative strategy.

However, as has been discussed in earlier chapters, consumer concerns are often dynamic, pluralistic and ambivalent and do not require foodstuffs to be perfect and pure (in the positional sense) or quantitative and neutral (as in the informative sense). With respect to the dynamic, pluralistic and ambivalent features of consumer concerns, the causal, positional and perspectival strategies do not always seem to be fruitful for addressing consumer concerns, because these strategies go against these features: they are not objective but ambivalent (i.e. they are not informative), they do not exclusively appeal to one tradition or origin as ethically acceptable (i. e. they are not positional), and they do not address the consumer exclusively as gourmet and connoisseur (i.e. they are not perspectival). This leaves the fourth narrative strategy (multi-perspectival), which can also possibly comply with a communication strategy directed towards (potential) ethical consumers. But, as we have seen, hybridization can also be a useful strategy in implementing consumer concerns in food networks and chains.

Conclusion

In this chapter we have presented several narrative strategies based on empirical research of various journals appearing in the Mediterranean, which show that food and nature are differently framed on the bases of different images of food, nature, landscape and society. The empirical research was inspired by our discussion of the work of Fischler, Douglas, Falk and Ferraro. These authors develop categories like purity, danger, order and naturalness, which comprise the possible relations that people have with nature and food. In reflecting upon our empirical material collected during the research we found two important issues. First we could prove that very often advertising strategies are subject to 'hybridization', and that they often confound different strategies that according to the authors we referred to cannot be confounded. Secondly, we showed that the various strategies when used in traceability programmes can have a different effectiveness.

References

Douglas, M. (1970) *Purity and Danger. An Analysis of Concepts of Pollution and Taboo.* Harmondsworth: Penguin Books.

Douglas, M. (1996) *Thought Styles, Critical Essays on Good Taste.* London: Sage.

Falk, P. (1994) *The Consuming Body.* London: Sage.

Ferraro, G. (1999) *La pubblicità nell'era di internet.* Roma: Meltemi.

Fischler, C. (1988) 'Food, self and identity'. *Social Science Information*, 27(2): 275–292.

Gorz, A. (2003) *L'immatériel. Connaissance, valeur et capital.* Paris: Editions Galilée.

Lévi-Strauss, C. (1978) *Myth and Meaning.* Toronto: University of Toronto Press.

Mason, R. (2005) 'Missing links: Product classification theory and the social characteristics of goods'. *Marketing Theory*, 5(3), 309–322.

Part II
Ethical Traceability in Three Food Supply Chains: Case Studies of Danish Bacon, UK Wheat-Bread and Greek Olive Oil

The purpose of this part is to focus on the current use of traceability in three different food chains and examine the actors' attitudes to ethical traceability and informed choice. The three chains described in the following chapters are: olives into olive oil in Greece, wheat into bread in the UK and pigs into bacon in Denmark. These chains all represent important cultural and economic elements of food production in the respective countries, with a long history of production. Olive trees were already cultivated in Crete by the Late Minoan period (1500 BC) in Crete, and perhaps even earlier, and the region today has one of the world's highest levels of olive oil consumption. Bread has been a staple food of the British throughout recorded history, and bread made from wheat flour is today eaten several times a week by almost the whole population. Since the 19th century, the production of bacon has played an increasingly important part in Danish agricultural production, illustrated by the fact that today approximately 25 million pigs are slaughtered every year – i.e. around five pigs per Dane. Today the negative impact of this production is often on the political (and the media's) agenda.

The case study approach enabled us to investigate many variables within a distinct context, drawing upon mainly qualitative information. The empirical data presented in the next three chapters was collected from 98 in-depth interviews with stakeholders and 59 in-depth interviews with consumers from the three chains. Stakeholder interviewees were senior actors in the chain, drawn from the major sectors: input suppliers, farmers, merchants, processors and retailers, as well as members of trade associations and other relevant stakeholder organizations. Before the interviewing began, initial 'scoping' surveys of the chains were conducted to identify relevant literature, and important sources, sectors, actors and potential interviewees. This 'scoping' survey formed the background for a common interview guide that was adapted to the specific supply chains (oil, bread and pork), and also to the context of the specific country (e.g. in Greece the term 'ethical food' had to be rephrased because in modern colloquial Greek, the word 'ethics' could not be connected to foods.

The interviews made use of the list of ten ethical consumer concerns identified as relevant to food production by initial work conducted by the project's philosophical partners (see Chapter 1 and Table 1.3). In all three chains, interviewees were asked to prioritize the ten ethical concerns, if possible, and to comment on their importance in the chain. The interviews also included questions about information: who holds it, how accessible is it, and how does it flow through the chain? The research took place between October 2004 and January 2006.

In the pork sector (discussed in Chapter 5) there is a long tradition of traceability systems. This tradition has gained extra momentum since recent outbreaks of food-borne illness (e.g. bovine spongiform encephalopathy [BSE]) and pathogens (e.g. *Listeria, Escherichia coli*, etc.). Two principal motives have driven the development of traceability systems for pork and pork products, namely: (1) *supply management* and (2) *safety and quality control*. Traceability systems enable slaughter plants and processors to track the flow of product and coordinate production more efficiently. The main objective of traceability systems in the meat sector is to trace and isolate all potentially affected pigs in the event of a disease outbreak. As with most other livestock traceability systems, it is *reactive* in nature and is not intended to transmit information on safety, production practices or the quality of the final product proactively downstream to firms or end-consumers. When we asked the actors in the pig-pork supply chain about the information 'stored' in the system, it was striking that a lot of information about various ethical aspects of production was already recorded by the system, but the information was not used actively – it never reached the next level in the chain. As for Danish bacon consumers, in general they feel that they are misinformed or even cheated, especially about the origin of the products. They tend to expect that when they buy bacon it is produced in Denmark and produced according to Danish regulations.

In the UK wheat-bread chain (Chapter 6), the ethical issues raised by the project were said to be increasingly relevant, but the routine industrial practice of blending wheat or flour to manipulate quality and price was widely seen as an obstacle to greater traceability and ethical traceability. There was also some evidence of a limited, 'sectoral' view of the ethical concerns, in which stakeholders did not see (or look) far beyond their own link in the chain: we referred to this as the 'field of ethical vision'. Stakeholders in the wheat-bread chain, which is predominantly highly industrialized, felt they had good access to information, and actively sought information from many sources. Consumers, in contrast, felt under-informed, and said that information was withheld and unreliable. This was found to be justified to the extent that some stakeholders did not want all information to be communicated to consumers, preferring some practices to remain invisible. However, consumers did not make much use of available information (e.g. none habitually read labels or checked web sites).

In the olive oil chain (Chapter 7), traceability and potentially ethical traceability were again said to be limited by the practice of blending. Olive mills and packing houses blend high-grade olive oil with cheaper grades of oil or with other, cheaper edible oils, in order to manipulate quality and price. This situation makes it difficult to trace the origin of the olive oil, and leads to a feeling among Greek oil producers that their product is being exploited, and among Greek consumers that they are being cheated. One result is that a high proportion of Greek olive oil consumers buy oil from a farmer or co-operative known to them personally, or buy by direct marketing.

The interviewees prioritized and interpreted the ethical concerns differently depending on their place in the chain. For example, in the wheat-bread chain, farmers stressed the importance of methods of production while consumers focussed on personal health. And whereas farmers interpreted 'methods of production' to mean the environmental impacts of farming, industrial bakers tended to think in terms of the impacts of mills, such as noise and energy use. Consumers in all three chains tended to prioritize substantive values over procedural values (see Chapter 1 for an explanation of these terms). In spite of widespread feeling that ethics are subjective – possibly too subjective to be 'captured' by standards or assurance schemes – there was in fact an agreement on which concerns were most and least relevant, but with some differences between consumers and stakeholders, and differences between the three chains. These findings seem to support the view that an 'overlapping consensus' can be reached on certain aspects of food production (see Chapters 9 and 10), but the differences illustrate what are termed elsewhere in this book the 'dynamic and dilemmatic' nature of consumers' ethical concerns (see e.g. Chapter 11), and raise important questions about how to determine which attributes should be traced by ethical traceability systems.

In all three case studies, we set out to examine both organic and conventional food chains, but we found that there was generally a clearer distinction between industrial and craft chains than between organic and non-organic chains. Ethical concerns were more explicitly relevant to craft-scale operators, who often had ethical motives for being in business. On the other hand, larger firms trusted other large firms to have staff and resources to ensure regulatory compliance. Whereas the industrial chain valued consistency and uniformity, the craft chain valued variation.

Consumers looked for proxies (such as brands, logos or retailers) for the quality attributes they were seeking, and tended to 'bundle' desired attributes together (e.g. organic certification was seen to imply fairer trade and fewer road miles as well as the absence of synthetic pesticides specified in organic regulations). This has implications for ethical traceability systems, in that time-pressed consumers, disinclined to search for product-specific information, look for bundles of attributes, and brands or logos to flag them.

Stakeholders did not always see traceability as a necessary or obvious vehicle for communicating ethical practice. Instead, the interviewees often mentioned other systems where 'ethics' resided, such as Codes of Practice. Some interviewees saw traceability purely as a tool to achieve certain ends (such as supply chain management); others saw it as an ethical activity in itself.

From our consumer interviews we found that when attitudes to food production are articulated in a place that is disconnected from concrete, everyday life and consumption, it is easier to formulate ideals about production. As a citizen, you can formulate your claims to the good life and good production. It is different when you are a consumer and the context is constituted by everyday life.

The economic and structural development of the chains has an important impact on levels of interest in traceability and the potential for tracing foods. This was most clearly evident in the cases of olive oil and the pork-bacon. In the pork case, traceability was made difficult because the monopoly company optimized its profits by placing its production where the wages were lowest, or in the case of olive oil optimized the quality: price ratio by blending oil of many different grades and origins.

Chapter 5
Ethical Traceability in the Bacon Supply Chain

Thorkild Nielsen and Niels Heine Kristensen

The consumers don't know what we are doing; they live in a fantasy about modern pig breeding. I had a visit from a school and the teacher and the children had never seen a piglet

Conventional farmer

Introduction

This chapter presents the analysis and findings of empirical research in the Danish pig-pork-bacon supply chain. The focus is on ethical concerns and traceability systems.[1]

While still increasing, pig production in Denmark has consolidated in recent years, although environmental regulations limit farm size. More than 95% of pig production is slaughtered through two producer-owned co-operatives, with the largest, Danish Crown, accounting for 90% of the slaughter (DS, 2005). The Danish pork industry incorporates several inconsistencies. While it remains an almost wholly farmer-owned and -controlled supply chain, the slaughtering and processing sector is highly concentrated, giving farmers little alternative to membership in one of the two dominant co-ops.

A traceability system has been introduced in the pig-pork sector, and the main objective of this system is to be able to trace and isolate all potentially affected hogs in the event of a disease outbreak. The traceability system is reactive in nature and is not intended to convey information proactively to end consumers on safety, production practices or the quality of the final product. At slaughter, the carcass grade and any veterinary marks are electronically connected with the producer ID number, and the information is sent to the farmer. It is therefore possible to trace each carcass from the cooling room back to the farm. Once the carcass is cut up, however, final cuts cannot be traced back to the farm of origin. This is the situation with bacon today.

---

[1] The research involved a literature study and qualitative interviews with stakeholders from the pig-pork-bacon supply chain and 12 bacon consumers.

C. Coff et al. (eds.) *Ethical Traceability and Communicating Food*,

The Danish slaughterhouses are following a strategy of 'diversification'. This means that the slaughterhouses, together with some of their customers – mainly large retail chains – have developed a number of 'special pigs' with well-defined attributes. These attributes are mainly connected to animal welfare, but a few focus on meat quality. Except for the organic label, consumers were not able to tell the difference between these different labels, although some of them were perceived positively.

In this chapter we will first of all provide an overview of the supply chain and its development into its present form. We also present some of the main ethical concerns in the sector. Obviously, animal welfare is a central ethical concern in the pig-pork-bacon chain, but the use of antimicrobial growth promoters and working conditions will also be discussed. The findings of the project's qualitative research are then discussed, first in relation to the project's ten identified ethical concerns and then to the broader issues involved in the use of information in the chain. Finally, the chapter will discuss the implications of the research for traceability and ethical traceability.

The Danish Pork Chain

In the 19th century, the Danish economy was predominantly agrarian and linked to the world economy mainly through its grain exports to the UK. In response to the entry of new competitive grain producers from Russia and the USA, Danish agriculture was reoriented from grain to pigs and cattle and especially butter. Up to the 1970s the dominant form of agricultural production was cows, pigs and feedstock cultivation on the same farm. After the 1970s, specialization and structural development speeded up. From 1975 to 2000, the number of so-called full-time farms declined from 91,000 to 13,000. The same development could be observed in the food processing industry. In the slaughter industry, there were 62 farmer-owned co-operatives in 1962, while today the number is 2. Table 5.1 shows the change in the number and ownership of slaughterhouses since 1970. The two remaining co-operatives handle approximately 95% of all Danish pigs. From 1992 to 2002 the number of slaughtered pigs rose from 17.7 million to 21.4 million, a growth of 20% in a decade. In 2004 the total production was 24.7 million pigs. Besides this, approximately 2.5 million weaners (small pigs aged about 7 weeks) were exported alive, mainly to Germany (Danish Crown, 2003; Danske Slagterier, 2004).

Danish Crown has 16 slaughter facilities spread over the country, but is still closing down older units and building newer and larger slaughterhouses. Besides the

Table 5.1 Development in the number and ownership of Danish slaughterhouses. (www.danskeslagterier.dk)

	1970	1980	1990	2004	2005
Coop slaughterhouses	50	18	5	2	2
Private-owned slaughterhouses	4	2	7	10	10

two co-operatives, there are about ten smaller, private slaughterhouses handling the last 5% of the pigs. None of these private companies has a licence to export the pork (Danish Crown, 2003; Danske Slagterier, 2004).

By all conventional input measures, Denmark appears to have a significant cost disadvantage when compared to major competitors: land is scarce and high-priced, manure disposal regulations are strict, wage rates in farming and processing are well above those of other major pork-producing countries, feed costs are high, line speeds in processing plants are slow and the growing markets in East Asia are distant. The relative success of the Danish hog and pork industry appears to be related to its structure and technology, enabling it to achieve strategic linkages along the marketing chain by means of its co-operative structure. This co-operative structure is probably the most striking characteristic of the Danish hog and pork industry. Perhaps one of the most unusual features of the sector is that because the co-op represents many stages of the pig marketing chain, adversarial relationships between buyer and seller, which are common in the marketing chains of meat industries in many other countries, appear to be largely absent (Hobbs, 2001; Danish Crown, 2003; Danske Slagterier, 2004).

The Danish hog and pork industry is highly export driven, and Denmark is one of the world's largest pork exporters with over 75% of its production going to some 100 countries. The Danish pork system is highly integrated, from the primary producer to the processing companies.

To ensure greater stability in the market and to make it easier for farmers and abattoirs in their management activities, prices are set by a committee each week and are based on market conditions in export markets. Using a predetermined processing margin, the price to be paid to pig farmers is then set; consequently each producer receives the same price for a given specification. This saves on transportation costs as there is no price incentive for producers to send live pigs over long distances, and it eliminates marketing costs such as the need for markets, middlemen or a network of buyers employed by the abattoir.

Development in the Danish pork industry has been shaped by changes in demand, consumption and demographics, with attention shifting towards 'softer' product quality traits, such as animal welfare, 'ethical' products and product origin. In the areas of productivity and technology, in particular, there has been an increasing focus on specialization in production and other production aspects such as monitoring technology and biotechnology. Over the years, pork production has become more knowledge intensive, with more and more advanced technology, which creates a demand for a skilled and flexible workforce, but at the same time we have seen some tendencies to move the production to countries with lower costs (wages, environmental standards, etc.), especially the new members of the EU. The pig sector itself has been pushing a strategy of expansion, planning for a total growth in the number of slaughtered pigs of 2% per year. But new actors are also entering the sector. Monsanto has filed an intellectual property application on a number of techniques for pig breeding. If this comes through, it could affect the economy in breeding and in the pig sector in general (Hobbs, 2001; Lemoine et al., 2002).

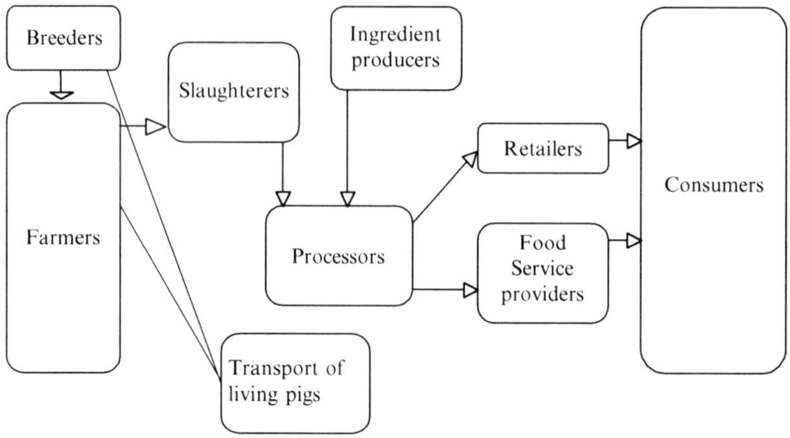

Fig. 5.1 Simplified diagram of the bacon supply chain

Today, the Danish bacon chain is well-established, intensive, industrial and globalized. The development is towards fewer and bigger holdings, with intensification in slaughterhouses and processors in Denmark paralleled by a growing transport of live pigs to slaughterhouses and bacon processors abroad. Figure 5.1 shows a simplified diagram of the bacon supply chain. The transport of live pigs is included in this diagram, although the transport distances and methods are not mapped here.

An interesting facet of today's Danish bacon supply chain is that it is not possible to buy Danish sliced bacon in Danish shops. It is sourced from production units outside Denmark, such as in Poland or Germany, although marketed as Danish bacon. On the other hand, what are specifically called 'UK pigs' are produced and processed in Denmark – and sold on the UK market. Only the slicing and packaging is performed in the UK.

Today's production of a pig, for slaughtering at 100 kg, takes roughly 225 kg grain and 75 kg ground soy. The way these two types of fodder are used varies, depending on market prices and production systems. Typically the feed is constituted of barley, wheat, rye, oats, maize, fishmeal, potato protein, molasses, skimmed milk powder, fat, whey, etc. A number of other nutrients, vitamins, minerals, synthetic amino acids, etc., are also added to the fodder. These form the typical, formally recommended nutrition for pigs. Besides these, other ancillary substances are used, such as the so-called technical, sensory, nutritional and zoo-technical substances (growth hormones, antibiotics, enzymes, etc.). A radical growth in the use of fodder produced by genetic modification techniques has been reported. These are regulated, but incidents have been reported where banned substances have also been used in fodder for pig production. Fodder production today is the object of a lot of scientific attention, being highly documented, tested and optimized to ensure maximum economic benefit. Another input to pig production is medicine. Pigs are treated with medicine by veterinarians and also by the pig farmers themselves after instruction from the veterinarian (Damm, 2004).

Today, two breeder organizations dominate Danish pig breeding: PIC and Danavl. Danavl is the dominant company, supplying about 90% of the marketed genes (Damm, 2004). In 2004, 24.7 million pigs were produced on 10,000 Danish pig farms, with more than half of these produced on 10% of farms – the farms with a production of more than 4,000 pigs per year. A variety of more specialized systems coexist with the dominant model of conventional pig production: Free-range pigs, UK pigs, organic pigs, Antonius, heavy pigs, etc. Fattening pigs are the most important, but a significant number of sows and boars go into the supply chain too.

With a flow of more than 80% of the volume of fresh and frozen pork products, supermarkets are the predominant outlet for pork products for Danish consumers today. Direct sales and butcher's shops count for a very limited amount. Sliced bacon for the Danish market is now mainly processed and produced in Poland and Germany. If consumers wish to buy Danish bacon they have two options: they can either go to butchers producing specialty meat products or buy bacon in whole, unsliced pieces at the supermarket.

Technologically, Danish pig production is highly developed. At Danish pig farms, many husbandry practices have been taken over by advanced stables, which have automatic feeding, automatic cleanout, etc. Different stable systems have been introduced, but the major systems are highly technologically developed.

The growing distances between farm and slaughterhouse have also had an effect on the use of transport technology. Today the pigs are transported over a longer period and it is often necessary to mix different litters (which can cause trouble with fighting over hierarchies among the animals). The killing of animals at the slaughterhouse is done according to various methods, depending on the age of the slaughterhouse and the technology used there. The most modern technology used in the stages up to the killing of the animal is more careful, and pays more attention to animal welfare. The technology used for production, transport and slaughtering for the specialty production systems (e.g. organic or free-range pigs) are especially designed to take animal welfare concerns into account (Hobbs, 2001; Lemoine et al., 2002; Damm, 2004).

Strategy of Differentiation

The two dominant meat companies in Denmark use a strategy of differentiation. The reason for this is that the retail trade wants to differentiate the products it supplies to consumers. This requires the farmers as well as the slaughterhouses to differentiate their production to a greater extent. Some of these 'special pigs' have been subject to stricter animal welfare requirements, or have been produced with the object of improving the eating quality. The most common specialty pigs are described below.

Antonius and Vitalus were established in 1976 with the idea of creating a quality pig. Having initially being targeted at improving eating quality, other goals have since been integrated into this production system, such as giving the sows straw in their stables. The two brands are sold in different retail chains (Danske Slagterier, 2005).

Free-range pigs (Frilands gris) are produced under the supervision of standards developed by the Animal Welfare Association. The standards specify, for example, that piglets are born on free-ranges, where the sows live, and that slaughtering pigs must have access to the open pastures. The second biggest supermarket chain in Denmark owns the brand (Dansk Supermarked; DS, 2005).

UK pigs have more space than the law requires. The UK pig is produced in accordance with standards set by some of the large retailers in the UK. Farmers rearing these pigs produce them under contract, following specific production practices that conform to the animal welfare and food safety requirements of the UK market. For example, sows must be loose-housed and not tethered; feed containing meat and bone meal may not be used; farmers must keep mandatory records of feed composition; and withdrawal periods for medications are regulated (Hobbs, 2001).

The *EU-heavy pigs* have a higher weight than UK pigs, which results in higher fat marbling in the meat. There are also requirements concerning slaughtering. Some German retail markets require slaughter weights of 83–105 kg as opposed to the 67–79.9 kg that is the standard weight rewarded by the Danish carcass-grading system (Hobbs, 2001).

These special production systems are only used to a limited extent in processed products. The exceptions are UK pigs, which are used in bacon production, and ham from the EU-heavy pig, which is used to make Spanish Serrano hams. Bacon using the brand *Danish* is from Danish pigs and is partly processed in Denmark, then sliced and packaged in the UK. The farms producing specialty pigs are audited to verify that the specified practices are being followed. There is an additional external 'audit of the auditors' to ensure that the system is being implemented uniformly in all areas. A farmer wishing to produce hogs under one of the above-mentioned systems must apply for a contract through his/her co-operative slaughter company (Hobbs, 2001; Danish Crown, 2003; Danske Slagterier, 2004).

Organic pigs are differentiated from other specialty pigs by the fact that there is public and legal regulation of the organic production system. The production of organic food in the EU is regulated by Council Regulation 2029/91. In relation to animal welfare, the EU organic legislation has some requirements that go beyond conventional animal welfare regulation. For example, it is stated that the management of farm animals must take into account their physiological, social and behavioural needs and be carried out in a 'natural' way. Thus, organic pigs must have access to grassland, and to daylight in the stables. Genetically modified (GM) fodder is banned in organic products, and the transfers of organic pigs up to the slaughter process are designed to minimize harm to the animals.

Two companies dominate Danish organic pork production: Friland Food and Hanegal A/S. Organic production still represents a very small proportion of total pig production in Denmark (just 0.2%), but the growth rate has been high in the last couple of years. There is now an export trade in organic pork meat to England, Germany and Italy, with approximately 60% of organic production exported. Demand is growing, in both domestic and export markets. The largest organic pork processor, Friland Food, which is organized as an independent company within the largest conventional processor, has a market share of 90% of total organic pork production in Denmark. The risks and the profits are thereby borne by the organic pork farmers, but at the same time they have the opportunity to use the organization and marketing channels of the 'mother' company, Danish Crown. Friland Food has evolved from an autonomous organic farmers' association which experienced major challenges in establishing steady distribution systems that could supply the retail marked with quality organic meat. The first steps to introduce organic pig meat to customers in FDB (Danish Co-op) were taken back in 1989. Organic pig producers involved in these first efforts created Økokød (Organic Meat Co-op). In 1992 this organization joined forces with producers of free-range pigs, and together they established Friland Food Ltd. In 1999 Friland Food Ltd. joined Danish Crown; in other words the largest organic slaughterhouse is owned by the largest conventional slaughterhouse (Claudi-Magnussen et al., 2000; Danish Crown, 2003; Danske Slagterier, 2004).

Other organic slaughterhouses are also actively involved. Another major one is Hanegal A/S – one of the first organic farmer and slaughterhouse initiatives. It introduced a fair trade system for the Danish pig farmers supplying the slaughterhouse. Today Hanegal A/S is especially well known in supermarkets for processed organic meat products. Hanegal A/S is the only purely organic slaughterhouse in Denmark. It was founded in 1994 by a couple living on a farm near Silkeborg, in Jutland. They have farmed organically since 1980, and pigs have been the main part of their business for a long time.

The slaughter of organic pigs is covered by a number of regulations intended to ensure careful treatment of the animals both during transport and in the slaughterhouse. The initial vision of the company was to organize a local or regional slaughterhouse – connecting organic farmers and pig producers with local consumers and catering. As it was not possible to establish enough interest and sufficient volume of business in the region, another strategy was developed. At this stage the company started supplying a broader range of supermarkets, health food shops and institutional kitchens (catering). The company established a fair trade concept for the farmers supplying the pigs. This concept resulted in a premium price that was not connected to the price of conventional pigs. Instead it reflected the costs of running an organic pig production unit.

In 1996 Hanegal A/S introduced a range of cold cuts for supermarkets. The Danish Co-op was the major distributor, and the main markets were found to be in major cities in Denmark (Copenhagen, Århus, Silkeborg, etc.). In 1998 the capacity was expanded by the establishment of a new slaughterhouse in Southern Jutland. This region of Denmark is known for its traditional and artisanal skills in the meat sector. After some turbulent years, Hanegal A/S has now reduced its own production to a minor level, concentrating on processing activities. Today most of its organic products are sourced from certified organic slaughterhouses in different European countries. The chickens are raised and slaughtered in Hungary, and labelled with their country of origin. This is not possible with bacon production because the system is unable to handle the frequent shifts in deliveries, due to the fact that the pigs come from many different countries. The company is very innovative and has recently introduced organic fresh chickens and organic fish into supermarkets.

Danish Bacon

In medieval times, Denmark was already exporting live pigs to Germany (Christensen, 1983), but it was not until the late 19th century that pig production became significant in terms of sales. The first recorded shipment of Danish bacon to England was in 1847, and Danish pig farmers quickly recognized a market opportunity, especially welcome in the light of the closure of the German border to live pigs, which until that time had been the main outlet for Danish pigs. The British Industrial revolution meant that agricultural production was unable to supply sufficient food for the UK's growing population. At this time, bacon was considered a luxury product. Records show that a worker on a good salary ate bacon and cheese daily, whereas a worker on average wages maybe consumed bacon 2–3 times a week, while the lowest paid workers could not afford to eat meat at all (Christensen, 1983).

By the end of the 19th century about 90% of Danish bacon exports went to the UK, and Denmark had replaced the USA as the main supplier of bacon to the UK. In the 1960s processing developments such as vacuum packing and mechanical slicing began to change the traditional supply of bacon products and this was indirectly the reason why the Danish co-operative slaughterhouses opened processing factories in the UK. During the 1980s, the Danes still supplied more than a quarter of the UK bacon market (Gray et al., 1981).

Bacon is a cut of meat taken from the sides, back or belly of a pig, cured and possibly smoked. Curing typically means preserving with salt. The traditional dry-cure process involves rubbing the meat, over a number of days, with dry salt or a mixture of salt, sugar and spices. Today it is difficult to find bacon produced in the 'traditional' way. The more usual production method, the wet-cure process, involves immersing the sides of meat in brine for 2–3 days. Modern mass-produced bacon uses the wet-cure process but also involves pumping additional water, sodium nitrite and phosphates directly into the flesh to speed up the process and add bulk. Mass-produced bacon is held for curing for 6–24 h before being cooked. Smoking is used to impart more flavour to the bacon and also to speed up the curing process. Smoked bacon is traditionally produced by hanging the cured meat in a room over smoking wood chips. Smoke flavour is imparted to the bacon either using natural smoke obtained by burning wood chips, or by spraying the bacon with a liquid smoke extract. In mass production, cooking is much quicker than traditional cooking due to the use of convective heat transfer, and can be completed in as little as 6 h. After cooking and smoking, the bacon must be chilled before being pressed and sliced.

Much of the bacon consumed in Britain is produced in Denmark, and marketed as Danish bacon with the word 'Danish' stamped on the rind. At the end of the 1990s, a UK ban on Danish bacon was called for. The then leader of the British conservative party, William Hague, argued that Britain's strict animal welfare regulations meant that British pig farmers could not compete with Danish producers, who use a sow stall system. The result of this controversy was that some of the UK retailers requested some additional animal welfare standards. The Danish pig meat sector responded quickly to these 'ethical' demands and the 'UK pig' described above was introduced. These pigs enjoy some welfare improvements, especially sows, which are allowed to move about freely. Today 1,800 farmers produce UK pigs and they receive a supplement of 0.05 €/kg (DS, 2004).

Relocation of Danish Bacon Production

In 2001 the import of bacon to Denmark comprised only 5% of consumption. This situation has changed radically in the last few years, because in 2004 and 2005 the two dominant slaughterhouses in Denmark moved their bacon production to Poland and Germany. The background to this development is that in 2004 one of the largest retail chains in Denmark decided to replace Danish-produced bacon with Polish-produced bacon. Only 6 months after this decision, imported bacon accounted for more than three quarters of total sales. The price was lowered to one third of the price of Danish bacon.

As far back as 1994, Danish pig farmers had started to buy closed-down state-farms in Poland, through the company Poldanor. Today this company is owned by a group of 60–80 mostly pig farmers. On their 15 farms and 2 feed companies they produce approximately 500,000 pigs annually (Heinemann, 2005). In 2004 Poldanor and the Danish co-operative slaughterhouse Tican started a co-operation that resulted in the establishment of the meat processing company Prime Food. Prime Food consists of an abattoir and a processing plant, and is situated in a small city in the northern part of Poland. This is where a large part of the sliced bacon sold in Denmark is produced today.

The Danish farmers and the co-operative slaughterhouse are not the only ones to invest in Poland: they have been joined by the largest pig producer in the world. In 1998, the president of Smithfield Foods[2] announced that he wanted to make Poland the 'Iowa of Europe' (National Hog, 2000). Poland is attractive to investors because land and labour are cheap, and it is seen as a bridgehead to lucrative EU markets. Smithfield Foods today operate in Poland through the subsidiary company Animex. The Danish meat co-operative Danish Crown has also invested in Poland. Together with the Finnish meat concern HK Ruokatalo, Danish Crown owns 80% of the Polish meat company Sokolow. Sokolow has six slaughterhouses and a production of one million pigs a year. Last year Swedish Meats decided to start producing bacon in the Polish port city of Swinoujscie. The plan is to close the factory in southern Sweden and produce 60t of bacon every week at the Polish plant (Landbrugsavisen, 2006).

In fact, the entry of these big industrial hog producers into the Polish market has resulted in overproduction and a drop in pig prices (National Hog, 2000). This has

---

[2] Iowa has led the USA in swine production and pig inventory since the 1880s, measured in hogs per square mile.

been disastrous for small farmers who no longer have a market for their pigs, but has proved a boon for the large companies because it lowers the cost of the raw material, without affecting the retail price of pork. However, some Poles are trying to fight off this invasion. The nationalist politician Andrzej Lepper, head of the Polish Farmers' Union, is fighting to oppose foreigners' attempts to take over the Polish hog industry. The Polish environmental organizations are still very weak, but there has been some resistance to intensive pig production in the northern part of Poland (National Hog, 2000).

As these global meat companies buy local companies to front their operations, it is very difficult to trace their products. For instance in Denmark bacon from Smithfield is marketed under the name Animex, while in the UK the bacon products are marketed under the brand name PEK (Fig. 5.2). However, it is more than likely that the bacon from these processors is ending up in mass-produced products such as pizzas.

Tulip Food Company, a subsidiary of meat giant Danish Crown, took over a meat packing plant in Oldenburg, Germany, in 2004. This meant that the luncheon meat plant in Viby, as well as a bacon facility in Vejle, would be consolidated at the newly acquired Oldenburg factory. More than 300 Danish workers have lost their jobs.

> *Profitability is important for a company like ours to be able to take part in a competitive global food market. And we must acknowledge that meat industry wage levels in this country are substantially higher than in Germany.* (Manager from a Danish meat processing company)

The meat industry in Germany is largely dependent on foreign workers, especially from Eastern Europe. The German food workers union, NGG, expects that 26,000 jobs in the meat sector will be taken by foreign workers at lower wages. The workers, mainly from Poland, are employed through subcontractors, who are hired to carry out specific tasks. The Polish workers are officially employed in Poland, which means that the Polish workers in Oldenburg get their wages paid in Poland, on Polish conditions. The hourly rate for the Polish workers is 5 €, while the German workers receive 10 € for the same work. In Denmark the minimum hourly rate for workers in slaughterhouses is 21 €, and it is forbidden to use 'column workers'.[3] (Column workers are formally self-employed. An intermediary signs the column workers up and pays them for a pre-arranged job. The use of column workers is not

Fig. 5.2 EU factory-numbers from two bacon packages. This is where most of the bacon sold in Denmark is produced: Oldenburg in Germany and Prezechlewo in Poland, indicated by the D and P respectively

[3] Interview with a person from the German food-workers' union Gewerkschaft Nahrung-Genuss-Gaststätten (NGG).

allowed in Denmark, but is legal in Germany. Today approximately 60,000 column workers are employed in the German meat sector, mainly from East Europe.)

Ethical Concerns in the Pork Sector

Based on our literature research and preliminary interviews, in the following paragraphs we discuss some of the areas most often identified as giving rise to ethical concern in this food chain. These areas are: pig breeding, animal welfare, feed, medicine, the environment, processing and working conditions.

Pig production is about producing pigs for slaughter. The *breeding of pigs* focuses on optimizing the production of pigs for slaughter, so sows must deliver the maximum possible number of piglets, which should be able grow as quickly as possible to be ready for slaughter. Progress in pig production is achieved through targeted breeding work to achieve more rapid growth and several piglets in every litter. Desirable traits include, e.g. faster growth on less feed – the sort of features which, according to Damm (2004) also affect the sows and their welfare. Until June 2004 it had been the aim for sows to give birth to a growing number of piglets no matter whether they survived or not. Ten years of breeding for this trait had raised the number of pigs per litter to an average of 12.8 (though in many holdings the average size is even higher and litter sizes of 13–14 live pigs seem to be the rule).

The average number of pigs reared per sow is approximately 25 each year. Sows produce around 4–7 litters before they become exhausted and are slaughtered, after 3–4 years, for sausages, pork pies and other low-quality products. In June 2004, breeding criteria for the number of live pigs per litter were changed to aim for the highest number of pigs surviving after 5 days. Since the number of stillbirths increases with the number of births per litter, the change is intended to mean that fewer piglets will be born, but more will survive. But the change will not come through in the stocks until after 3–5 years and not fully until after 5–10 years.

But whether, breeding is about having more piglets in the litter or more living pigs after 5 days, the development will result in more surviving pigs per litter, and it is harder for the sows to bring up the individual piglets when the litter is bigger. The weight of the piglets is lower when there are more in the litter, and there are simply not enough teats for all the piglets. There is a significant risk that the piglets will die from hunger or lack of warmth, or from being crushed by the sow; combined with minimal space for the pigs and insufficient time on the part of the farm worker; these are the main reasons why sows are fixed during farrowing. Fixation of the sow during farrowing provides for the production of more pigs (Damm, 2004).

Big litters also mean that it is necessary to 'average out' the litters and use suckling sows. By this means, piglets are moved from sows with large litters to sows with fewer piglets. Among suckling sows, there are one-step suckling sows and two-step suckling sows. A one-step suckling sow has her own piglets weaned off after 21 days, after which the sow takes over supernumerary piglets from one or more litters. A two-step suckling sow has her own piglets weaned off after 21 days and receives in their place a litter of maximum 1-week-old piglets, which have been moved from another suckling sow. The sow that gave up these week-old piglets is then assigned supernumerary piglets. The use of suckling sows is very extensive and has the effect of extending the suckling period, with associated risk of weight loss and with a longer stay in the farrowing stable, where they are fixed in a farrowing crate. A suckling sow may bring up her own pigs for 3–4 weeks and thereafter alien pigs for the next 2–3 weeks, with the result that the sow has been fixed in the farrowing crate for up to 7 weeks. Large litter sizes and the extended use of suckling sows is one of the main reasons why the farrowing stable in many production units causes a bottleneck.

It is a principle of organic production systems that the animal's integrity should be preserved. Organic pigs must have space to move freely and access to grass. The feed must be organic, which means that feed from conventional farms cannot be used. There has been no scientific research on the potential impact of this feeding strategy on fertility. On the other hand, organic pig production can pose dilemmas between ethical and economic interests. The major breeding concern in organic pig production, according to Claudi-Magnussen et al. (2000) and LandsUdvalget for Svin (2006), is to support the pig's overall constitution: its strength, temper (pigs naturally have low aggression and live together gregariously) and maternal instinct; more specifically, the number of coloured hair follicles is rated highly in organic pig breeding.

Animal welfare is overwhelmingly the main reason why consumers are sceptical about modern husbandry and meat production. Local veterinary reports[4] state that the pigs suffer from sundry health problems directly related to the production system (Baadsgaard, 2003). Some of the problems mentioned were:

- At least 24% of the sows live in constant pain
- Autopsies have revealed broken limbs, sores, worn joints, chronic bone inflammation and other leg abnormalities and severed ligaments
- Mortality among the sows has been rising for several years. So has fertility
- Forty-five percent of the sows were killed before they could deliver their second litter
- Thirty-three percent of all slaughtered pigs received adverse comments by the public veterinary inspectors in the slaughterhouses, referring to, e.g. tail bites, lung scars, bedsores
- Ulcers were found in one third of the sows
- A Danish sow produces an average of 25 piglets a year, which is double the rate of 20 years ago. Yet still the industry is striving to increase fertility to 30 piglets per sow per year
- The pigs reach their slaughter weight after 4–5 months: half the time it took 20 years ago
- Overcrowding is a major problem in many pig herds, especially with weaning piglets (Baadsgaard, 2003)

[4] Quotes from 16 vets' report published 2002 by Danish Pig Veterinary Society (Baadsgaard, 2003).

In recent years the transportation of young piglets has raised considerable concern in Denmark. This is an example of the changes that have taken place in the pig-sector. A few years ago, almost all Danish pig farmers were members of a co-operative slaughterhouse, and guaranteed to sell all their pigs to this company. Today, a few very large farmers have expanded their operations to the whole of Europe, and they sell their pigs where the prices are highest. Furthermore, the developments in East European pig production have made it attractive to sell live piglets to these countries. The number of piglets exported was expected to rise from 1.7 million to 2.8 million from 2004 to 2005 (Danske Slagterier, 2004).

Denmark imports one million tonnes of soy annually directly from Argentina for use in *pig feed* (accounting for more than 93% of total soy imports). In Argentina almost the entire soy production is genetically manipulated.[5] According to the EU rules it is permissible to use GMOs in animal feed without declaring it to the end-consumer, but is mandatory for the feed mill to indicate on the package that the feed contains ingredients produced by means of gene technology.

In 2002 one of the largest retailers in Denmark tried to promote non-GM pig meat in the shops. The attempt lasted for only half a year. The retailer argued that consumers were not willing to pay the additional charge of 15%.[6] Consumers, for their part, using calculations from the environmental organization NOAH, said that the price premium was too high. According to these calculations, the extra cost for using non-GM soy in pig production would be 0.5 € per pig.[7]

In 2004 The Danish Plant Directorate sampled pig feed containing soy and corn. It turned out that 45% of the samples that did not say they contained GM soy did in fact contain it.

The main concern connected to *medicine* is the use of antibiotics. In 1999, pig producers were pressured into ending the use of antimicrobial growth promoters. The pressure came from the media, the large retail chains in the UK, politicians and consumer groups. Despite this voluntary agreement, antimicrobial consumption in animals increased by 9.7%, from 102.5 to 112.5 t, in 2004. However the total consumption of antimicrobials in animals in Denmark is still considerably lower than before the termination of the antimicrobial growth promoters. The increase in 2004 was due to a 13.3% increase in antimicrobial consumption in pigs, from 81.8 t in 2003 to 92.7 t in 2004 (DANMAP, 2004).

Antimicrobial consumption varied depending on the age of pigs. The increase was mainly observed in weaners (19%), and less in finishers (12%) and sows/piglets (5%). A number of factors may have contributed to the increase. The administration of medication via water pipes has increased in recent years and individual water pipes often supply more pens than individual feed pipes. Therefore, in addition to the diseased pigs, a relatively larger number of healthy pigs will be medicated

[5] Information from interview with manager from feed mill.

[6] Information from interview with retail buyer.

[7] Information from interview with person from the environmental organization NOAH.

when using water pipes rather than feed to administer drugs. In this way an increased amount of antimicrobials are needed for medication. The occurrence of Post-weaning Multisystemic Wasting Syndrome (PMWS) in the Danish pig population may also explain part of the increase in antimicrobial consumption (Hassing and Bækbo, 2004).

When conventional pig farmers are confronted with these figures they typically mention that the antimicrobial consumption in Denmark is still far below consumption in the USA. Indeed, according to official registrations (DANMAP, 2004) the use of antibiotics per kilogram of meat was considerably larger in most other countries playing a role in the world market: the Netherlands used three times as much and the USA more than four times as much as Denmark (DANMAP, 2004).

Pig production has several implications for the *environment*. A project called 'Environmental Information in the Pig Production Chain' (Kolind Hvid et al., 2005) used so-called life cycles analysis (LCA) to look at the effects of pig production on four environmental problems: global warming, acidification, nutrient enrichment and land-use. A Danish slaughterhouse and ten farms supplying pigs for slaughter participated in the project. Emissions of nitrous oxide (N_2O) are the most important contributors to global warming from the production of pigs. Ammonia from manure contributes to the total acidification of water courses and lakes. Leaching of nitrates from the pigs' stables contributes to nutrient enrichment. The problem with excess nitrogen is that it leaches into groundwater and contaminates the sea, causing algal bloom and 'dead fjords'. (From the point of view of ethical traceability, it is worth noting that the Environmental Information project also demonstrated that it is possible to prepare and communicate product-oriented environmental information reasonably easily in a product chain involving pig meat.)

The environment is also an issue for the slaughterhouses and meat processing companies. The authorities increasingly demand that the companies must comply with stronger environmental legislation. New protocols, such as environmental management, life cycle assessments and Best Available Technology (BAT) are becoming more common (Kolind Hvid et al., 2005).

From a welfare point of view, *stunning* the animals before the slaughter is of great significance. Pre-slaughter stunning is used to ensure that animals do not suffer needlessly and are unconscious and insensible to the slaughter procedure. The stunning method itself should be painless and as close as possible to instantaneous in its effect. Furthermore, it should guarantee that the animal remains unconscious long enough to ensure that slaughter intervenes before the recovery of sensibility (Cook et al., 1992). In the EU there is a legal requirement that all animals destined for meat consumption must be rendered insensible instantaneously and remain insensitive to pain until there is a complete loss of brain responsiveness due to exsanguinations (Council Directive 93/119/EC). Under commercial conditions, stunning of animals for slaughter may be achieved by a mechanical instrument (penetrating captive bolt pistol, non-penetrating percussion stunner or free bullet) or by means of an electrical current (head-only or head-to-back stunning). In addition, anaesthesia with carbon dioxide may be used to render pigs insensible to pain.

When the meat is further *processed*, one of the ethical concerns is with 'invisible' processing methods. For instance, meat removed from bones by machines (mechanically de-boned meat) could contain bones or sinews, which may then be present in minced meat or sausages. Injections of phosphates or proteins are also examples of 'invisible', processing methods in that the consumer cannot see what he or she is getting.[8]

The slaughterhouse industry in Denmark is in general known for strong structural rationalization, a tayloristic organization of work and a hazardous *working environment* with a high frequency of accidents, repetitive strain injuries and lower back pain.

A comparative study of the psychosocial work environment in different industries identified slaughterhouses as a 'problem industry'. Compared to other industries the slaughterhouse workers had low influence on their own work combined with a high work pace, partly resulting from the use of task-based contracts. The problematic working environment in the food sector is indicated by the fact that the food processing industry contributes 4.1% of early retirements, even though it employs only 3.5% of the workforce. Another indication is that illness is 25% above normal (Arbejdstilsynet, 1997). However, studies have shown that there are large differences between the slaughterhouses, in terms of sickness-related absenteeism and labour turnover, and furthermore the working environment for slaughterhouse staff has been improved. In the newly established slaughterhouses the heavy lifting, dangerous cutting and uniform, repetitive working operations have been taken over by robots.

At the farm level, too, statistics suggest that there are several problems. The main problems mentioned in the literature are accidents, dust, noise and heavy lifting (Arbejdstilsynet, 1997). Dust is a large problem, especially in indoor pig farms; the number of pig farmers developing asthma because of dust is above the average in agriculture. In free-range pig production, the main problems are ergonomic, e.g. lifting the piglets from the outdoor farrowing houses and typically more manual feeding.[9]

Traceability and Ethical Traceability in the Chain

Companies balance their benefits and costs to determine the breadth, depth and precision of the traceability system they establish (Golan et al., 2004). *Breadth* refers to the amount of information the traceability system records. Given the huge number of attributes that could describe any food product, full traceability is a theoretical goal, if only because of the volume of record-keeping it would entail, although some of the new technologies will make it easier to record more attributes. Food companies decide which attributes to trace based on their own needs, their interpretation of consumer interest and the cost of keeping the necessary records.

---

[8] Interview with a consultant in food processing methods.

[9] Interview with specialist in working condition in farming from Landbrugets Rådgivningscenter

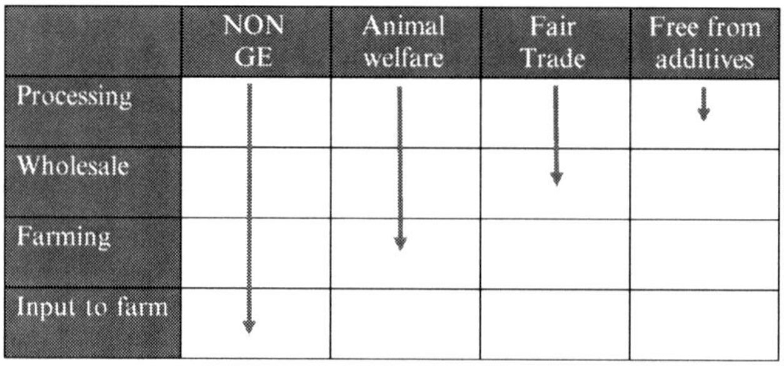

Fig. 5.3 The depth of traceability refers to the number of attributes the traceability system records. The information shown is for[some] animal products

The *depth* of a traceability system refers to how far back or forward the system is able to track an item. For pigs and pork, the question is whether it goes back to the specific farmer, the abattoir, the processing plant or the retailer (Fig. 5.3). Current EU law says that companies only have to trace 'one back and one step forward' (see Chapter 2). If the objective were to provide full information to consumers, the system would go all the way back to the farm.

Precision is about how precisely a food can be traced, meaning how small a unit can be individually traced. Slaughterhouses typically have low precision, because they usually trace the meat back to a whole day's production, which might involve pigs from many farmers. The degree of precision is closely connected to the cost of recall and the frequency of recalls.

The EU's regulation on traceability (Regulation 178/2002) is not very detailed, and food businesses have for quite some time been unsure of how the requirements should be fulfilled. Likewise, the EU's controlling authorities have been unsure of how the provisions should be supervised. According to the definition of traceability in Article 3, 15 of the regulation, food businesses must be able to 'trace and follow' every 'substance' included in the production process. Article 18, however, setting out the requirements for traceability, is more general, requiring 'systems and procedures' for identifying 'any person from whom they have been supplied with a food, a feed, a food-producing animal, or any substance intended to be, or expected to be, incorporated into a food or feed' and also 'the other businesses to which their products have been supplied', i.e. one stage forward and back to/from the food business (OJL, 2002).

The interpretation of this traceability requirement by the Danish authorities has not been very strict. According to the Danish authorities, a strict interpretation would burden food businesses with extra costs, compared to their foreign competitors, without enabling a Danish product to be marketed as 'safer'.

In the opinion of the Danish Veterinary and Food Administration, the traceability requirement does not directly apply to the internal processes of food businesses, but requires only one step forward, one step back traceability between businesses – a view supported by the authorities in Sweden and the UK. This has been one of the most controversial areas of interpretation of the Regulation, however.

In the view of the Danish Veterinary and Food Administration, the majority of food businesses in the Danish food industry comply with the general traceability requirements stipulated in Article 18 of the Regulation. It is generally felt that traceability procedures should be incorporated into the food businesses' internal control programmes. The internal controls should at a minimum control 'who has delivered what and when' to the business and 'what was sold to whom and when' by the business.

The principle currently in force is that the businesses' internal control systems define the size of the consignment/batch/lot which, if relevant, has to be withdrawn from the market in connection with a risk to food safety. In other words, the businesses' own risk analysis and liability will provide the basis for the volume/consignments demanded by the Danish Veterinary and Food Administration to be withdrawn from the supply chain/retail sector if the Administration suspects, or has reasonable grounds to suspect, that a product poses a risk to food safety.

Two primary motives have driven the development of traceability systems for meat and meat products: supply chain management and safety and quality control. Traceability systems enable abattoirs and processors to track the flow of product and coordinate production more efficiently. Traceability systems also help plants to minimize the extent of safety or quality failures, thereby minimizing damages.

One of the main driving forces for implementing these systems is not 'traceability' but rather the management's interest in improving the efficiency of the company, making the production 'lean', faster and enabling more precise responses from the production level to the management. In other words, traceability is a tool for management to improve efficiency by making production more flexible, minimizing waste, etc.

> *Today we are working to combine the management system with the production system, and traceability [to consumers] is not the driving force for this development.* (The responsible system developer of a large slaughterhouse)

A number of major outbreaks of food-borne illness and heightened awareness of food safety issues have led many producers to adopt increasingly precise traceability systems. These systems reflect not just the fact that awareness of the benefits of traceability is rising, but also the fact that technological innovations are reducing the costs of traceability. These trends are expected to continue as retailers and importers demand safer food.

Slaughter plants and processors have developed a number of sophisticated systems for tracking the flow of production and monitoring quality and safety. In accordance with quality guidelines (e.g. ISO 9000), most track inputs by batch or lot and then assign new batch or lot numbers to track product as it is transformed. To control food-borne pathogens such as *Salmonella*, a number of processors have established very precise sampling, testing and tracking protocols.

In choosing lot size, companies consider a number of factors, including procedures for accounting, production technologies and transportation. As these factors vary within and among industries, the lot size also varies from plant to plant. Consequently, there is no standard traceability unit. Furthermore, a company is likely to have a different lot size for incoming and outgoing products. Firms add value in their production and marketing practices by commingling, transforming and processing products. The size and shape of a lot is therefore likely to change at each processing juncture. As mentioned earlier, some of the slaughter facilities only have traceability for a 'lot' consisting of a single day's production. This could entail more than 8,000 pigs from more than 100 farms. The lot size of one such company is more than 7,000 pigs. Other slaughter facilities have different lot sizes. For instance a smaller, private, organic abattoir has a lot size below 50 pigs.

Table 5.2 outlines the various traceability systems and track *ethical* aspects of production in each sector of the pork-bacon chain. The terms in parentheses in the final column link the ethical concerns addressed with the project's ten ethical concerns.

Stakeholders' and Consumers' Response to Ethical Concerns

This report is a case study trying to unfold the concept of ethical traceability in relation to a specific type of food production, namely bacon and the pig-pork chain. To explore what meaning *ethical traceability* has in a concrete product chain we proposed a framework of qualitative investigation, based on semi-structured, in-depth interviews, supplemented by observation at companies and plants and by documentary research. The analysis uses a predominantly qualitative approach, based on interviews with stakeholders from all sectors of the chain as well as consumers, and as far as possible puts the actors themselves at the centre of the study.

The interviewees were shown the list of ten ethical consumer concerns relevant to food production, used by all three case studies (see Chapter 1 for a discussion of these concerns, and Table 1.3 for a list of concerns). Some of the stakeholders said that the concerns were not comparable. Others thought that some of the listed concerns (notably trust, voice and quality) were not really about ethics. Some of the consumers mentioned that the concerns were not mutually exclusive, in the sense that some of the concerns covered several of the others: '*I think that transparency covers most of the others*' or '*For me animal welfare is a central part of quality*' were sample comments.

We identified some differences in the way the ten concerns were perceived throughout the chain. In particular, we saw differences between the stakeholders in general and the consumers (e.g. while all the consumers mentioned animal welfare, the main concern among the stakeholders was quality). In general we found that priorities varied depending on the interviewees' place in the chain.

When we contacted the headquarters of one of the large slaughterhouses and asked for an interview about ethical traceability, we were immediately asked to call the section dealing with organic and free-range pigs, as if this section of the company was the place where the ethics of the company were deposited.

When we asked conventional farmers for their opinion of ethical traceability, their first association was typically that it had something to do with the consumer or media criticism of Danish pig production. The farmers and the retailers referred

Table 5.2 The relationship between traceability, information and ethical concern along the pig-pork-bacon chain

Level in chain	Traceability systems with ethical aspects	Information traced	Ethical concern addressed
Input to farm Feed:	Special emphasis on GMO and *Salmonella*. Before GM crops or a processed GM product can be used in feed, a risk assessment must be carried out and approved according to EU regulation 1829/2003 about GM organisms in food and feed. Feed containing GM crops should be labelled and have an accompanying document. *Salmonella* samples are taken from ships, feed stores and at feed processors.	Where does the raw material come from (origin) Documents telling that the feed is free from GMO	GMO Dioxin, heavy metal, etc. Ghost acres Bacteria (Human health, Methods of production, Transparency)
Breeding	In the Danish breeding system, 'Danavl', a set of common breeding objectives is set, and the result is disseminated to Danish pig producers through sales of animals and semen.	Documentation of the specific breeding characteristics, combination of pig races, health level, etc., accompanies each breeding animal, and is gathered at breeding livestock	Mainly economic traits (litter size, weight gain, meat percentage, etc.) Dilemmas: e.g. between low-fat pigs and the emergence of shoulder wounds. Ownership of genetic material (Animal welfare, Transparency)
Breeding	Ear tags, The central database (CHR). The same central database is used for the management of registration of holdings and herds, all head of cattle and all movements of pigs, sheep and goats. The database is called CHR (Central Husbandry Register). CHR is owned by The Ministry of Family and Consumer Affairs and was launched in 1992. In the attempt to reduce spread of diseases all information in the CHR – with a few exceptions – is available to the public on request. Since 2002 movements of pigs have been recorded in CHR. (Bækstrøm Lauritsen, 2005)	The number of pigs moved, the date and time of movement, places involved in the movement and registration number of the vehicle used for the transportation.	Mainly economic traits (litter size, weight gain, meat percentage, etc.) Surplus of nutrients Change in landscape and socio-cultural environment ('pig factories') Medicine (e.g. antibiotics, resistant bacteria) Working conditions (dust, noise, heavy lifting)

(continued)

Table 5.2 (continued)

Level in chain	Traceability systems with ethical aspects	Information traced	Ethical concern addressed
Input to pig fattening and breeding Medicine	The veterinarian is not allowed to prescribe antibiotics before the disease has been diagnosed. The antibiotics are available by prescription only, and the farmer receives a written specification on how to use the medicine and retention time for the treated pigs.	The veterinarian maintains exact records of his farm visits, and the farmer must register his consumption of medicine, and specify which animals have been treated.	Continued spread of resistance means that treatments for common infections will become increasingly limited and expensive – and, in some cases, non-existent
Pig fattening for slaughtering	Ear tags/tattooed number on the gammon with the number of the farmer. Approximately 20% of the trade in piglets takes place via a pool arrangement in which a purchaser receives piglets from different pig producers (whose herds may also be identified by means of ear tags). The piglets are sold before they leave the herd, ensuring that the purchaser is always known. The purchaser will also know the health status of the piglets.	Farmer Date of transport Number of animals	Animal welfare (space, stress) Health (use of medicine) Methods of production Working conditions
Transport Export Local abattoir	Ear tags/tattooed on the gammon with the number of the farmer Transport (haulier) is authorized by the Food Administration	Name and number of the pig farmer Number of pigs	Space in wagons Time spent in wagon Food and water for animals Veterinary check Biting in case of 'mixed' litters Energy Methods of production (CO_2 emissions)
Abattoir	The farm number is registered either from the ear tag or the tattooed number. The weight is recorded at the slaughterhouse no later than 45 min after sticking. Any observations made by the veterinary inspector are registered with the weight and the farmer receives this data at the same time as he receives his payment. (Bækstrøm Lauritsen, 2005)	Weight Number of pigs Fat content Veterinary comments Date for processing	Animal welfare (is the hog killed decently (stunning)) Working conditions (Repetitive work, stress, heavy lifting, accidents) Health (*Salmonella* program)

	Before the carcass goes to the cooling room, it is automatically branded with an EEC health label, trade grade mark and sub-grade mark. In addition to the oval EEC mark, a coded week and day mark is applied to the carcass, which is used in the quality control system to establish product durability. The carcass is either sold after the cooling or it will be processed further at the abattoir. If meat products are manufactured in a separate plant, they must be marked with the authorization number of the plant in accordance with EU Directive 77/99/EEC. If slaughter, cutting and production are performed at the same plant, only one registration number need be applied. The meat is packaged either as a whole carcass or cut into smaller pieces (belly, ham, fore end, etc.) in boxes or jars with identification (RFID, barcode, etc.		
Processing	The processors typically receive boxes with batch numbers from the abattoir. From the final product (e.g. sliced bacon) it will be possible to trace it to the abattoir.	Temperature, fat content, processing time	Working conditions (repetitive work, stress, heavy lifting) Health (nitrosamines) Transparency (Phosphates to 'bind' water and hide 'bad meat')
Slicing, packaging	Meat cuts and meat products must be accompanied by information identifying the specific lot number. If a date of minimum durability or the 'use by' date is labelled on a product, this information may be used as lot-ID provided that the date consists of at least the day and month.	Retail packed meat must be labelled with the name of the distributor, or the company packing the product, or the manufacturer of the product in conformity with EU Directive 2000/13/EEC.	Transparency (labelling) Working conditions (repetitive work, stress, heavy lifting)
Retail	In accordance with EU Directive 2000/13/EEC, the name of the manufacturer or the distributor must appear on the packaging. The latter is then able to trace back the product on the basis of information on the product type. Promotional material	Ingredients Provenance Name of manufacturer Shelf life	Transparency (price setting, labelling)

to existing traceability systems, and found it difficult to see the difference between these systems and the concept of *ethical traceability*. Others thought that it had something do with organic or free-range production, as if these production systems, in opposition to traditional production, were the representatives of ethical production.

At the processing level, especially, ethical traceability was associated with security, i.e. the withdrawal of meat products from the market in the event of a risk to food safety.

The interviewees' responses to the ten separate ethical concerns are discussed in more detail in the following paragraphs.

Animal Welfare

Animal welfare was collectively seen as the most important of the ten concerns, discussed by stakeholders throughout the chain, as well as by consumers. From the processors' angle, animal welfare was typically associated with the 'welfare pigs' or 'special pigs' described above. The increasing number of pigs that are specially adapted to the demands of both domestic and export markets were mentioned, mainly by the farmers. These special pigs still only account for a small part of the total Danish pig production, and it was pointed out that consumers' concerns about animal welfare often disappeared when they were choosing meat in the supermarket.

Although conventional farmers agreed that the animals had a better life in free-range production systems, some of these farmers argued that free-range or organic pigs also had welfare problems (such as a high mortality rate in piglets, or late treatment in the case of illness). One of the conventional farmers, who had formerly been a free-range pig producer, had to give up this type of production, because he could not charge more for his pigs, due to lack of consumer demand.

The largest slaughterhouse referred to a publication called 'Code of Practice' (Danish Crown, 2003). This publication sets out a number of requirements regarding animal welfare in primary production. These requirements are mainly based on current legislation, and could be extended as necessary for special production.

A consumer poll from 2001 carried out by the Association of Danish Slaughterhouses (*ds-nyt*, 2001) showed that 50% of respondents wholly or partly agreed with the statement that pigs are not well treated in modern hog production. According to the survey, modern, industrialized hog production was viewed negatively because it infringes the pigs' freedom through the use of tail docking, immobility, lack of access to the open air and cramped and crowded pens. This contrasts with perceptions of the 'good old days' when farmers had time to scratch pigs behind the ears and when pigs were more unrestrained in their 'pursuits'.

> *I have this picture in my head of a typical, romantic farm in the country with pigs wandering around snorting, rolling in the mud and seeming very contented. And then a terrifying sight appears: you drive past these huge, stinking buildings and wonder if the pigs are having a good time.* (Consumer)

Our consumer interviews show that these contradictions are symbolized in two ways. One is the pig with a docked tail as opposed to one with a curly tail. The other is the free-range pig as opposed to a pig living under industrialized conditions. Based on the awareness that the good old days are not a realistic alternative, the free-range pig with the curly tail becomes the epitome of the preferred type of production.

One of the topics often mentioned by consumers was the transportation of living animals. A month before the interviews, an episode was reported in the newspapers in which a truckful of piglets died because the driver did not give them water. The minister tried – without success – to strengthen the EU rules for the transportation of living animals.

> *I think that the transport of living piglets in Europe is a large problem. I don't know how to avoid it.* (Consumer)

Consumers know little about new initiatives to safeguard animal welfare on Danish hog farms and respondents were actually sceptical when told about innovations such as firm floors and straw bedding. Animal welfare in modern hog production, including transportation, is primarily perceived as the farmers' responsibility.

On the other hand, people understand that farmers are involved in cut-throat (international) competition that leaves little room for more minimal animal welfare. There is some recognition that the consumer's focus on the price of meat/bacon also locks the farmer in an unacceptable form of production.

As suggested before, the issue of freedom was an important topic for participants who place a high priority on free-range pigs. The issue here is the respondents' perception that pigs bred in modern, industrialized herds do not have enough space to move around in. This is a problem of which many are already aware. The perception of industrialized pigs living in confinement is also often included in the argument about the pigs' integrity. According to this view, modern pig farming is so far removed from the pigs' natural behaviour that it violates the pigs' integrity. Their living conditions are described as 'unnatural' referring to, for example, the lack of opportunities for natural behaviour such as wallowing in mud, rooting through the soil or the 'right' to have a curly (undocked) tail. Similarly, the docked pig's tail is a symbol to many people of something gone wrong. This is predominantly expressed as (a symbol of) a violation of the pigs' integrity: the pig is born with a whole tail and this is natural.

> *All animals should have the right to unfold natural behaviour.* (Consumer)
> *Farmers keep building bigger and bigger hog houses, and filling them with more and more pigs. In my opinion, it's not worth it. The pigs can't be happy with living in the buildings they're using today.* (Consumer)

The participants did not perceive that pigs were thriving in today's pig breeding system, either during the growth period or in relation to slaughtering (i.e. transportation to the slaughterhouse and the actual slaughtering). Overall, an ideal picture of 'the happy pig' emerged in contrast to a bogey of 'the sad pig'. Some participants saw organic farming with outdoor pig breeding as the solution to the perceived problems of modern, intensive pig breeding.

There are two kinds of production: organic production and ordinary production where the pigs are behind bars and all that. (Consumer)

Pollution, medicine, the pigs are victims of inbreeding. Bone meal as fodder, medicine that give sores. They get injections, hormones. It is all a matter of money. It's business. (Consumer)

Others doubted that consumers were willing to pay a premium for organic meat.

Nevertheless, these participants also believed that more decent production methods were needed in today's pig breeding, but they regarded the perceived ideal as unrealistic:

If pigs were bred the way we would like to see it, they would cost a fortune. (Consumer)

These participants did not perceive free-range pigs to be better off than pigs in intensive systems. Among other things, it was said that the free-range sows might squash their free-range piglets by lying on them.

A few of the interviewees viewed the issues from the farmer's side as well. The farmer was perceived to be subject to requirements from authorities and consumers and perceived to be in tough competition with other producers:

The farmer has to follow the development. It is also though-provoking that consumers spends so little money on food, compared to other goods. (Consumer)

In general, both intensive and extensive pig breeding methods were seen to be applied in modern pig breeding in Denmark, and a number of undesirable elements were associated with intensive pig breeding. The participants were uncertain about how common 'the unhappy pig' was, but they suspected the worst and realized that they tried to suppress this when they went shopping for food:

I must say that I have a bad conscience, because I think they are treated really horribly. I ought to buy the pigs that run around in the fields. But I never get down to doing it. And I eat the other meat anyway. (Consumer)

Well, the negative sides also have to be there, but we are a bit undecided, because it tastes good and it is nice to be able to get inexpensive bacon. (Consumer)

In general, stakeholders in the chain felt that the media and consumers were ignorant about modern pig production. A farmer expressed this frustration this way:

They think we like to give medicine, trim the piglets' teeth or castrate the animals. Even our own family thinks we mistreat our own animals. (Conventional farmer)

This farmer also agreed that pig producers had failed to communicate with consumers in a trustworthy way:

We market our pigs as lying in straw, but realistically 95% of our pigs live in stables with slotted floors. (Conventional farmer)

The interviewee from the breeding company focused on the fact that breeding decisions influence which characteristics animals born in the future will have, and was aware of the ethical consequences this could raise:

The traits we seek to optimize when we change the genetic composition are mainly chosen for economic reasons. (Breeder)

The consequences have caused several ethical conflicts. For instance, the breeding of a slim pig has had the effect of considerably increasing the number of shoulder wounds.

The organic farmers mentioned that organic pig farmers are co-operating with an animal welfare organization about some of the areas where the two groups had different rules. One of these areas concerned the common practice of nose-ringing outdoor sows to prevent them from rooting and damaging pastures. In the UK and the Netherlands this is prohibited in organic farming. The practice of castrating organic male pigs has also been discussed.

Human Health

This was widely cited as an important concern, and especially in the processing part of the chain. The comments that related to human health were mainly divided into three discussions: (1) health related to bacteria (*Salmonella* and other zoonoses); (2) health related to antibiotics in meat; and (3) health related to the content of fat in the meat.

The problem related to antibiotics is the potential issue of resistance to antibiotics in human beings. The fact that there has been a rise in the use of antibiotics in recent years has caused some discussion, especially among the farmers we interviewed. It was clear that the animal welfare organizations and farmers had different views on this question. The conventional farmers focused on the fact that it would be unethical not to treat the animals when they were ill:

> *The reason for the increased use of antibiotics in recent years is that we have seen a rise in the disease PMWS [Postweaning Multisystemic Wasting Syndrome] in pig farms, and the only way to cure this is antibiotics. Anything else would be mistreatment of the animals.* (Conventional farmer)

The animal welfare organizations, on the other hand, argued that many of these diseases were preventable, and that it is important to look for reasons for this disease.

> *The piglets are weaned off already after four weeks when their weight is only 7 kg. This, combined with lack of space and general stress in pigpens, is the overall problem. PMWS is a symbol of the problems in modern pig production ...The veterinarian should replace the use of antibiotics with advice to farmers on animal welfare improvements.* (Animal welfare organization)

The fact that the official figures show that Denmark and the Scandinavian countries generally have a lower consumption of antibiotics compared to other countries with a considerable pig production was stressed by the farmers. The use of antibiotic growth promoters in feed was prohibited in Denmark 6 years ago, no matter if the feed is for sows, piglets or slaughter pigs. An agreement has been made with the feedstuffs industry that antibiotic growth promoters may not be added to pig feed. Shortly before the stakeholder interviews a TV documentary programme revealed that the Danish farmers producing pigs (and bacon) in Poland for the Danish market actually added growth promoters to feed. This was mentioned by several of the stakeholders:

> *We live in a global world, and we compete with farmers in the US, Brazil and Poland, and they produce under very different conditions.* (Conventional farmer)

The fact that pork in general was perceived to be fat meat has for many years been an incentive to breed pigs with lower fat, and these initiatives have in fact been successful. In the last 30 years the fat content has been reduced by one third. The representative from the breeding company agreed that this has also caused some dilemmas, because: '*modern pigs suffer from shoulder wounds, due to this development of slim pigs*'.

The consumer organizations drew attention to genetically modified (GM) livestock and transgenic pigs. American researchers have already produced pigs cloned with omega-3 genes. These pigs are not in the market, because cloned livestock are still not allowed in the food chain. The interviewees from the pork sector did not expect this technology to be of major importance for many years. They were reluctant to pursue the technology because of divided public opinion on GM foods (Biello, 2006).

The health issue has for many years focused on the safety and danger aspects, specifically and narrowly in terms of the production method, because it has essentially concentrated on the relationship between the production system and the spread of undesirable bacteria such as *Salmonella*, but also on the use of antibiotics, and the inherent risk of developing resistant bacterial strains. In public forums, this is manifested by discussions derived, in particular, from media 'exposures' of industrial production systems whose very form necessitates the use of antibiotics as preventive medication, connecting this use to the prevalence of *Salmonella*.

The debate about health seems to be split in two: on the one hand, a slightly narrower debate dealing with the risk of zoonotic diseases related to handling and eating pork – a debate that in many aspects parallels the discussion of broilers and eggs; on the other, a broader theme discussing whether the widespread use of antibiotics involves the risk of forming (multi-) resistant bacterial strains, thus constituting future obstacles to the treatment of human disease.

The fact that the health issue was so relatively far down on the agenda in our interviews was surprising. The reason could be that this problem has recently attracted less media attention, partly because of the successful Danish *Salmonella* action plan. Another reason for the lack of attention to zoonoses could be the fact that *Salmonella* and *Campylobacter* are not regarded as an ethical concern.

When directly asked, it became apparent that some of the consumers perceived pathogenic bacteria to be an inherent risk of eating meat. A few stated that they regarded *Salmonella* as a natural phenomenon and, thus, not much of a problem. Most consumers forgot this topic during the first round of the interview and did not become aware of it until it was mentioned by the researcher midway through the interview.

> *I think the use of antibiotics in pig production is very important. It is a threat to human health. Even though Danish pig producers uses less antibiotics, I think it still too much.* (Consumer)

Several consumers mentioned the fat content of bacon as a particular problem. Apparently, many people have a deeply ingrained perception that pork is high-fat meat. The fat content is regarded as both positive and negative: positive because the fat adds taste, negative because it is unhealthy. When this point was mentioned it was typically in relation to their actual choice at the refrigerated counter in the supermarket.

A few of the consumers referred to better veterinary control systems and more decent production methods in Denmark:

> *I buy Danish bacon to support the Danish producers. In this country, we have a fairly good control system and I believe that there is less control in other countries. In this country, control is more in focus and they have to live up to the requirements.* (Consumer)
> *It has to do with disease. The standards are significantly higher in Denmark than in the countries we are surrounded by. Veterinary control gives security.* (Consumer)

The media are generally an important player in the formation of attitudes towards pork. However, the behavioural consequences of the attention created by television features on food scares and pig production were often insignificant:

> *When there is something up about, let's say, growth enhancers (red: on TV), I lose my appetite. It is psychological and influences me for the next couple of days. When I go to the supermarket I pick another kind of bacon. But it doesn't last long, after a few days I am back to the old habits again.* (Consumer)

Participants sharing this opinion felt guilty that they took it so lightly and tried to defend their actions:

> *I don't have the time or the knowledge to get qualified advice, for instance whether there are medicine residues in the bacon.* (Consumer)

The retail outlets were not perceived to know enough to give trustworthy advice about these issues anyway:

> *No matter whether it is a supermarket butcher or an independent butcher's, he doesn't know how much penicillin was in the bacon he bought. He has to trust the people that he buys from, and it is like that the whole way through.* (Consumer)

Healthiness was an important dimension of quality in relation to bacon. In general, however, healthiness was perceived as an invisible factor, which was difficult to evaluate. Therefore it often only played a minor role in purchase situations. Nevertheless, as indicated above, factors perceived to affect the healthiness of bacon, such as antibiotic residues, had high level of awareness. Part of the reason why conventional pork did not engender complete trust among the respondents was because they remembered incidents when residual antibiotics were detected in bacon.

Methods of Production and Processing and their Impact

Several of the stakeholders thought that this concern covered animal welfare, quality, working conditions and the environment. From the (conventional) farmers' side it was argued that Danish environmental regulations were much too strict in comparison to other countries, and were considered to be the main obstacle to further development of the sector in Denmark. In other words, the regulations were accused of being the principal cause for the relocation of Danish pig production to eastern European countries in recent years.

In the last 10 years, several of the slaughterhouse companies have been approved in accordance with the EU's Environmental Management and Audit Scheme (EMAS). This is probably the reason why many of the slaughterhouses highlighted this concern.

The use of GM feed was also mentioned under this concern. In general, the pig farmers were sceptical about the use of GM feed, because of the general scepticism about GM products among consumers today. Some of the farmers also mentioned that they were split between being a consumer and a producer. As a consumer they felt insecure about GM products, but as a producer they knew that it would be difficult to separate GM feed from non-GM feed. This also raised the question of labelling. At present it is not compulsory to disclose that animals have been fed with GM feed, and organic pig production is the only production system that uses non-GM feed. The conventional and organic farmers expressed different opinions about GM feed. One of the fundamental principles of organic production, namely the precautionary principle, was emphasized by an organic farmer:

> *The consumers don't have a real choice with the new EU regulation. You can't see if the pig got genetically manipulated feed. We think meat from these animals should be labelled. Ecology is about the precautionary principle.* (Organic pig farmer)
> *Why should we mention that our pigs have eaten this feed. You could not find GM in the meat. Besides I think that consumers are well aware of this.* (Conventional pig farmer)
> *I find it very problematic that the European regulations, which govern GM crops and food, do not cover animal feed. Consumers have not been consulted and have, once again, been denied the opportunity to opt out of the GM experiment.* (Consumer organization)

The retailers said they would prefer to label the meat, if the animals were fed with GM feed. At the same time, they did not believe that consumers were willing to pay for the additional cost of such labelling. In fact they already tried to market meat reared with non-GM feed some years ago.

Terms of Trade (Fair Price etc.)

This concern was mentioned only by one farmer and the consumer organizations. The farmer was one of the relatively few Danish farmers who are not members of the co-operatives and therefore are not guaranteed a fixed price for their pigs. This meant that the farmer had to find his own slaughterhouses round Europe and negotiate an individual price. In his view fair *terms of trade* mean that you keep your promises, which was not what he had experienced.

The consumer organizations mentioned that the *terms of trade* were important because of increased pressure on small- and medium-sized farmers and processors. Fair prices were seen to have a close relation to transparency. This was not mentioned explicitly in any of the consumer interviews, but some consumers referred implicitly to this when they talked about (organic) farmers' prices and the retailers' pressure backwards in the chain. The most common way this concern was understood was in relation to Max Havelaar products,[10] and this was not considered to be relevant for bacon.

Terms of trade are important for me. I typically buy Max Havelaar coffee. (Consumer)

Working Conditions

Several of the interviewees felt this area was well covered by the labour markets' voluntary agreements. Again, interviewees tended to focus on working conditions within their own sector, or even their own company, rather than the chain as a whole.

Not surprisingly the union of the food workers regarded this concern as very important. The employees from the slaughterhouses realized that the improvements that have been won could turn out to be a pyrrhic victory, because many of the Danish jobs in the pig-pork sector are now being transferred to countries where the wages are much lower and working conditions poorer.

We know that the work at the slaughterhouses is hard, both physical and psychological. Therefore we still see that our members are worn out already when they are in their forties, and the mortality among our members at the slaughterhouses is one of the highest in the country. (Food workers' union)

There is still a long way to go, but we have improved the conditions for the slaughterhouse workers by increasing their wages. But the market is under severe pressure from globalization and the conditions will get worse. A few companies have tentatively abolished work by contract. (Slaughterhouse manager)

The concern about working conditions was mentioned unprompted by only one of the consumers. More consumers found that working conditions were relevant when they had watched the brief presentation on the bacon chain that was made during the consumer interviews, and especially been told that all bacon production had been moved outside Denmark. One of the consumers found it striking that he had mentioned the welfare of the animals and had not considered the welfare of the workers involved. Other consumers also reflected on the paradox that we are more concerned about animal welfare than the welfare of the people who produce our food.

We often discuss conditions for the animals, but maybe we should start to look at the conditions for the workers on the farms and factories. (Consumer)

[10] The Max Havelaar Foundation fosters fair trade for products from economically disadvantaged regions in the southern hemisphere under conditions that ensure the livelihood of the local population. These products are cultivated according to ecological principles and increasingly follow organic farming methods.

Quality (Taste, Composition, etc.)

Almost all the stakeholder interviewees said that producing a 'quality' product was their goal, but the perception of quality varied along the supply chain. The farmers saw quality as meaning they had an effective, highly productive unit with no diseases. More specifically, quality was closely related to the payment agreement with the slaughterhouse, i.e. the precise weight of the animal and a high meat percentage.

The discussion of quality often included a discussion of producing bulk versus highly processed products, and here too there were different perceptions of the modern Danish pig sector. A representative from the farm input sector suggested that the Danish pig sector should continue with bulk production, as he explained:

> *It is the 'bulk' that Danish pig farmers shall earn their living from. The ethical market is only 5–10% of the total market.* (Breeder organization)

But several of the farmers, not only the organic farmers, were not satisfied with this argument stating instead that:

> *We should make a good taste and be proud of our production. We should not be produce for the bulk market.* (Conventional farmer)

When the large processing companies mention quality they emphasize three things. Firstly, the uniformity of the products, achieved by slaughtering the pigs within a narrow weight range, and by detailed sorting of the carcasses at the slaughterhouses. Secondly, the veterinary approval secured by independent inspection; and thirdly, dependability of supply formulated as the need to '*assure the customers of delivery of the right quality and the volume ordered at the agreed time*'.

Some of the slaughterhouses mentioned that the large number of different types of special pigs, and correspondingly large number of labels, is confusing to Danish consumers and is a weakness of the Danish quality-labelling system. This was confirmed in our consumer interviews, where consumers really did not know what the different labels meant, except for the organic label, which is relatively well known by consumers.

A few of the consumers mentioned that quality actually covered most of the other concerns, because health, animal welfare, trust, etc., form part of their perception of quality. A variety of attributes that are not intrinsic or visible aspects of the bacon were also mentioned as indicators of quality. These were: country of origin, whether the product was on offer, whether the product is organic, or brands and quality marks.

Further indication of the quality was obtained when the bacon was prepared/fried. Here, the following elements were observed: shrinkage, water content, texture and structure (crispy), and the smell.

Origin and Place

When the farmers mentioned origin and place, it was quite often in recognition of the fact that international competition is seen as a threat to their livelihood. Farmers stress that costs in Denmark are considerably higher than in competing countries. Besides the relatively high level of salary, they mentioned stricter environmental and animal welfare regulation. The problem is that these benefits are not visible to the consumer. Bacon is mentioned as an example where, e.g. Polish bacon is produced under different conditions than in Denmark, where environmental regulation and rules governing the use of medical drugs are stricter, and farmers have entered into a voluntary agreement not to use growth enhancers.

In the processing and retail sector, it is generally assumed that consumers are not interested in the origin of the meat. The companies argue in different ways when they are questioned about the lack of information about origin:

> *Danish bacon belongs to a bygone age, and not in a modern, globalized time. It is years since Danish consumers were willing to pay for Danish products.* (Manager from medium sized processing company)
> *We acknowledge the European common market and don't label our products with information on the origins of the meat.* (Large processing company)
> *In practice it is impossible to label the country of origin of the pig meat on the bacon packages. It would complicate the packaging process because we would have to change packaging several times. It takes time and would increase the cost of production considerably.* (Organic processor)
> *If you are interested in the country where the bacon is produced, salted and sliced you can read it on the packaging. It shouldn't be of importance for the consumer, because you can't taste the difference.* (Manager from slaughterhouse company)

The fact that *Salmonella* is found three times as often in foreign pig meat compared to Danish meat is an incentive for Danish pig/pork producers to provide the country of origin labels, but the fact that this is not compulsory has discouraged the slaughterhouses from labelling all their products with thecountry of origin. Some of the processors mentioned that their analyses did not show that consumers were willing to pay more for *Salmonella*-free products. This was also confirmed in our consumer interviews, where *Salmonella* was not mentioned at all when they were asked the reasons for their pork and bacon purchases.

Almost all the consumers interviewed were very surprised when they heard that all the sliced bacon they could find in supermarkets was produced not in Denmark but probably in Germany or Poland. Most of the consumers said that they felt cheated, especially when they were presented with a picture of well-known bacon packages and told the bacon was not produced in Denmark. There was a strong belief that Danish bacon was better than foreign bacon, but substantive reasons for this could usually not be given.

> *Somehow, I just have more confidence in Danish bacon [than in foreign bacon] There is no special reason – it is just something I believe.* (Consumer)

Trust

The farmers mostly discussed the opposite of trust, namely distrust in modern agriculture. In their analysis, one of the reasons for this distrust can be found in the media's coverage of agriculture, and especially the way they cover animal husbandry. A farmer mentioned:

> *They [consumers] have their knowledge about agriculture from the front page of Extrabladet [a tabloid newspaper].* (Conventional farmer)

Generally the farmers agreed that they and their organization have for many years neglected to communicate with consumers. One farmer mentioned that in his area the farmers had organized a scheme in which they open their farms for schoolchildren.

One of the processors said that this concern actually should be placed above all other concerns, because: '*if we lose the consumers' trust it doesn't matter how we produce*'. The organic farmers and slaughterhouses, in particular, also mentioned trust in relation to labels, and said that the extremely high level of consumer trust in the Danish organic label was due to the fact that the state guaranteed that the products actually were from organic farms. When consumers mentioned trust it was typically in relation to a feeling of 'lack of trust' in the pig farmers. The perception of the pig-pork sector as being solely driven by the profit interest is very prevalent, and is argued to be the main reason for lack of trust in stakeholders in the sector.

> *Roughly speaking it is the pig producers and the meat companies that have deceived us. They [meat companies] want to produce as cheaply as possible and sell as expensively as achievable.* (Consumer)

Regardless of whether organic farming and/or products are specifically discussed, in relation to pork or in more general terms, one of the most recurrent themes by far is the issue of credibility. The central issue is the extent to which a consumer believes the assertions regarding a specific organic product and, to a considerable extent, it is a question of whether the consumer relies on the system of authorities, farmers and food manufacturers behind the product. Several of our consumer interviewees emphasized that organic products are credible, by contrast with other labelling schemes (e.g. private labels such as 'Antonius pork' or 'free-range pigs'), which are subject to somewhat more consumer scepticism. The very system of state inspection and control connected with the organic system engenders credibility in the consumer because the product has to comply with the law. This contrasts with private labelling schemes controlled by manufacturers or retailers, both of which have obvious economic interests.

> *The only real, genuine and certified product here is the organic bacon. And the only certified aspect of this product is that the pig was raised according to organic rules. And the state guarantees this.* (Consumer)

Voice (Participation)

The conventional farmers, especially, formulated this concern unprompted. It was seen as the only concern where it was possible for them to place or express their frustration about the strict environmental and animal welfare policy with which they have to comply, and which imposes costs on them. The farmers felt that there was no place for them to make their voice heard, and that the media only wanted to present pig farmers in general as criminals. They expressed a strong wish for recognition and appreciation.

Some of the non-governmental organizations also mentioned voice as an important concern, and had an overall wish that all stakeholders, including consumers, would in future have influence on food production in general.

This concern was not well understood by consumers, and they found it difficult to explain its meaning. The most common comment was that they felt that the consumers had too little 'voice' today and that the big companies and retailers decided what we should eat anyway.

Transparency

Typically transparency was confused with traceability, but during the interview many of the interviewees turned back to it and found it crucial for this topic. Some mentioned that this concern was an essential prerequisite for the whole idea of following the food chain backwards.

> *If we don't have transparency we can forget all the other things we have talked about.* (Consumer)

However, this concern was not mentioned much, other than by some of the organic farmers and processors. They saw transparency as an extension of the organic principles, and also as an essential prerequisite for obtaining trust from consumers.

Communication in the Chain

The interviews with the stakeholders and consumers also focussed on information flows through the chain. The discussion concentrated on the types of information that flow along the chain, the means by which information is communicated, and barriers to communication. Initially the stakeholders were asked what they understood by the term 'ethical traceability'. Typically the stakeholders said that they were well informed about traceability but not in connection with ethics. Ethics was seen as something very different from traceability. The processors, especially, stressed the safety argument, notably the need to withdraw products from the market in connection with a food safety risk, but typically they did not connect the safety argument with ethics.

The most common argument from the stakeholders was that they found it difficult to see ethical traceability being realized in practice, mainly because they did not believe that consumers were willing to pay the considerable costs these systems entail.

The interviewees were asked for their thoughts on information and communication at their stage in the chain. Only a few of the stakeholders thought of information as a mean of increasing consumers' knowledge about the product. Information was predominantly seen as a way to secure some basic quality standards (e.g. weight, fat content, cuts) or as mandatory certificates showing origin of the product. Finally, information about how to improve production or solve problems was important for the stakeholders. Farmers typically mentioned the agricultural advisory system that seems to function relatively efficiently, but also mentioned colleagues and magazines. Other information sources mentioned by stakeholders were trade associations, the Internet, customer feedback, trade publications and newspapers.

The organic farmers also highlighted research on organic production as an important information source. The larger processing companies and retail chains typically had their own research and development centres, to assist them with new information, but external consultants were also used. The organic meat company had used government technology centres when they had some technical problems with production. They could not afford to hire some of the private consultancies.

When the stakeholders were asked if they thought their ultimate customers had a precise picture of their production system, the majority of stakeholders said that the consumers did not have a precise idea of how the food industry in general, and the pig-pork sector in particular, worked. A farmer expressed it in this way:

> *The consumers don't know what we are doing; they live in a fantasy about modern pig breeding. I had a visit from a school and the teacher and the children had never seen a piglet.* (Conventional farmer)

When the stakeholders were asked to mention barriers to achieving ethical improvements, the one mentioned most often by far was the lack of interest from consumers, as expressed by their (lack of) willingness to pay for these improvements. Typically this argument was substantiated by the fact that the organic meat sector still lagged behind other organic product groups.

> *The main barrier is the price. Consumers don't want to pay as much as they say they will.* (Conventional farmer)

The consumer interviews revealed that consumers did not have much knowledge of bacon production, and most of their information came from the media. Immediately before the interviews were done, one of the media's main stories concerned 90 pigs that had died due to overheating and lack of water in a truck heading for southern Europe. In the middle of the consumer interviews, a short presentation was given on the Danish pig-pork-bacon chain and some of the ethical concerns it raises. The typical initial reaction from the consumers to this presentation was a lot of surprise at the fact that the bacon they purchased was not processed in Denmark.

> *I was surprised that TULIP [a well known Danish brand] was not produced in Denmark. I knew they had activities in foreign countries but not that they had closed the whole production in Denmark.* (Consumer)

Although most of the consumers were surprised and found it unsatisfactory that they could not find Danish sliced bacon in the refrigerated display counter, some consumers thought that animal welfare or other attributes were more important than origin, and this was actually highlighted by consumers buying organic bacon:

> *I don't mind whether the organic meat comes from Dutch or Swedish organic pig farmers.* (Consumer)

The interviews also revealed that consumers found it difficult, at least for bacon, to find any information on the ethical attributes, and especially information regarding provenance. To these questions the bacon packages did not give any answers, and the sparse information available was not understood, e.g. none of the interviewees had heard about the mandatory label indicating where the product was last processed. The reason for the discrepancy between the consumers' perception of modern pig production and the 'reality' was discussed, especially among the stakeholders. The discrepancy was partly blamed on the media, but also on the sector itself, because their products often were connected to pictures of pigs very far removed from reality – either in unrealistic, idyllic surrounding or showing pictures of mistreated or suffering animals. The general answer throughout the chain was that stakeholders wished to send a message, that although there '*is a black sheep in every flock, we produce decent products and we care for our animals*'.

An organic processor differed from this when he said that he was convinced that '*at least my consumers have a realistic picture of my production*'. He had made a great effort to communicate a realistic picture of his production system via the web site and by packaging. Some of the larger meat companies were more focused on their direct customers, namely the large retailers or processors, and here information about the companies' capability to fulfil the requested service, in terms of capacity, skills, level of security systems, etc., was highlighted. The retail sector mentioned the fact that the systems they use nowadays are only able to operate with very large costumers, which is a problem for some of the more innovative and smaller supply chains.

> *It is a barrier that we need to have meat in all our shops. We cannot establish a 'farm to shop' concept because we can not find big enough farmers.* (Retailer – purchasing agent)

The consumer organization mentioned a more general problem of how to prioritize the information to be communicated, and also the problems connected to information overload, and especially the credibility of this information.

> *A special problem is that consumers don't have knowledge about existing problems in food production. What is the most important animal welfare problem? Is it the space, the feed or the fixation of the sows?* (Consumer organization)

Many different types of bacon were purchased, but often the consumers could not remember exactly what type of bacon they had bought last time. When they were shown pictures of the different bacon brands, they often recalled the brand. Many consumers said that they did not think much when they bought bacon, because it was 'only' regarded as supplement to a larger meal. Some of the more common reasons for purchase are mentioned below. They can be split into three groups, namely price, quality and animal welfare:

> *I bought the cheapest. Price is my main argument when I choose bacon. Quality is not at stake.* (Consumer)

I look for fat-free products. If the bacon slices are mainly fat I'll avoid it. (Consumer)
To be honest, it is only taste and price I look for. I like Antonius pigs, not because it is free-range – that's what I think it is – but because of the taste. (Consumer)
I bought organic bacon, because it isn't stuffed with drugs you can't see and are hazardous to your health. I'll support that kind of production even it is more expensive. (Consumer)

Consumers were asked if they had ever avoided certain products for ethical reasons, and if so, why. The most frequent answer was related to pesticides and the subsequent the health argument. The reason for this focus on health could probably be sought in the fact that most Danes today choose organic products because of the content of pesticides in conventional products. (Pedersen and Obelitz, 2001 and Kristensen et al., 2004) Below we have presented some common answers to the question of why they had avoided a food product due to ethical concerns:

I don't buy eggs from cage hens, but organic or free-range eggs. It is too unethical to buy eggs from cage hens. Of course if the organic eggs are sold out we buy the others. (Consumer)
You cannot see the content of pesticides on the product. I often read the analyses when they are published. Generally vegetables from the South have higher pesticide content. That is why I prefer Danish fruit and vegetables – if I can afford them. (Consumer)

Finally, the consumers were asked if they thought they needed specific information on bacon production. Many of their comments had already been mentioned in answer to earlier questions. Here is a sample of the most cited answers.

The amount of antibiotics should be mentioned on the package. (Consumer)
As a minimum I would like to know where the pigs have been bred and slaughtered. It could influence my purchase. (Consumer)
I would like to know the water and fat content. When you buy cheap bacon it often disappears when you fry it. (Consumer)

Discussion of Findings

A couple of hundred years ago, Danish pigs were taken to Germany and sold in markets in northern Germany. Today, the transport of living pigs is on the rise after more than 100 years of *local* breeding, fattening and slaughtering of pigs. Due to changed market conditions, the export of live pigs from Denmark to most of Europe has increased considerably, both for piglets and fattening pigs. Denmark has been successful at breeding and the demand for live piglets has increased, while transport costs are low. Growing pig production in the new European member states is another underlying factor in this development. This example illustrates how market conditions have direct implications for ethics, in this case animal welfare.

The Danish pig and pork sector has managed to adapt production to changed market conditions. An example is the 'special pigs'. These production systems meet special requirements from different customers, typically the retail trade. The requirements are mostly connected to improvements in animal welfare. Today you can choose between organic pigs (corresponding to the EU organic standards), free-range pigs or UK pigs (with minor welfare improvement for the sows), all inside the same dominant meat company. Also a few pig types have 'add-ons' with regard to eating quality, for instance Spanish ham producers demand hams from 'heavy pigs'.

Even though Denmark has a long tradition of bacon production, it is impossible to buy sliced bacon produced in Denmark. The majority of the bacon sold in Danish supermarkets today is produced in Poland and Germany. The background for this shift 2 years ago was that the leading retailers in Denmark said that according to their analysis consumers did not mind where their bacon was produced. Our analysis indicates that consumers actually do care where their bacon is produced and the pigs are raised, and that they are convinced that bacon bought in Denmark also is produced in Denmark.

The objective of the Danish traceably system is to be able to trace and isolate all potentially affected pigs in the event of a disease outbreak. As with most other livestock traceability systems, it is *reactive* in nature and is not intended to transmit information proactively to downstream firms or end consumers, whether about safety, production practices, or quality of the final product. One of the managers from a large slaughterhouse admitted they did not made use of the information that was embedded in the system.

Some of the stakeholders interviewed, especially farmers and processors, argued that consumers are responsible for the extent to which food production today is ethical. But paradoxically, the consumers stressed that they expect the authorities to ensure ethically acceptable conditions. One of the reasons for this discrepancy is the fact that in the case of sliced bacon consumers are disconnected from any information about the product except the price. What is at issue, therefore, is not the isolated responsibility of a consumer/citizen, but the fact that a consumer makes his/her choice in a space and context essentially different from the context in which he/she presents his/her own opinions. The purchase situation is structured by the way the distributive trade organizes its sales, by the sets of rules that apply to labelling and advertising and other factors. To a great extent, consumers must apply their own knowledge in the purchase situation, enabling them to make qualified choices at the refrigerated food counter. While product prices may be directly comparable, the degree of respect for animal welfare associated with various meat products, for instance, is not.

In general, the pig farmers and the associated industry have a very defensive or reactive way of arguing. This is especially the case with discussion about animal welfare, but also in the discussion of environmental regulation or other issues that could imply increased cost. One of the problems is that several of these ethical issues are complicated and difficult to communicate to consumers through added value. Examples are breeding traits and GMO-free feed.

The organic sector has to some degree managed to establish a dialogue with consumers. In the discussion about the problems with free-range hens and the above-average dioxin contamination in free-range animal products, the organic sector has managed to present relatively complicated dilemmas to consumers.

From the interviews with Danish bacon consumers it is clear that they feel misinformed or even cheated, especially about the origin of the products. Danish consumers in general expect that when they buy bacon it has been produced in Denmark according to Danish regulations. In this project we have tried to follow the trace from bacon packages found in Danish supermarkets backwards. From many of the bacon products it was very difficult to follow this trace, especially to get the accompanying information on ethics.

Among consumers there was widespread criticism of modern industrial pig production. There seems to be an idea of the good and the evil, represented by two production forms. The good is of course a small-scale pig farmer where the animals have access to free movement in the open air. Organic and free-range pig production[11] seem to be the production systems that are close to this perceived 'good' production system. The evil is of course represented by the dominant large, intensive hog production units, often called 'pig-factories'. Production on these farms is characterized by rationality, effectiveness and economic optimization.

In our interviews with conventional hog farmers, we have found that they generally express frustration about this discrepancy between the consumers' idealistic perception of modern farm production and the reality, and because of this lack of knowledge, they were afraid of the consequences of total traceability in the food sector. First, the farmers had to educate the consumers, or as it was expressed by one of the farmers: 'The time is not ripe to tell the consumers that our animal production is an industry'.

The question is whether consumers really are interested in a deeper knowledge of the conditions of pig and pork production. Most of the interviewees from the retail sector questioned this, and did not believe that consumers would change their purchase habits if they had access to information on ethical attributes. From the stakeholders' point of view, it might look as if it is the consumers who should solve the problems of pig and pork production through their consumption of pork produced under ethical conditions.

From several investigations (e.g. Bredahl and Poulsen, 2002; Lassen et al., 2002) we know that consumers are ambivalent when it comes to their actual buying behaviour and their ethical concerns. Many of the stakeholders used this ambivalence to crystallize the discussion about ethics and food to an oversimplified question, namely either to have a guilty conscience and buy cheaply, or buy expensively and have a clear conscience.

In our interviews with both consumers and stakeholders we were presented with several structural barriers affecting the pig and pork supply chain. Many of these barriers arise from the economic and competitive conditions the pig farmers are dealing with. These barriers contribute to maintaining the existing form of production, in this case the intensive industrial pig and pork production system.

From a sociological perspective an important conclusion is that there are different contexts or rooms where you articulate your attitudes and where you consume. When attitudes to pig production are articulated in a room that is disconnected from

---

[11] Often consumers confuse organic and free-range products because they believe that 'organic' is equivalent to 'free-range' food (See also Zanoli, 2004 and Kristensen et al., 2004).

concrete everyday life and consumption, it is easier to formulate ideals; you are a citizen and can formulate your claims to the good life and good food production. It is different when you are a consumer, and the context is constituted by everyday life. In this case the barriers to achieving the ideal of pig and pork production will be visible, and practice will look very different from the attitudes expressed elsewhere. When we asked consumers to reflect upon the last time they bought bacon they mainly spoke about taste, health and economy; but when we asked the consumers to reflect upon pigs and pork in a more abstract way, the themes spoken about were the environment, animal welfare and criticism of industrial pig production.

One important barrier is the fact that the export market is dominating the home market. In the case of the Danish pig/pork sector this is crucial because the majority of the production is sold to export markets. If consumer power were absolute, it would not be the Danish consumer who decided which products should be produced, but the Americans, the Germans, the British, etc. One of the farmers expressed this dilemma in this way: '*The Danish consumer should not be the only one to decide how we should produce. He only buys 5–10% of our total production*'.

Seen from the consumers' point of view there are several reasons to be sceptical about the assumption that consumers have absolute power over the production sector. In general, the pig and pork sector is not much swayed when pig production is criticized by environmentalists, animal welfare organizations, consumer organizations, etc.; but when their main customers – in Britain, Germany, etc. – comment, they really listen. The 'UK pig' is a result of this.

One of the representatives from the retail sector questioned whether consumers in general were interested in more information about the products: '*Everybody asks for information but nobody uses it. It is typically consumer organizations that talk about ethical traceability. It is not the interest of the consumers*'.

In order to make the right choices, it is of course necessary to have accurate knowledge about the products, and here labels are often mentioned as one way to provide the knowledge in a condensed form. Nevertheless, many of the consumers in our survey reflected that they did not have much confidence in the systems behind some of these labels. In Denmark this is not the case for the organic label. This label indicates that the Danish state guarantees that the product has been produced according to EU organic laws. The public food authorities control all organic products.

The labels are also often criticized because they have no real information value, and often it is difficult to tell weather it is manipulation or real information. The question when we talk about ethical traceability is how to prepare the consumers' foundation of knowledge in order to make choices when they buy pork. One way could be to give factual information directly in the stores where pork/bacon is sold. For pork from traditional pigs, this information would say: '*pigs raised on slatted flooring*' if this was the case; and it would prevent the use of pictures that misrepresent the reality of the conditions in which pigs have been raised. As one of the farmers said: '*In our commercials we give the impression that the pigs live in straw, but the truth is that 95% live in housing units with slatted floors*'.

The average Dane spends less than 10 min on his daily shopping. He does not have time for much reflection or to read a lot of product information. And the average consumer does not know all the details about what is good or bad when it comes to the multiplicity of themes that fall within our '10 ethical concerns'. It does not improve matters that products, production processes and product contents are in the process of changing fast. The development of consumers' knowledge and competence as regards environmental, health and ethical matters is by no means making similar progress.

References

Arbejdstilsynet (1997) (The Danish Working Environment Authority) *ADOS-ANALYSE – Anmeldte arbejdsskader i landbruget*, 1993–1997.

Baadsgaard, N.P., P.H. Jørgensen, Aa. J. Jørgensen, P.H. Rathkjen (red.) (2003) *Rapport vedrørende velfærd i svineproduktionen*. Dansk Veterinær Hyologisk Selskab.

Biello, D. (2006) 'Scientists Engineer Pigs with Heart-Healthy Meat' in *Scientific American*, March 27, 2006 (http://www.sciam.com/article.cfm?articleID = 00095050-1EB7-1423-9EB783414B7F0000).

Bredahl, L., and C.S. Poulsen (2002) 'Perceptions of pork and among Danish consumers', Project paper no 01/02. The Aarhus School of Business, – Århus, June 2002.

Bækstrøm Lauritsen, H. (2005) Danish Quality Guarantee. DS, København.

Council Directive 93/119/EC of 22 December 1993 on the protection of animals at the time of slaughter or killing. *Off J Eur Commun*, No. L 340/21.

Claudi-Magnussen, C., S. Jensen and L. Hansen (2000). Produktkvalitet, in Hermansen J E (ed.) Økologisk svineproduktion – Udfordringer, muligheder og begrænsninger, FØJO 2000.

Cook, C.J., C.E. Devine, A. Travener and K.V. Gilbert (1992) 'Contribution of amino acid transmitters to epileptic activity and reflex suppression in electrical head stunned sheep', *Res Vet Sci*, 52: 48–56.

Christensen, J. (1983) *Rural Denmark 1750–1980*. Copenhagen: The Central Co-operative Committee of Denmark.

Damm, B.I. (2004) Velfærdsproblemer hos danske søer [Welfare problems in Danish sows]. Dyrenes Beskyttelse.

Danske Slagterier (2005) (Danish Meat Association) http://www.danskeslagterier.dk

Danske Slagterier (2004) (Danish Meat Association) Statistik 2003. Danske Slagterier, København.

Danish Crown (2003) Code of Practice – rules for Danish Crown's pig farming co-operative members, Randers.

DANMAP (2004) Use of antimicrobial agents and occurrence of antimicrobial resistance in bacteria from food animals, foods and humans in Denmark.

Golan, E., B. Krissoff and F. Kuchler (2004) *Food Traceability – One Ingredient in a Safe and Efficient Food Supply*. Economic Research Service, USDA, Vol. 2 April, 2004.

Gray, J.I., B. McDonald, A.M. Pearson and I.D. Morton (1981) 'Role of nitrite in cured meat flavor: a review'. *J Food Protect*, 44: 893–896.

Hassing, A., P. Bækbo (2004) *PMWS Manual*, Dansk svine produktion, FAGLIG PUBLIKATION, published 29.11. 2004.

Heinemann, T. (2005) *Dansk svinefarm i Polen undviger miljøkrav* in Fagbladet 3F, marts 2005.

Hobbs, J.E. (2001) *Against All Odds*. Centre for the study of Co-operatives, University of Saskatchewan, Saskatchewan, Canada.

Kolind Hvid, S., P. Nielsen, N. Halberg, and J. Dam (2005) Miljøinformation i produktkæden. Et case studie af produktkæde med svinekød. Miljøstyrelsen, Miljøprojekt nr. 1027.

Kristensen, N.H., T. Nielsen, and M.W. Hansen (2004) *Understanding Consumers Attitudes to Organic Food in Denmark.* Results from focus groups and laddering interviews, Technical University, Denmark.

Lassen, J, E. Kloppenborg and P. Sandøe (2002) Folk og Svin. En interviewundersøgelse om danske borgeres syn på den danske svinesektor og svinekødet. Center for Bioetik og Risikovurdering.

Landbrugsavisen (2006) http://www.landbrugsavisen.dk/ LandbrugsAvisen /2006/8/5/Svensk + bacon + til + Polen.htm

Lemoine, W., Ragus L.C., Christensen J.M. (2002) Forsyningskæder for ferskvarebrancher – Danish Crown et case studie. Institut for Transportstudier og SDU.

LU (Landsudvalget for Svin) (2005) http://www.danskeslagterier.dk/smcms/Landsudvalget_Svin/Videnscenter/Forsoeg_og_udvikling/Stalde_produktsystem/Index.htm?ID = 1402

National Hog, Farmer Magazine, (2000), no. 5.

OJL (2002) 'Regulation (EC) No 178/2002 of the European Parliament and of the Council of 28 January 2002 laying down the general principles and requirements of food law, establishing the European Food Safety Authority and laying down procedures in matters of food safety', *Official Journal of the European Communities*, 1.2.2002, L31/1–24.

Pedersen, S. and M. Obelitz (2001) Kvalitativ undersøgelse af danskernes holdninger, Analyserapport, København: Danske Slagterier/ Tranberg Marketing Rekommandation.

ds-nyt (2001) The Association of Danish Slaughterhouses' newsletter, Issue 6.

Zanoli, R. (ed.) (2004) *The European Consumer and Organic Food*, OMIaRD, Vol. 4, Aberystwyth.

Chapter 6
Ethical Traceability in the UK Wheat-Flour-Bread Chain

Rosalind Sharpe, David Barling, and Tim Lang

Introduction

This chapter presents the findings of an investigation into the role of ethical concerns in one of the UK's most important food supply chains, covering the production of wheat and its transformation into flour and bread. Bread has long been a staple food in the UK. Today, the British consume the equivalent of around nine million large loaves a day, almost all of it made from wheat flour. Wheat and bread are thus linked by a complex chain that must be both flexible and reliable. This chain presents considerable challenges both to traceability and to the tracing of ethical aspects of production.

In the UK, most bread is the product of intensive cultivation and an industrial manufacturing process, and is sold by supermarkets rather than specialist bakers. Beyond the farm gate this chain is very concentrated and there is considerable vertical integration between millers, ingredient suppliers and bakers. Coexisting with this dominant chain is a comparatively small 'craft' chain, characterized by smaller production units, less mechanized and more time-consuming manufacturing methods, and less use of inputs or additives. However there is trade (e.g. in ingredients and services) between the two chains. Beyond the farm gate, it is hard to discern separate organic and conventional chains, because the organic chain depends on the conventional chain for some supplies and distribution facilities, and the conventional chain also deals in organic goods. Around 85% of the wheat used for non-organic UK bread is grown in the UK, but less than 50% of the wheat for organic bread is home-grown. Less than 2% of flour and bread are imported or exported.

Reflecting its central importance in the diet, bread was one of the earliest foods to be regulated (e.g. the Assize of Bread, which was established in 1266 to set the weight of bakers' loaves, was itself based on earlier rules). Today, the chain is regulated in one way or another for most of its length, with rules governing the development of new wheat varieties, the purity of the seed supply, the handling of agricultural chemicals, the addition of substances to flour, the ingredients permitted in bread and the information that must be declared on the labels of wrapped loaves. However, these rules are contested (e.g. some claim that the selection of substances that must be added to flour is anomalous) and also subject to change (e.g. labelling

C. Coff et al. (eds.) *Ethical Traceability and Communicating Food*,

law is currently being reviewed by the EU). In addition to these regulations, some sections of the UK wheat chain (mainly between farm and mill) have within the past decade been covered by 'assurance schemes' (in effect private regulatory systems defined and policed by stakeholders in the sectors concerned), which aim to ensure that the supply base consistently meets specified standards of food safety and quality. Further along the chain, and usually working backwards from retailer to baker and/or baker to mill, the chain is additionally 'regulated' by means of detailed contractual specifications, which prescribe ingredients, processes and procedures, and which aim to ensure that stipulated standards of food safety, quality, consistency and product differentiation are met. This regulation and quasi-regulation covers some of the issues identified by this project as ethical concerns (see Chapter 1 and Table 1.3 for a discussion and list of these ethical concerns).

Many of the regulations and assurance schemes mentioned above depend on traceability systems to identify consignments, confirm compliance or verify attributes. However, although bread has a symbolic status as the 'staff of life' it is not currently perceived by consumers or treated by the industry as a risky food requiring highly visible traceability (unlike meat). Nor is it marketed on the basis of place of origin or plant variety (unlike, e.g. cheese, wine or coffee).

A significant barrier to traceability in this chain is presented by the longstanding and ubiquitous practice of blending wheat or flour so that it can be conveniently stored and transported, and to manipulate consistency, quality and cost. Blending the output of several fields in one barn, or several farms in one silo, or several silos into one milling line, or several flour bins into one type of flour, makes traceability back to farm or field difficult and costly (though examples are to be found). A response to this challenge, in the face of pressure throughout the food industry for increased traceability, has been to develop the assurance schemes described above. By ensuring that the whole supply base meets specified standards, the schemes reduce the risk attached to being unable to trace back to individual farms. As a corollary, 'Identity Preservation' (IP) systems, which do maintain separation of wheat through the system by source or type, are available at extra cost. It must, however, be asked whether the goal of more precise or narrower traceability for ethical reasons – e.g. to guarantee locally sourced bread to the consumer – might not undermine practices such as blending, which have been developed to improve other aspects of quality, such as consistency and reliability of supply.

Ethical traceability – meaning traceability which records ethical aspects of production – is limited and patchy in this chain. Bread made from organic wheat is an example of visible ethical traceability spanning the whole chain. Our research found that firms which felt that their sourcing and production practices were more ethical than the industry standard often went to great lengths to communicate these practices to customers, via labels, web sites or in-store notices, both to attract customers who shared these values and also in some cases as a critique of the rest of the industry. These claims were not always independently audited, however. On the other hand, industry assurance schemes, such as the Assured Combinable Crops Scheme, which sets standards for growing, harvesting and storing wheat, are independently audited, but in the case of bread are not communicated to consumers.

The remainder of the chapter describes the findings of qualitative research conducted among actors and consumers in the wheat-bread chain, looking both at ten identified ethical concerns (see Table 1.3) and at the wider use of information in the chain. It then discusses the implications of the research for traceability and ethical traceability. An overview of the supply chain and its historical development is provided to set the research in context.

Wheat into Bread: Overview of a Mature, Complex Supply Chain[1]

Worldwide Wheat

Wheat was one of the earliest domesticated plants, first grown in the 'Fertile Crescent' of what is now the Middle East around 8,500 years ago. Cultivated varieties of the genus *Triticum* now represent the world's most important food crop, grown on more than 240 million hectares worldwide. Wheat's success can be attributed to its nutritional properties (it is an easily digested source of starch, protein, vitamins, minerals, oil and bran), its agronomic adaptability (it grows in many climatic zones) and the ease with which the dried grains can be stored. The raised loaf of bread is possible because the wheat kernel contains gluten, an elastic form of protein that traps bubbles of carbon dioxide when fermentation occurs in leavened dough, causing the dough to rise (Curtis, 2002).

There are now 30,000 varieties of wheat, of which only a few are commercially grown in the UK. It can be planted in either autumn or spring, for harvest the following summer. It has been selectively bred over millennia to favour characteristics desired by growers (such as vigour, yield and straw strength) and increasingly by processors (such as protein quality). In experiments, biotechnology has been used to improve wheat protein quality, but there is currently no genetically modified wheat in commercial cultivation anywhere in the world (David, 1979; Curtis, 2002; Pena, 2002).

World wheat production increased dramatically in the second half of the 20th century, due mainly to yield increases, and has remained above 500 million tonnes per year since 1986. Major producing regions include the EU, China, India, the USA, Russia, Canada, Australia, Pakistan, Turkey and Argentina. The UK is Europe's third largest wheat producer, behind France and Germany, accounting (in 2005) for 12% of the EU total. Production can be adversely affected by weather, but it is unusual for all producing regions (which span all continents and both hemispheres) to be affected in the same year, so trade can usually compensate for

[1] In addition to the sources cited throughout, many details in this chapter are drawn from research conducted with stakeholders from the supply chain, as described in the Preface to Part II and later in this chapter.

Table 6.1 World wheat production, consumption and trade, 2003–2006 (USDA Quarterly International Trade Report Feb 2006)

	2003/04	2004/05	2005/06	Year to year change 04/05 to 05/06	
	(million tonnes)	(million tonnes)	(million tonnes)	(million tonnes)	(%)
Production	555	627	616	−11	−2
Consumption	589	609	624	+15	+2
Trade (July/June)	105	113	110	−3	−3
Ending stocks	132	150	142	−8	−5

Table 6.2 UK average prices, milling and feed wheat, and milling premium 1996–2006, £/t (Defra, 2006)

	1996	97	98	99	2000	01	02	03	04	05	06
Feed wheat	111	89	75	73	66	75	63	75	76	66	78
Milling wheat	122	101	84	81	74	82	71	83	86	73	83
Premium (milling over feed)	11	12	9	8	8	7	8	8	10	7	5

regional shortfalls. However, climate change presents an unquantifiable threat to the world's wheat supply (Curtis, 2002; ABARE, 2007; Defra, 2007a).

Although most wheat is consumed in the country where it is produced, roughly one sixth is traded. Table 6.1 shows recent global production, consumption and trade figures. The major exporters are the USA, Canada, Australia, the EU and Argentina. Major importers include Egypt, Japan, Brazil and China. As a globally traded commodity, wheat fluctuates in price depending on forecasts of supply and demand around the world, and an increase in speculative investment has made prices more volatile. In recent years, as markets have expanded, demand has often exceeded supply, with wheat increasingly used as animal feed and a new market emerging in the form of ethanol, a biofuel (it has been estimated that the latter could consume 10% of UK wheat production by 2010 if EU targets are met). This growing demand, coupled with uncertainty over supply, has put pressure on world stockpiles, reported in 2006 to be at their lowest level for 25 years, triggering fears of a crisis in supply and driving up prices. In the UK, it remains to be seen whether the recent recovery in prices will prove to be a temporary blip or the reversal of a downward trend (Table 6.2) (Curtis, 2002; USDA, 2006; *FT*, 2006; Finch, 2006; ABARE, 2007; Defra, 2007a).

Historical Context in the UK

The UK is currently a net exporter of wheat, but this has not always been the case. Wheat was intermittently exported from Britain from the days of the Roman

Empire until the 18th century, when imports began to increase gradually, reflecting population growth. Then, in the mid-19th century, the repeal of the UK corn laws removed import tariffs, and the opening up of the American interior by railways brought a flood of American wheat to Britain, which even with transport costs could be sold in the UK for half the price of home-grown grain. Britain became a centre of the growing world grain trade, while domestic cereal production by acreage rapidly reduced, almost by half from 1869 to 1891. Imports rose from 0.86 million tonnes in 1857 to 2.5 million tonnes in 1877, and by 1880 English bread was for the first time made almost equally with home-grown and imported grain. By 1914, less than a quarter of the wheat used was home-grown, and this dependence on imported bread wheat lasted until the 1970s. By then, changes in baking technology allowed a higher proportion of British wheat to be used in commercial bread-making, and rising import tariffs following the UK's accession to the EU in 1973 provided an incentive to the industry to choose home-grown wheat. As a result, the proportion of domestically grown wheat in the milling 'grist' (the mixture of grain fed into the mill) has doubled since the 1970s (Fig. 6.1). UK wheat production has also expanded, mainly through yield increases (Fig. 6.2) (Burnett, 1968; Morgan, 1979; Montague, 2000; Nabim, 2006a; Cauvain and Young, 2006; Defra, 2007a).

The wheat grain is made up of an outer husk (the bran), a starchy centre (the endosperm) and the germ, which is the embryo of the next wheat plant (Fig. 6.3). Milling reduces the grain to uniform particles of a desired size. From ancient times,

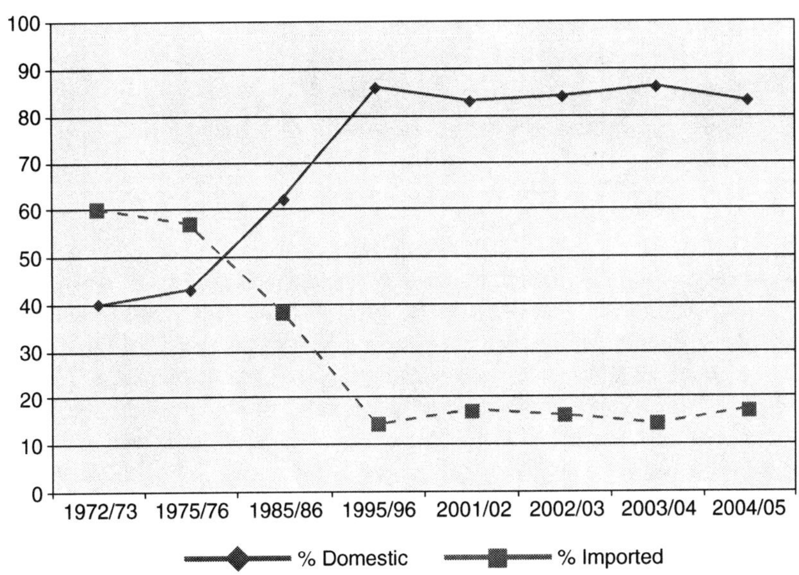

Fig. 6.1 UK milling grist: percentage imported and domestic wheat, 1973–2004 (Nabim, 2006b)

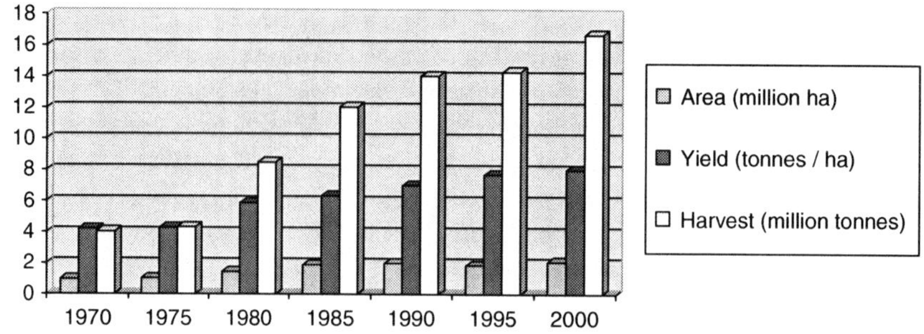

Fig. 6.2 UK wheat area, yield and harvest, 1970–2000 (adapted from Defra, 2006)

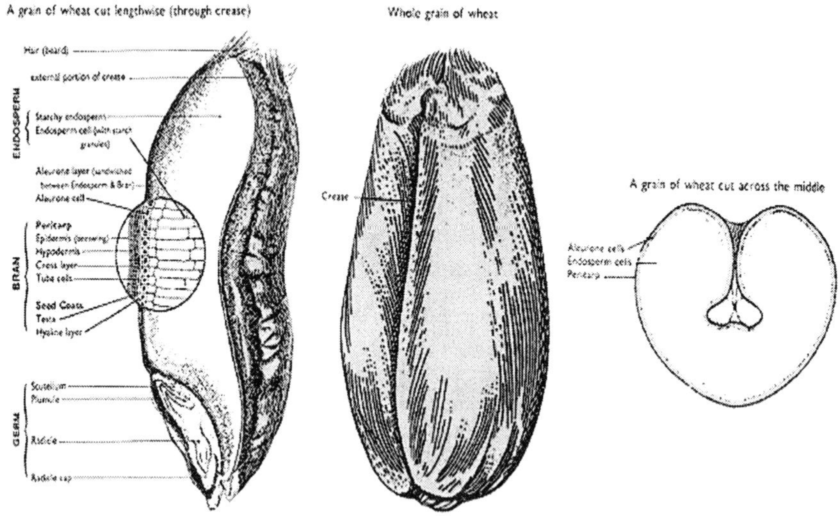

Fig. 6.3 Wheat grain (Nabim, 2006a)

this was done by grinding the grain between stones. Traditional mills, powered successively by water, wind and steam, used large, circular stones, one of which was rotated above the other. Grain was fed down a hole in the centre of the top stone and the resulting flour driven to the rim of the stones by centrifugal force, and then collected. The coarseness of the flour could be controlled by adjusting the distance between the stones, and some of the coarser particles could be sieved out to give a finer, whiter flour (traditionally the preference of the wealthy). This flour, however, would still incorporate all parts of the grain (bran, germ and endosperm), resulting in a flour and bread with (by modern standards) a coarse texture and greyish colour (David, 1979; Nabim, 2006a).

From the 1870s, stone milling was rapidly replaced by roller milling, which uses a series of rotating steel cylinders to break open the wheat grains, separating the bran and germ and then reducing the size of the particles. This system offers significant commercial advantages over stone-milling – the flour is whiter, it keeps better because the oil-bearing germ is removed, and the bran and germ can be sold as animal feed, adding to the miller's profits. When roller milling was adopted, however, it was not understood that the germ and bran contain vital nutrients, and that removing them impoverishes the flour. Burnett (1968:140) comments, 'It was fortunate from this point of view that by the time roller milling had come into general use, bread alone was ceasing to be the principal food of the English people' (Burnett, 1968; David, 1979; Nabim, 2006a).

Although wheat's main food value is as a carbohydrate, the quantity and quality of protein in the grains (especially the elastic protein gluten, which allows dough to stretch and rise) have come to be seen as key determinants of use and price. The protein content of wheat varies depending on the variety and the conditions in which it is grown, and for climatic reasons wheat grown in the UK tends to have lower protein levels than wheat grown in more extreme climates, such as North America. For this reason, around 15% of the wheat used for flour milling continues to be imported high-protein wheat, sourced mainly in Europe and North America. Only a handful of bakers use exclusively UK-grown wheat (David, 1979; Curtis, 2002; Nabim, 2006a; FoB, 2006).

The Contemporary UK Wheat-Flour-Bread Chain, by Sector

The dominant UK wheat-flour bread chain is intensive, industrial and, beyond the farm gate, very concentrated (Fig. 6.4). Coexisting with this dominant chain is a comparatively small 'craft' chain, characterized by smaller production units, less mechanized and more time-consuming manufacturing methods, and less use of inputs or additives.

Of the average annual wheat harvest of around 15 million tonnes, the largest portion (more than 40%) is used for animal feed. Around 33% is milled into flour, and around 15% exported, mainly to Spain and other southern European countries. Some is reserved for use as seed, and some used by distillers. Table 6.3 provides a breakdown of UK wheat supply and use for the year 2004. Within the food industry, the flour-bread chain is characterized as 'low risk', involving stable, ambient goods. It is an indication of the strategic importance of the wheat-bread supply that the government requires millers to file monthly returns detailing wheat stocks and flour production (Defra, 2007b).

Farming

Wheat is the UK's most important cereal crop, with an average annual harvest of around 15 million tonnes. Around 28,000 UK farmers grow some wheat, with between 8,000 and 10,000 growing substantial quantities. The harvested area is

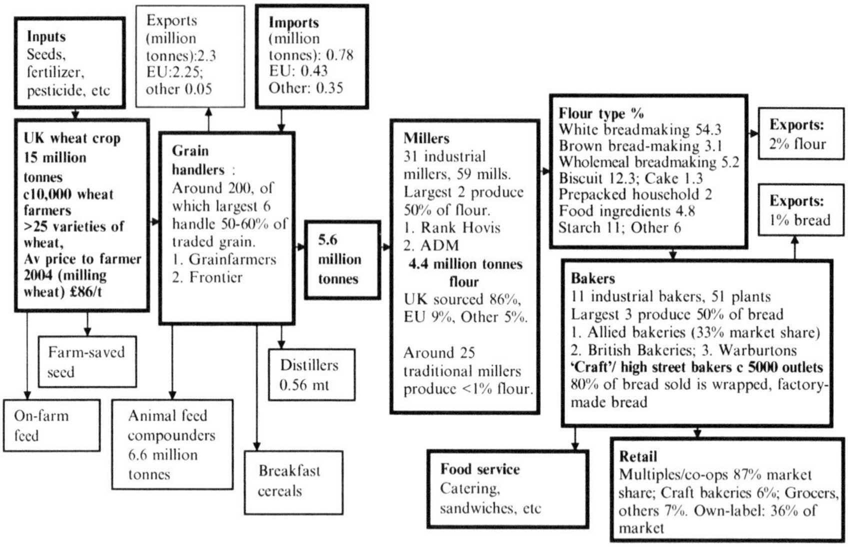

Fig. 6.4 UK wheat-flour-bread chain (Defra, 2006; FAB, 2006; FoB, 2006; Mintel, 2005; Nabim, 2006a; Nabim, 2006b)

Table 6.3 UK wheat supply and use, 2004 (Defra, 2006)

Production	
Area (000 ha)	1,990
Yield (t/ha)	7.8
Volume of harvested production (000 t)	15,473
Prices	
Milling (£/t)	85.8
Feed (£/t)	76.1
Imports/exports (000 t)	
Imports from: the EU	432
The rest of the world	352
Exports to: the EU	2,250
The rest of the world	43
Total domestic uses (000 t)	13,287
Of which: Flour milling	5,576
Animal feed	6,633
Seed	280
Other uses and waste	798
% home-grown wheat in milling grist	86

around 1.8 million hectares (more than 40% of the total arable area); almost all of it intensively cultivated winter wheat. The wheat has a relatively low protein content but very high average yields of 7.5–8 t/ha (compared with around 2.5 t/ha in North America), achieved through intensive use of fertilizers and pesticides. Depending on local growing conditions and prevailing prices, farmers may choose

to grow wheat for milling, feed or other uses. Milling wheat sells for a higher price (see Table 6.2), but feed wheat tends to give higher yields. The UK farmer's share of the price of a white sliced loaf was 15% in 2005, down from 23% in 1988 – a fall of 35%. Recent CAP reform, which partially decouples subsidy from production, may affect UK wheat production. Arable Area Payments in the UK in 2004 (excluding set-aside) amounted to £900 million, of which £447 million were for wheat. The area of the organic wheat crop is comparatively tiny, at 15,000 ha, with the result that more than half of the organic milling wheat is imported (Curtis, 2002; Defra, 2005a, 2007a).

Inputs (Seeds and Agrichemicals)

The agricultural inputs sector does not break down its figures for separate crops, so it is difficult to assess the size of the sector as it applies specifically to wheat. According to the sector's largest trade association, the Agricultural Industries Confederation (AIC), the UK pesticides market is worth £370 million and the fertilizer market worth £500 million, with the multiplication and distribution of agricultural seeds worth a further £200 million. This puts the total inputs market at £1.07 billion, of which wheat, as the UK's largest arable crop, will account for a significant portion (AIC, 2006).

There are currently eight seed-breeding companies in the UK with wheat programmes, only one of which is UK-owned. Seed supply in the UK has been regulated since the early 20th century to control quality and consistency, with regulation latterly taken over by the EU. Seed varieties can only be marketed if they are listed in the EU Common Catalogue of Agricultural Plant Species or its national equivalent, the National List. To be listed, new varieties must undergo several years of testing to demonstrate that they are 'distinct, uniform and stable' (DUS) and have 'value for cultivation and use' (VCU). In the UK, there are further tests for inclusion on the Recommended List, produced by the UK cereals advisory body, the Home Grown Cereals Association. Varieties are selected for the Recommended List by a committee of growers, millers, bakers and academics, who look for characteristics desired by growers and processors, such as yield or suitability for breadmaking. Recommended List varieties account for 95% of UK wheat production. For winter wheat for 2006–2007, the Recommended List includes nine varieties judged suitable for milling into bread flour (BSPB, 2006; HGCA, 2006).

To be sold, the seed itself must be 'Certified', meaning it meets statutory quality specifications relating to variety, germination capacity, disease and impurities. Most seed varieties now in use are protected by Plant Breeders' Rights, and royalties are payable, including (at a reduced rate) when the seed is saved by the farmer for use the following year (Farm Saved Seed) (AIC, 2006; BSPB, 2006; HGCA, 2006).

Almost all UK wheat is gown intensively. The average crop is treated with three herbicides, three fungicides, two insecticides, one growth regulator, plus fertilizer. A 500-hectare wheat farmer might spend £75,000 on pesticides and £60,000 on fertilizers annually and use 50 different products, though not necessarily all in the same year. Agrichemicals are subject to an elaborate testing and registration process, involving both UK and EU regulation (CSL, 2005; CPA, 2006; AIC, 2006).

Merchants

Most wheat in the UK is traded by merchants, who buy from farmers and sell to millers or other users, with only a small amount sold directly by farmers to millers. According to the AIC, the UK market in merchanting arable crops is worth £3 billion. Interviewees estimated that of around 200 firms in the sector, the largest six handle 50–60% of traded grain. Although wheat is traded as a global commodity, and increasingly traded speculatively, the process of trading between farmers and merchants remains relatively personal, usually conducted by telephone, between parties known to each other. Farmers can check current wheat spot and futures prices on web sites tracking commodity prices (e.g. www.farmersweekly.co.uk/Prices). The UK's relatively small area, and the consequent proximity of farms and mills, has produced a system in which most wheat is stored on farm until needed, then delivered directly from farm to mill (unlike in geographically larger production areas, where grain may be pooled several times between farm and mill) (AIC, 2006).

Milling

According to the milling sector's trade association, the National Association of British and Irish Millers (Nabim), to which most industrial millers belong, the UK currently has 31 industrial millers, operating 59 mills, with the 2 largest companies (Rank Hovis McDougal and Archer Daniels Midland) accounting for 50% of production. In 2004, the millers processed 5.6 million tonnes of wheat into 4.4 million tonnes of flour, with domestically produced wheat accounting for 86% of the total and the rest imported, mainly from the EU and North America. The sector's total turnover is just under £1 billion per annum. Grain is transported from farm or port to mill by road (imported wheat is transported by sea and by road). A large mill produces around 200,000 t of flour per year (Nabim, 2006a; Nabim, 2006b).

There are also 25 members of the Traditional Cornmillers' Guild (TCMG), whose members operate watermills or windmills. Their output is probably less than 1% of total flour production (TCMG, 2006).

The millers stringently test and analyze grain as it arrives at the mill, to avoid contaminating their grain stocks by the addition of mould or impurities, and to identify its characteristics in order to be able to use it appropriately. Grain samples may be tested for, among other things, the presence of mycotoxins, moisture levels, protein content, specific weight and the amount of the enzyme alpha-amylase present (tested by measuring the Hagberg Falling Number). Based on the information so obtained, millers blend wheat from different sources to achieve the properties they are looking for, and also to control costs. Before being milled, the grain is 'conditioned', during which process it is cleaned and its moisture content stabilized (Nabim, 2006b; HGCA, 2006).

Of the 4.4 million tonnes of flour produced by the industrial millers, the bulk (53% in 2005) is white bread-making flour. The other main types are biscuit (12.3%), wholemeal bread-making (5.2%), food ingredients (4.8%), brown bread-making (3.1%) and cake (1.3%). Packaged flour for household use accounts for 2% of production. Within these categories, a large mill may produce 60 or so different blends of flour, with minute variations to suit the requirements of individual customers (Nabim, 2006a; Nabim, 2006b).

In the UK, flour is classified according to its 'extraction rate' which refers to the percentage of the original grain which is incorporated in the finished flour. Wholemeal flour has an extraction rate of 100%, meaning that the entire grain is turned into flour (though in a roller mill its constituents will have been separated and reassembled during processing). Brown flour is a generic term for flours with an extraction rate of between 80% and 85%. White flour has an extraction rate of between 70% and 75%, meaning that 25–30% of the grain, mostly consisting of the bran and germ, has been removed (FAB, 2006; FoB, 2006).

UK law requires that flour with an extraction rate of less than 80% must be 'fortified' with a number of nutritional supplements, with the aim of restoring the nutritional value to the level of 80% extraction flour. These regulations, introduced in the early 1950s, recognize that roller-milled white flour has a depleted nutrient content. (They were introduced because during World War II, an extraction rate of around 85% was imposed on all UK millers, to make most efficient use of wheat supplies; the resulting 'National Loaf', though unpopular, was understood to have health benefits, which post-war ministers feared would be lost when the milling restrictions were lifted and the population once again expressed a preference for white bread.) The nutrients that must be added are thiamine (vitamin B1), nicotinic acid (vitamin B3), iron and calcium carbonate, and these 'statutory additives' need not be listed on product labels. However, the nutrients chosen reflect the level of nutritional knowledge of the 1950s, and the list has been criticized for being anomalous, as well as for being 'invisible' to consumers. For example, Whitley (2006:23) lists 19 vitamins and minerals which are lost to varying degrees when flour is milled to a 70% extraction rate. The UK government is currently considering adding another nutrient, folate, to the list of statutory additives, to reduce the incidence of congenital neural tube defects. The additives are usually delivered to millers premixed, and dosed into the flour at the end of the milling process (David, 1979; SI 1998:141; FAB, 2006; FoB, 2006; FSA, 2006; Whitley, 2006).

Baking

The UK bread and bakery goods market is worth over £3 billion, and is one of the largest sectors of the food industry. The total volume of production is 2.9 million tonnes per year. Of this, 80% is produced in bread factories, also known as 'plant bakeries'. A further 17% is produced in an estimated 1,500 'in-store bakeries' (ISBs). These are located in retail outlets, such as supermarkets, and may either bake bread from scratch or use factory-made dough, supplied frozen, partly or fully baked, which is then given a final baking on-site. Just 3% of bread (by volume) is made by traditional methods in 'craft' bakeries (see Table 6.4 for a comparison with other EU countries) (Mintel, 2005; FAB, 2006; FoB, 2006).

The UK industrial bread sector is dominated by 11 companies, operating 51 factories. The three biggest companies (Allied Bakeries, British Bakeries and Warburtons) account for half the bread market by volume. There is strong vertical integration between millers and bakers, with the two largest plant bakers, Allied Bakeries and British Bakeries, owned by millers Associated British Foods and Rank Hovis MacDougal respectively. Less than 1% of bread consumed in the UK is imported, and around 2% of total output is exported (Mintel, 2005; FoB, 2006).

There are two main methods of making bread mechanically: the Bulk Fermentation Process (BFP) and the Chorleywood Bread Process (CBP). The BFP, which is used in smaller, 'craft' bakeries, broadly follows traditional methods, in which ingredients are mixed together to form a dough and then left to ferment for anything from an hour to overnight. During this time, the dough becomes elastic, the time taken depending on the amount of yeast used and the temperature. Bread made by this method typically consists of flour, yeast, salt, water and (sometimes) fat (FoB, 2006; Dove's Farm, 2006a).

The Chorleywood Bread Process is used in plant bakeries and accounts for more than 80% of UK bread production. Introduced in 1961, having been developed by the Flour Milling and Baking Research Association (now part of the Campden and Chorleywood Food Research Association) at Chorleywood, in southern England, it fundamentally transformed the method of bread-making. Sometimes known as the 'no-time' method, it eliminates the need for slow fermentation of the dough in bulk. Instead, the dough's elasticity is produced by a few minutes of high-speed mechanical mixing at a precisely controlled energy level, and the development of the dough depends on the presence of a number of specially developed additives to stimulate and regulate the action of the main ingredients. Bread made by this method typically includes higher proportions of yeast, salt and water than are likely to be found

Table 6.4 Market share of plant and craft bakers, selected EU countries: (FoB, 2006)

Country	Plant bakeries	Craft bakeries	Others (e.g. ISBs)
Austria	20	80	
Belgium	36.1	58.8	5.1
Denmark	51	49	1
France	23	68	9
Germany	35	65	
Italy	10	90	
Netherlands	73.7	21	5.3
Spain	34	66	
UK	80	3	17

in traditionally made bread. Although originally devised as an aid to small-scale producers, the CBP was widely adopted in plant bakeries, because it enabled mass production of loaves of uniform quality, and reduced costs. The system has been exported around the world, and has recently been adapted to allow the production of different types of bread, including croissants, French sticks and naan bread. However, the system is not without critics, including David (1979:195), who asked 'Could it be that the technological achievements of the bakery scientists have removed bread just that much too far from the image of a real loaf as it once was?' Whitley (2006), another critic of the CBP, lists six categories of additives that may be used in the process, including emulsifiers (to control the size of gas bubbles), reducing agents (to create more elastic dough) and enzymes from fungal, cereal, bacterial and animal sources (used for a variety of purposes, including to extend shelf life by enabling bread to remain soft for longer after baking) (David, 1979; Cauvain and Young, 2006; FoB, 2006; Dove's Farm, 2006a; Whitley, 2006).

A separate sector now exists to supply bakers with their ingredients, including yeast, salt, fats, improvers, emulsifiers, preservatives, relaxants, raising agents and dough conditioners. For example, the web site of the ingredients manufacturer Puratos lists 30 different bread improvers, such as 'Primavera clean label improver for no-time doughs', the term 'clean label' indicating that because the ingredient is classed as a 'processing aid', which is destroyed in the baking process, it does not have to be listed among the ingredients on the label. The Association of Bakery Ingredient Manufacturers (ABIM) has 15 member companies, representing a turnover of £500 million and employing 3,000 people (Puratos, 2005; ABIM, 2006).

Retail and Consumption

The British buy a small and declining proportion of their bread at shops dedicated to selling bread (currently just 6%). Although the press reports a resurgence of interest in artisanal bread, more than 80% of bread is sold through chain retailers (supermarkets, co-ops and convenience stores). Independent grocers and others account for the remainder. Bread manufactured by plant bakeries but sold under retailers' own brands accounts for 36% of the market in wrapped bread (Mintel, 2005; FoB, 2006; Lepard, 2006).

Bread is eaten several times a week by 96% of the UK population, although consumption has been declining for more than a century, and now averages 0.75 kg per person per week (compared with up to 5.5 kg per person per week in the mid-19th century). But although volume sales of bread for home consumption fell by 5% between 1999 and 2004, in terms of value the bread market grew by 9% over the same period. This is partly explained by a trend away from low-cost 'economy' loaves towards 'premium' (higher-priced) or speciality breads. The sector showing greatest growth is 'plant-baked speciality breads'. Other growth areas include functional breads (e.g. containing fibre or folates) and sandwiches (Burnett, 1968; Mintel, 2005).

More than 76% of bread sold in the UK is white (28% of it pre-sliced). A survey by the market research agency Mintel (2005) found that 84% of respondents were white bread users and 60% were brown bread users, varying by social class and family composition. Young adults and families with children were heavier users of white bread, as were people in poorer socio-economic groups. More affluent groups were heavier users of brown bread. Sales of organic breads and cereals increased from £29 million in 1998 to £75 million in 2003, equivalent to an annual growth rate of 21% (Mintel, 2005).

The overall decline in bread consumption is a cause of concern to the industry. It is partly caused by the availability of alternatives and partly by the decision of some consumers to avoid bread for dietary or health reasons, such as the avoidance of gluten (which is an allergen), or the avoidance of carbohydrates in order to lose weight. According to Mintel, 16% of the population may be actively reducing their bread intake, and the industry promotes the nutritional benefits of bread to counter what it terms 'media-induced mistrust'. Bread manufacturers are also major advertisers, spending £15 million in 2004 (Mintel, 2005).

Traceability in the Chain, and its Ethical Dimensions

The UK wheat-flour-bread chain has various systems in place for tracing goods or procedures. In some cases, these systems predate and exceed the 'one step back, one step forward' traceability required by EU law since 2005 (EC178/2002) (see Chapter 2 for a description of EU law). Examples include the traceability of organic wheat into flour and bread, or a system whereby wheat and flour for kosher bread are kept separate through the milling system, or a company which sources wheat from contract growers in Canada and traces it from farm to bakery (via a chain of dedicated lorries, trucks and tanker-holds) under an IP system (see below).

While most traceability systems arguably have ethical origins (such as the prevention of fraud or food poisoning), many now have primarily commercial objectives, such as legal compliance and stock control. But several systems in the wheat chain also use traceability as an instrument for tracking and communicating ethical aspects of production. Of the above examples, the organic and kosher traceability systems could be said to track ethical attributes, whereas the system for tracing Canadian wheat is driven by commercial considerations, because the baker judges this wheat to be necessary for its production system. One UK mill produces a brand of flour – 'Dove's Farm white all-purpose' – which carries the Soil Association's pilot 'Ethical Trade' logo, signifying that the company and product meet standards of ethical trading specified and audited by the Soil Association, the UK's largest organic certifier (the standards relate to price paid to farmers, treatment of workers and community support) (Soil Association, 2006; Dove's Farm, 2006a; Dove's Farm, 2006b).

Traceability systems are also used to implement the 'assurance' schemes which now cover the bulk of UK wheat production from farm to mill. These schemes (which have been introduced into most UK commodity chains over the past

15 years or so) are in effect private regulatory systems, in which groups of actors within the chain agree to set standards for specific attributes or procedures. Participants in the schemes are subject to third-party audit, to verify compliance. The schemes were introduced following food scares (notably in the meat sector), which highlighted the hazards of an untraceable supply base, and also as an industry-wide response to the 1990 Food Safety Act, which introduced the requirement that actors in the food chain must be able to demonstrate that they have satisfied themselves as to their suppliers' standards, in the event of prosecution for a food offence. According to a review of the schemes by the Institute for Grocery Distribution (IGD), they are now 'the main baseline standards required of modern farming (legal compliance plus good agricultural practice)' (IGD, 2003:2). The schemes were also either intended to be, or became, a means by which processors could regulate production to suit their own requirements. They have also been used to communicate aspects of production histories to consumers, to allay or pre-empt anxiety.

In the wheat chain, the main schemes are the Assured Combinable Crops Scheme (ACCS), which covers cultivation methods, the handling and application of agri-chemicals, and grain storage; and the associated Trade Assurance Scheme for Combinable Crops (TASCC), which covers transport, storage and laboratory testing of grain. The schemes cover all wheat grown for human consumption. Most of the provisions relate to food safety, although some affect worker safety or address the environmental impact of farming. To some extent, therefore, they can be seen to have ethical dimensions. These schemes fall within the umbrella organization of Assured Food Standards (AFS), which uses the Red Tractor logo to communicate to consumers that farm produce has met specified standards. However, the Red Tractor has rarely been used on bread products, mainly because of the stipulation that the products must contain 95% UK-assured ingredients, thus ruling out most UK-produced bread (which contains more than 5% imported wheat).

The detailed contractual specifications supplied from retailer to baker and from baker to miller, which prescribe most processes and practices, are likewise primarily commercial instruments, but may also have ethical dimensions, for example by specifying that assured wheat (with its ethical aspects) must be used; or by incorporating the Ethical Trade Initiative questionnaire relating to labour standards in the supply chain (see below) (IGD, 2003; USDA, 2004; AFS, 2006).

The widespread practice of blending different types of wheat or flour so that it can be conveniently stored and transported, and also to control costs and manipulate quality, is a commonly identified barrier to traceability in this chain, and particularly to internal traceability. (Internal traceability is the tracking of goods as they are transformed or commingled during processing; it is encouraged but not required by EU law EC178/2002.) In the UK, wheat is typically stored on farm (where the harvest of several fields may be mixed in one barn) then transported to the mill as required. At the mill, the contents of a 20-tonne lorry may be added to a bin with a capacity of 100 t, which already contains wheat from other farms. The wheat may be blended again as it is drawn through the production line, to achieve the desired 'functionality', and the flour may be blended again. The 'lot' size, for traceability and recall purposes, is thus very large, and the difficulty of segregating small lots in such circumstances is a compelling reason for developing farm-level assurance schemes which set standards for the whole supply base, as a pre-emptive measure to reduce risk.

Nevertheless, some examples were found in which relatively small lots wheat were tracked by lot from farm to bakery. Two mills had systems for tracing lots of French wheat through the mill, to make French bread. A supervisor from a kosher firm producing matzos (unleavened wheat bread) explained that a mill had facilities to keep selected wheat, designated as kosher, separate from other wheat through the whole process of storage and milling. Moreover, in order to select the wheat specifically for Passover matzos, he was able to accompany the farmer on the combine harvester, and the wheat then gathered was kept separate from the time of harvest until it was needed for milling some months later, when it was processed through a special line at the mill. A grain trader said his company would be able to supply a mill with, say, south-east England wheat, or wheat of a particular variety, if that was required, a system already used for baby food contracts.

Indeed, a small number of brands of flour and bread are already sold on the basis that they are identified with a particular place of origin. A few (mainly small) mills produce flour made from entirely British or local wheat and brand it as such, for example the Watermill, in Little Salkeld, whose packaging states that the flour is made from 'Organically grown English wheat'. Some of these mills also sell bread, but an industrial baker pointed out that committing to produce bread from regional or British flour on a large scale could be a liability in years like 2004, when the UK wheat harvest was of such poor quality that extra milling wheat had to be imported.[2]

Some stakeholders drew a distinction between 'traceability' and 'Identity Preservation' (IP). In the latter system, procedures are put in place to maintain separation between different consignments of wheat or flour, where the customer has contracted to buy goods from a specified source or of a specified variety. For example, the baker mentioned above, which sources wheat from Canada, uses an IP system. (In other chains, IP systems are used to separate GM from non-GM produce.) Given that it fulfils the definition of a traceability system as set out in EU law (see Chapter 1), IP can be seen as a type of traceability. In practice, in this chain it offers an 'optional extra' level of deeper or narrower traceability, at additional cost to the client.

Table 6.5 outlines the various traceability schemes that track *ethical* aspects of production in each sector of the wheat-bread chain, and the schemes are summarized below. The terms in parentheses in the final column link the ethical concerns addressed with the project's ten ethical concerns (Table 1.3).

At the inputs stage (seeds and agrichemicals) traceability is required by legal registration and quality-control regulations (BSPB, 2006; CPA, 2006; PRC, 2006).

---

[2] In November 2007, a major UK retailer, Sainsbury's, announced that it would become the first supermarket to use flour fully traceable to a specified co-operative of UK farms in its 360 in-store bakeries (Sainsbury, 2007).

Table 6.5 Ethical traceability in UK wheat-flour-bread chain

Level in chain	Traceability systems with ethical aspects	Information/attributes traced	Ethical concerns addressed
Inputs: Seeds	Farmers buying seed must choose seed from National or Recommended List (BSPB, 2006)	Variety, attributes (e.g. yield, straw strength)	To ensure wheat supply meets agronomic and process requirements, and also meets policy goals, e.g. reduced use of pesticides (Quality, Methods of Production, Human Health)
	Only Certified seed can legally be sold (BSPB, 2006)	Variety, purity, seed producer, growing conditions	To safeguard quality of seed supply (Quality)
Inputs: Agrichemicals	Pesticide registration process covers worker, user and bystander exposure; also environmental impacts (CPA, 2006)	Active ingredients; toxicity effects; safe handling, storage and transportation procedures; usage and dosage instructions; residues, environmental effects; hazard warning.	To safeguard health of users and bystanders, and limit damage to environment (Human Health, Methods of Production)
Farm	Assurance schemes cover various aspects of cultivation and on-farm practice – Assured Combinable Crops Scheme (ACCS) England and Wales, Scottish Quality Cereals (SQC) Scotland (AFS, 2006)	Use of agrichemicals (seed treatment, storage, date of application, reasons for use, weather conditions and dosage details); grain storage and handling (condition of buildings, use of post-harvest treatments, vermin control, cleaning procedures); transport (condition of vehicles, previous loads); insurance	Environmental impacts of cultivation (e.g. water pollution, harm to wildlife); operator hazard; food safety (e.g. contamination after harvest); to restore consumer trust (Methods of Production, Human Health, Trust)
	Organic certification systems (e.g. Soil Association, Demeter)	Compliance with certifier's standards and EU organic regulations	(Methods of Production, Human Health)
Farm to mill	Trade Assurance Scheme for Combinable Crops (TASCC) covers storage, analysis, merchanting and transportation (AIC, 2006)	Storage (condition of buildings, previous contents, cleaning procedures, vermin control, pesticides/fumigants); haulage (condition of lorries, previous loads); testing (facilities, staff training, sampling, analysis methods,);	Food safety, standardization and verification of testing procedures to eliminate variation from lab to lab; transparency in trading process (Human Health, Methods of Production, Terms of Trade, Transparency)

(continued)

Table 6.5 (continued)

Level in chain	Traceability systems with ethical aspects	Information/attributes traced	Ethical concerns addressed
		trading (marketing, specifying, sampling, rejection procedures, record keeping).	
	Assurance via contract: merchants negotiate individual contracts to supply wheat with specific attributes (including ethical attributes) and traceability	Traceable provenance from a specified region or cultivation method	Specified concerns (e.g. Methods of Production, Origin and Place)
Milling	Assured grain delivered to mill must be accompanied by a Post Harvest Treatment Declaration' or 'grain passport' and a unique farm identification number (AFS, 2006)	Unique number identifies farm. Passport includes information on post-harvest treatments and haulier's details, including declaration of previous three loads carried by vehicle	Food safety (Human Health, Methods of Production, Origin and Place, Quality)
	Tested at mill to assess quality and conformance with specifications	Specific quality parameters for, e.g. protein content, protein quality, moisture content, specific weight – predictors of milling and baking quality	(Quality, Terms of Trade)
	Assurance via contract: millers negotiate individual contracts to keep supplied consignments of wheat (e.g. French, kosher) separate during milling process	Traceable provenance, quality characteristics affecting milling process (e.g. variety, moisture content	(Quality, Origin and Place)
Baking	Detailed specification to miller	Specifies 'functionality' of flour – characteristics required for different baking purposes	Food safety; quality characteristics suitable for purpose (Human Health, Methods of Production, Quality)
	ETI audit (ETI, 2006)	Working conditions along supply chain, if company uses ETI Code of Practice	(Working Conditions)

	Traceability required on additional ingredients	Composition, provenance; possibly also working conditions in supply chain, if company uses ETI code of practice	Food safety, working conditions (Human Health, Working Conditions, poss Origin and Place)
Retailing	Detailed specification to baker	Specifies quality characteristics of bread, e.g. ingredients, provenance of ingredients, physical characteristics, shelf life, etc.	Food safety (Human Health, Methods of Production, Quality)
	ETI audit (ETI, 2006)	Working conditions in supply chain, if company uses ETI Code of Practice	(Working Conditions)
	Labelling, promotional material, websites, point-of-sale information	Ingredients, nutritional information, name of manufacturer, use-by information, provenance	Food safety, nutritional value, other ethical attributes (Human Health, Methods of Production, Origin and Place, Terms of Trade)

On farm, all wheat destined for human consumption is covered by assurance schemes, which set standards for the handling and application of agrichemicals, grain storage and vermin control. The principal scheme is the Assured Combinable Crop Scheme (ACCS), part of Assured Food Standards (AFS). The certifying organisations accredited to audit the scheme are CMi, EFSIS and Genesis QA. Farms are audited annually, with a visual inspection of the farm and an inspection of the detailed records kept by farmers. Membership of the schemes, which farmers must pay for, is in theory voluntary, but in practice essential for farmers selling wheat into the mainstream UK market, as the major millers demand assured wheat and require proof of assurance as part of their specification (AFS, 2006; Genesis QA, 2006; EFSIS-FABBL, 2006, CMi, 2007).

The farm assurance scheme dovetails with the TASCC, which operates between farm and mill, with standards relating to how grain should be handled, stored, transported, tested and traded. It is managed by the trade association of UK agricultural suppliers, the AIC. The scheme is made up of four Codes of Practice covering Storage, Haulage, Merchants and Testing Facilities (AIC, 2006).

Wheat merchants may participate in the TASCC scheme, or contract to supply wheat with specified, traceable attributes, which may be ethical, such as organic, or coming from a specified location.

For the bulk of flour production, wheat is tested on arrival at the mill for quality and functionality and is then blended, processed and marketed on the basis of its inherent characteristics, not on the basis of origin, variety or production system. The millers work to detailed specifications supplied by their clients, the plant bakers. The bakers, in turn, work to specifications supplied by their clients, the retailers. These specifications, which may run to 30 or 40 pages, cover most aspects of production. In compliance with EU law and due diligence obligations, these specifications require ingredients to be traceable, and some aspects of ethical traceability (such as human health and methods of production) may be included. Where firms participate in the Ethical Trading Initiative (ETI, described below), this will elicit information on labour conditions of supplier companies.

The legal obligation to provide 'one step forward' traceability ends with the retailers. However, labelling regulations, which determine the nature and amount of information provided to consumers, clearly have ethical aspects. In the case of flour, the 'statutory additives', described earlier in this chapter, do not need to be included in ingredients lists. In the case of bread, as for other processed foods, 'processing aids', which are deemed to disappear during manufacturing, need not be listed. Some production systems use logos to communicate to consumers that goods have been produced to specified standards, which may have ethical aspects, such as the Demeter biodynamic logo or the Red Tractor logo of the Assured Food Standards organization. In practice, however, except when producers feel that information will help to sell a product, little of the information collected by traceability systems is transmitted to consumers. In some cases, traceability information is included on packaging but is intentionally indecipherable to consumers (such as barcodes).

Table 6.6 Base code of the Ethical Trading Initiative (ETI, 2006)

• Employment is freely chosen
• Freedom of association and the right to collective bargaining are respected
• Working conditions are safe and hygienic
• Child labour shall not be used
• Living wages are paid
• Working hours are not excessive
• No discrimination is practised
• Regular employment is provided
• No harsh or inhumane treatment is allowed

The Ethical Trading Initiative is an explicitly ethical traceability scheme focusing on working conditions in international supply chains. ETI is an alliance of corporations and civil society organizations set up in the UK in 1998 to extend and strengthen the use of Codes of Practice in this field. Table 6.6 sets out the provisions of the ETI base code. Some companies in the UK wheat-bread chain are already required by their customers, and/or require their suppliers, to conduct an 'ethical audit' based on this code, usually consisting of a questionnaire – there is no third-party audit of compliance. In the wheat-bread chain, where most ingredients are sourced within the developed world, it was felt that any supplier complying with national law would meet the criteria of the code. However, there was evidence that the ethical auditing process was highlighting problems within the UK (e.g. relating to the excessive hours worked by agency-supplied bakery drivers). The Suppliers Ethical Data Exchange (Sedex) is a web-based organization that enables companies to maintain and share data on labour practices in the supply chain, to avoid the need for suppliers to provide similar information repeatedly to different customers. It is supported by the milling-baking company RHM, as well as by major retailers including Marks & Spencer, Waitrose and Tesco (ETI, 2006, Sedex, 2006).

Perspectives on Ethical Concerns along the Chain

The qualitative research on which this chapter is based consisted of 34 semi-structured interviews with senior stakeholders from the chain, including input suppliers, farmers, merchants, millers, bakers and retailers, and 15 semi-structured interviews with consumers. (This is a distinction of convenience: in practice consumers are clearly stakeholders in the chain and actors in the chain are also consumers – a point which several interviewees made, and which informed their views of ethics in the chain. For example, an industrial miller said he worried about the suitability for his own children of a TV advertisement for a product for which his mill provided ingredients. And a consumer said that, as a worker, she sympathized with the right of others to be fairly paid for their work.)

The interviewees were shown the list of the ten ethical concerns (Table 1.3) and, as well as being asked to discuss them individually, were asked if they could list them in order of priority. Most indicated that they attached more importance to some concerns than to others. There was a tendency for interviewees' priorities to vary depending on their place in the chain, and for actors at the same stage to share priorities (e.g. consumers attached more importance to human health than millers did). Interviewees also interpreted concerns differently depending on their place in chain (e.g. consumers saw human health as personal health; millers more in terms of product and worker safety). We referred to these shifting perspectives as 'fields of ethical vision'.

On the whole, there was strong agreement that the ethical concerns most relevant to the wheat-bread system were methods of production and human health. Terms of trade were also important to stakeholders. The least important concerns were animal welfare and voice. This agreement emerged in spite of interviewees' comments that ethics are diverse and subjective, and that *'as a society we have no ethical consensus' (official from agrichemicals company)*, implying that it might be prohibitively difficult to incorporate ethical concerns into food standards.

It is striking that the concerns said to be most important are among the least abstract, and are already covered to some extent by assurance schemes or regulation, though still the focus of public (and media) controversy. It may therefore be the case that these were the concerns that interviewees were most aware of, or felt they were concerns that something could or should be done about. Some interviewees (especially stakeholders) said that some concerns (such as working conditions or human health) were 'not ethical issues in this supply chain', because in the UK they are adequately covered by law and regulation. Some of the concerns less often said to be important were nevertheless discussed at length (such as trust and transparency), suggesting they had an underlying significance, though they were perhaps not readily categorized as issues that could be traced. The lack of importance attached to voice reflects a widespread lack understanding of what the concern meant. Responses to the ten concerns are discussed individually below.

Animal Welfare

Most interviewees said this had no relevance to the wheat-bread chain. A typical consumer response was, *'I'm not aware of any problems with animal welfare and the production of bread'*. Although some farmers pointed out the interdependence of livestock and cereal production on mixed farms, there was little recognition that even where wheat farms and livestock farms are separate, there is still mutual dependence, with more than a third of the UK wheat crop being used as animal feed. Only one interviewee made the point that ethical concerns in the livestock sector could have knock-on effects on UK wheat producers: *'If animal welfare issues push livestock production abroad, then the market for feed wheat vanishes'* (Farmers' trade association).

Human Health

Most interviewees saw this as a major concern, and for consumers it was overwhelmingly the most important issue. There was, however, some difference in the way the concern was perceived – an example of shifting fields of ethical vision. Consumers tended to talk about it mainly in terms of their own and their families' personal health. Stakeholders, on the other hand, discussed it mainly in the context of food safety (i.e. the importance of keeping food free from contamination) and of worker safety (e.g. the importance of protecting mill workers from exposure to dust). Similarly, though both groups mentioned public health, they saw it differently: with stakeholders citing industry-wide initiatives to reduce salt levels in factory-made bread and consumers commenting that they felt cheap bread was less nutritious than more expensive bread, and that poorer people would suffer the consequences of this. Among consumers, there was widespread suspicion of 'additives', and pesticide residues were a concern.

In contrast with the consumers' rather generalized suspicions about the content and nutritional value of 'mass-produced' bread, stakeholders identified much more specific (and potentially more serious) health risks (e.g. the danger from mycotoxins, fungal infections of grain which can be fatal to humans). But there was a strong view among stakeholders that the risks were understood and well controlled. One stakeholder said '*By and large, in the UK all food is safe. This is not an ethical issue in this marketplace*' (Grain merchant). However, there was wide recognition among stakeholders that there is public anxiety about health in relation to food (which the interviews with consumers confirmed). Health was also identified by many stakeholders as a commercial opportunity – e.g. producers who made ethical claims for their flour or bread tended to associate them with health benefits; a miller pointed out that flours can be heat-treated to permit pastry to be made with less fat, in line with current public health goals, and a baker who specialized in producing gluten-free bread (for allergy sufferers) commented that baked goods that were both organic and gluten-free offered a 'double premium'.

Methods of Production and Processing and their Impact (Environment, Landscape, etc.)

This was seen by stakeholders as the most important concern, and it again illustrated how the concerns were interpreted differently by different groups of actors along the chain. Thus, input suppliers and farmers focused on agricultural impacts, bakers and millers on the impacts of their factories (notably noise and energy use) and the retailers on packaging and distribution. Consumers, for whom it was also important, mentioned the impact of wheat-farming on the environment, on landscape and on farmers' livelihoods, as well as wondering about hygiene in bread factories and shops, and the impact of 'mass production' (a term used pejoratively by several consumer interviewees) on the 'goodness' of bread. Again, and not surprisingly, the stakeholders' interpretations were more specific and informed than the consumers'.

Stakeholder interviewees pointed out that some concerns in this category are already covered by regulation or law. An interviewee from an assurance organization commented that he was increasingly expected to incorporate environmental concerns into the scheme's standards, and provided an insight into the process by which ethical concerns are converted into measurable standards:

> *We believe that controlling the negative impacts of farming is something we must take on board. But then you might say that taking 30 cm out of the hedge at nesting time is bad, or you might want to preserve footpaths – these are desirable. But these are not the same as limiting the negative impacts of farming itself* (Assurance organization)

Terms of Trade

Among stakeholders, this gave rise to emotional responses along the chain, with interviewees at various stages feeling that terms of trade were unfairly biased against them. At the beginning of the chain, a seed breeder said that payments lost to the seed breeders' association as a result of farmers failing to pay royalties on Farm Saved Seed was costing breeders 10% of their income, and that this threatened the UK's seed breeding programmes. An organic farmer said, '*All farmers get worked up about this, farmers struggle to make a living*', but a conventional farmer said that while he earned much less per tonne of wheat now than he did a year ago, '*I say that's not unfair, it's supply and demand, fairness doesn't come into it*'. An industrial baker described how the major retail chains attached various conditions to trading, such as imposing fines if specifications were not met or deliveries not made correctly. Several stakeholders made the point that 'a sustainable supply base' (i.e. a system in which the farmer could sell above cost), was a prerequisite for an ethical supply chain. But others distanced themselves from responsibility for unfair terms of trade. A merchant said '*This is not an ethical issue because there are established industry standards and we conform to them*'. The industrial millers acknowledged the pressure on farmers caused by low wheat prices, but blamed the global commodity system.

As for consumer interviewees, several said they did not understand the term, or felt they did not have enough information to comment. A common consumer response, to this and other questions, was to the effect that 'I am not aware that I should be concerned about that, but if I were told that it was an issue in this chain, then I would be concerned, and might act on my concerns', expressing a pervasive sense of wariness, a sense that food was a source of hidden risk, and that as consumers they felt under-informed. Some expressed a strong view that trading arrangements should be 'fair', but the issues involved were felt to be complex, and here as elsewhere the consumer interviewees commented on the need to make difficult trade-offs, having to choose between one priority or 'good' and another.

Embodying the often-reported tendency for consumers' behaviour and attitudes to be contradictory, one interviewee commented that although she felt that smaller, independent producers and retailers were likely to have '*taken a decision that we're not going to turn out cheap rubbish*', she nevertheless bought supermarket brands on the basis of value for money, knowing that supermarkets were '*taking absolute advantage of economies of scale...which is good for me but probably not good for... farmers*'.

Working Conditions

Stakeholder interviewees were confident that this concern was well covered by existing law, and also that working conditions in mills and bread factories had improved within recent memory (though it should be noted that interviewees were managerial, not workers on the production line). Consumer interviewees, on the other hand, felt uninformed about working conditions in the bread chain, but said that fair wages and the right to good working conditions were important. An interviewee from a farm assurance organization raised the issue of gang labour on UK farms: '*We cannot pretend these are not issues in this country.*'

Quality, Composition and Taste

This was discussed at length, but definitions varied. Consumers' definitions tended to be more subjective than stakeholders'. For consumers, quality was variously defined as good flavour and texture, nutritional value, healthfulness, freshness, keeping qualities, purity and 'naturalness', and generally incorporated the idea that bread should have been produced to reliably high standards. It was striking that all the consumer interviewees spoke about taste, whereas stakeholder interviewees rarely mentioned it. Composition was also a concern for consumers, in that they wanted 'good quality' ingredients and to avoid 'artificial stuff' that might have been added. The question then arose how consumers identified 'quality'. Most tended to buy from shops which they felt would reliably provide the quality they were looking for. Some trusted brands and some relied on the look or feel (tested by squeezing) and the freshness (determined by looking at the use-by date). None habitually read ingredients labels.

For many stakeholders, quality was at least partly defined as a checklist of attributes specified by the client – the goods could then be tested for compliance with the specification, providing an objective measure of quality for both parties. Thus, to farmers, it meant supplying wheat with the levels of protein and moisture specified and then measured by the millers (though it was noted by both farmers and millers that the superior testing equipment available to millers gave them more power in the transaction). To an industrial miller, it meant producing a flour with the '*assured functionality*' demanded by his industrial baker clients, who were '*machine minders*' and needed the flour to perform in a predictable way. To the industrial baker, it meant '*meeting customer requirements*', as set out in detailed and binding specifications negotiated with the retailer. A mass retailer distinguished between '*legal compliance and compliance plus*', explaining that quality attributes which were 'optional' and added cost, such as some ethical quality attributes, might be included in premium products but excluded from cheaper versions.

This tendency to standardize and specify quality was much more marked in the industrial chain than the craft chain. In the industrial chain, consistency was highly valued and producers tended to see it as their job to eliminate and mask variation. In the craft chain, variation was valued.

> *Homogeneity of the product is a very important principle in this country, so the job of the millers has been to blend and blend to give the desired consistency and uniformity* (Industrial miller)
> *In order to make a consistent product in as highly controlled a process as baking, and to give the uniformity our customers demand, then we need consistent raw materials* (Industrial baker)
> *The artisan process admits a range of variations, reflecting variations in the ingredients, temperature, handling – that is the joy of it. We wouldn't want perfect consistency* (Craft baker)

Commenting on composition, craft bakers strongly criticized the use of what they deemed unnecessary ingredients in the industrial baking process, including processing aids, the 'statutory additives' and the additives required by the Chorleywood Bread Process. Two craft bakers listed the statutory additives on their labels and one miller-baker refused to add them to his flour.

Origin and Place

The general response from stakeholders was that consumers are not interested in where the wheat in their bread comes from, and indeed several consumers said they had never thought about it:

> *Bread is just an ordinary, everyday commodity. Customers look for consistency, it doesn't matter whether it comes from Timbuktu or Tooting* (Industrial baker)

However, there was some sense that this might change, in the light of a perceived growth of interest in provenance and local sourcing, presenting an opportunity for farmers to add value to their product:

> *Every farmer in the UK is within 50 miles of consumers and can store almost his total production on farm. This gives UK farmers a great opportunity to differentiate themselves to customers* (Merchant)

Some consumers interpreted 'origin' to mean not where the wheat came from but where the bread was baked or bought. Three interviewees expressed a preference for buying local foods where possible, to avoid food miles; but although they used local shops to buy bread, they did not apply the same criterion to wheat and flour (perhaps because the information on the provenance of the wheat is not readily available).

Among farmers, there was a resigned acceptance that wheat and flour, being bulk traded and blended, cannot usually be identified with the primary producer: '*You just have to accept that your wheat goes into a bag and is marketed under the miller's label*' (Organic farmer). But this was regretted: all the farmers interviewed, from small-scale organic farmers to large-scale conventional farmers, said they would prefer to have some direct contact with the end consumers of their wheat, and wished that wheat could be more associated with its place of origin.

Paradoxically, in spite of the fact that Canadian wheat is regarded by many bakers to be of superior bread-making quality, a major British industrial baker that sources a higher than average proportion of its wheat from Canada, via an elaborate IP supply chain, does not promote this fact to customers, even though the higher than average price reflects the expense of the sourcing policy:

> *Origin and place is obviously important to us because it defines our quality. But it is not on our customers' radar screen. They trust us to put the best ingredients in. The UK customer may have the idea that UK bread is all-British ... We don't think there's really a positive message in highlighting the Canadian content or putting it on the wrapper. It might be a turn-off, rather than a turn-on. But for us, sourcing the material, it's absolutely vital* (Industrial baker)

Trust

This concern was discussed at length by stakeholders, and seemed to underpin some of their comments on other issues, but did not spark much comment from consumers, one of whom said, '*I've never associated the word trust with bread*'. Some consumers made the point that bread consumption required less trust than, say, meat consumption – implying that it is less associated with risk of harm. The stakeholders tended to share the view that this particular chain is relatively safe, but made the point that the food system as a whole was not trusted. (An interviewee from an agrichemicals company said that in his sector the stringency of the testing regime seemed in itself to undermine trust: '*People say, how can you say there's no risk when you've got to do all these tests?*') But although it is true that many mechanisms in the chain are designed to substitute for trust, or make it unnecessary – e.g., the audits, laboratory tests and specifications that formalize transactions – personal trust was still felt to play an important part, both between stakeholders and consumers and among stakeholders in the chain.

> *Trust is very important between me and the no more than six people I buy and sell from. I know them all* (Farmer)
>
> *Trust is so important. We have to honour verbal contracts while paper ones are processed, even though the price may change against us in the interim. [And] some of the more subtle quality aspects are subjective, a philosophy rather than a contract, and therefore you have to trust [the farmer]* (Merchant)
>
> *I think people trust the regulations. In a developed country they think, well, someone has looked into that* (Industrial baker)

Voice and Participation

Many interviewees either did not understand this concern or did not see its relevance. When prompted that it could mean, 'Do you feel your voice is heard in this supply chain?', the farmers almost unanimously answered 'No'. One major retailer asked, '*How could we deliver on voice or participation?*', illustrating the distance between this concern and the ones that can be more easily 'operationalized' into standards. Two consumers said that by making a choice from what they felt was a wide range of options when choosing bread, they were able to have a voice in the chain.

Transparency

For some interviewees, both consumers and stakeholders, traceability and transparency were inseparable: '*There's no point having traceability if there's no transparency*' (Farmer). In general, however, this concern sparked more comment from stakeholders than from consumers, and stakeholders' views on the desirability of transparency varied considerably, although there was a widespread view that the chain was having to become more transparent, in response to food scares, and at the instigation of the retailers, who were '*shining the torch upstream*' (Agrichemical industry).

Craft-scale enterprises tended to see transparency as a useful way of differentiating themselves from industrial-scale producers:

> *I was very happy to say as much as I possibly could about what was in the products, because by being so transparent we were poking a finger at others, who were less transparent. We had no advertising budget, but we always hoped that that when we said, 'we use no additives', it would prompt people to ask why others did ... when you're a small company, transparency is all you've got* (Organic craft baker)

Another interviewee, by contrast, felt that larger companies stood to gain more from transparency:

> *Transparency is good for us, because we are [large enough to] have systems in place. BSE and FMD were caused by people circumventing the rules, not by big companies abiding by the rules* (Merchant)

An interviewee from the millers' trade association implied that it was naïve to assume that all chain partners would wish for full transparency, and argued that it was useful to have industry forums where problems could be discussed frankly, and perhaps resolved, without making the issues public and thereby perhaps causing unnecessary alarm:

> *We do spend time talking about Methods of Production, but not necessarily with a view to communicating it to consumers* (Industrial millers' trade association)

Several interviewees mentioned labels, either as an example of transparency (because of the amount of information which must be disclosed on them) or its opposite. An organic craft baker described bread as '*the most untransparent product*', because of the number of things (e.g. statutory additives and processing aids) that can be omitted from them. Although traditional, craft-scale bakers tended to pride themselves on their transparency and went to some lengths to communicate their production processes and values to customers, wrapped, factory-produced bread is nevertheless more likely to be labelled and carry nutritional information than unwrapped bread from a craft baker.

Information and Communication along the Chain

Traceability is about information and how it is communicated. Interviewees were invited to discuss the types of information that flow along the chain, the means by which information is communicated, and barriers to communication.

Before being asked about information flows, however, stakeholder interviewees were asked what they understood by the term 'ethical traceability'. Most said they had not encountered the term before but could guess what it meant. A few said that although they were familiar with traceability and understood the term ethics, they didn't see how the two were linked. On the whole, however, they were quick to grasp the term and showed an informed interest in the challenges of putting it into practice. Several said that the term summed up an issue of growing importance in the agrifood industry. None said the idea of ethical traceability was irrelevant to their work, and for some it was fundamental. As the responses below indicate, however, there was some scepticism about how far ethical traceability could be put into practice. This was partly because of anticipated difficulty in reaching agreement about which (or whose) ethics to act on, and partly because it was felt that actors in the chain were not interested in achieving or communicating higher ethical standards, and also that consumers were not prepared to pay for them. There was also some puzzlement as to how ethical traceability could be realized, even if it were accepted as desirable. One of the more sceptical interviewees (from a major retail chain) commented that part of his previous doubt about ethical trading had been based on the practical difficulties of implementing it ('*How do I take on this great mushrooming thing?*'); and that it had taken him years to be able to see how it could be put into practice. Now, he thought it was feasible and would be '*the next big thing*'.

> *OK, fine, you're able to demonstrate to consumers that food has been produced to ethical standards. What percentage of consumers will say, fantastic, I'll pay a premium for it?* (Agrichemicals company)
>
> *I agree that quality standards have expanded to encompass not just production standards but also ethical standards* (Farm assurance organization)
>
> *I think it shows a kind of moral connection between the grower and the consumer* (Organic farmer)
>
> *On one hand, in my cynical mode, it means nothing, because the people we are dealing with, i.e. the retailers and their customers, can't spell ethical. On the other hand, intellectually, I understood the term completely. It means tracking from farm to plate. It means that foods have been produced in a way that means that neither human nor resource has been abused* (Conventional farmer)

When the interviewees were asked about information and communication, it became apparent that there was a gulf between stakeholders and consumers. While stakeholders felt they had access to plentiful, high-quality information, consumers felt that information was rationed, manipulated and untrustworthy.

Stakeholders were hungry for information, making use of a variety of sources, including trade associations, web sites, clients, suppliers, customer feedback, family knowledge, trade publications, paid advisers, government agencies, non-governmental organizations (NGOs) and the media. They also made use of forums such as trade associations to exchange information, mainly within but sometimes between sectors. For example a seed breeder said: '*There is good dialogue between the millers and the breeders, both at company to company level, and between trade associations.*' This shows one sector (the breeders) leapfrogging another (the farmers) to engage in dialogue with the millers, to learn what qualities they wanted from the grain. (Notably, both parties to this dialogue are areas of corporate concentration). Information did not flow uniformly along the chain, though, and some stakeholders had been surprised in the past at how ignorant actors at one stage were of the workings of another sector (e.g., farmers of milling and millers of farming).

Consumers, on the other hand, were hesitant and haphazard information gatherers. Labels and the media were their main source of information, but they were not seen to be full or reliable sources: '*The only information you get is what you're given on the packaging*', one interviewee said. There was a feeling that more information might be found if the interviewees '*spent a long time on the internet*', but few seemed inclined to investigate. Shops were also seen as sources of information, with supermarkets viewed as poorer sources than small, independent shops:

> *I don't for one minute think [the shop assistant in a supermarket] would have any idea at all. If you were to go down to the local baker and say what's in this bread, he would be able to tell you...but going to a supermarket, they would not know* (Consumer)

It was striking that although the consumers complained of being in the dark, they did not take advantage of the information channels that were available to them. (For example, one said, '*I think there should be a law to say exactly what's in products and what the additives are*', when in fact most packaged foods would already have fairly comprehensive labels.) But few interviewees read labels at all and even fewer looked at web sites or said they might do so in the future. On one hand, the consumers had a sense that they were not being told everything, but on the other, they did not make efforts to inform themselves – perhaps because they felt any information they might find could not be trusted.

For stakeholders, the requirement to record and transmit specific kinds of information to comply with traceability systems, assurance schemes and other contractual obligations formalizes the flow of information between stages in the chain. For example, a merchant would need to see proof of a farmer's organic certification when buying organic grain and a miller would need to see the delivery truck's TASCC documents before accepting a delivery. An industrial baker said:

> *We ensure we have a full production spec on every raw material and on packaging. With composites, we have to know the composition because we have to declare its components*

> *on the label. And we have to get confirmation of origin. We need to be able to supply the retailer with this information* (Industrial baker)

One test of the adequacy and veracity of the information being transmitted along the chain to consumers is to ask whether it conveys an accurate picture of the chain's operation. Stakeholders were unanimous in saying that consumers had a very inaccurate picture of food production. Indeed, there was a widespread, somewhat aggrieved, view that consumers had a false (*'romanticized'* or *'very inaccurate'*) view of how food is made.

> *The man in the street probably doesn't have a clue about the industrial process* (Industrial baker)
> *There is still a perception that farmers are getting money for nothing* (Organic farmer)

In some cases, this was accompanied by a view that consumers did not need or want to have an accurate knowledge of the food chain:

> *It can be a disservice to provide people with information if they are not equipped to understand it* (Seed breeder)
> *I suspect most of the time [consumers] don't even know about us. And [if the regulatory regime is doing its job] they don't need to know* (Agrichemicals company)
> *Most consumers don't want to be thinking exactly how their food has got to them* (Millers' trade association)

The consumer interviews tended to confirm the stakeholders' impression that there is widespread ignorance of how food, including bread, is produced. During the interviews, consumers were shown a short presentation on the wheat-bread chain. Most interviewees said they found the presentation told them things they hadn't known about, particularly about the sources of wheat, the scale of modern production, the extent to which the UK's bread is factory-made and the level of additives used. Some interviewees showed considerable ignorance of food production – for example, one asked whether the large mills were publicly or privately owned. Interestingly, although consumers generally said they resented the thought that food manufacturers withheld information, the 'statutory additives' – the nutrients which manufacturers are legally obliged to add to flour, but which need not be declared on the label – did not cause consumers much concern, although they caused angry comment among craft millers and bakers. Even consumer interviewees opposed to the use of 'additives' in general did not object to the hidden use of additives officially deemed beneficial, supporting the view expressed by some stakeholders that consumers 'trust the regulations'.

Given that an important reason advanced by stakeholders both for communicating or for withholding information was that it would affect consumers' behaviour, it was notable that only one consumer interviewee said that the presentation on bread production 'probably would' change the way she bought bread. On the other hand, several interviewees said the information would not change their bread buying, because they bought on the basis of convenience or price or taste, and these were not outweighed by anything in the presentation.

The information stakeholders sent out, in contrast to what they gathered in, seemed narrower in scope and used fewer channels. Confirming the consumers' suspicions, it was indeed manipulated to serve the producers' ends, whether this was to withhold information that might be damaging or to promote ethical or other attributes. The further back in the chain, the less need producers felt for communicating with end consumers. Thus, a company that produced the active ingredients for agrichemicals, which were then sold on to other companies for formulation and packaging, had no contact with end consumers at all – nothing on any packaging downstream indicated the source of these ingredients. But the 1990 Food Act, which created the duty of 'due diligence', was said to have caused a sea change in the way companies viewed communication and transparency (though not necessarily transparency to the consumer). Several interviewees said they were no longer able to 'keep their heads down', as they once could.

The main channels of communication with consumers were packaging, promotional material and advertisements, web sites, site visits and product sampling. Many interviewees mentioned the importance of educating consumers, in particular children, through school and site visits. In the industrial chain, many of the 'messages' interviewees wanted to send out reflected a sense that consumer perceptions are inaccurate or contradictory:

> *[The message is]: You don't need to worry about this product, it's OK. There aren't any big issues in the bread supply chain. If a problem arises, an answer is ready [thanks to internal dialogue within the sector] that closes it down* (Industrial baker)
> *It would be excellent ... to try to overcome this misty-eyed view of what food production is. If [consumers] want to flock to the supermarkets which offer the lowest prices, this will drive production to the lowest-cost producers in the world, who are not concerned about these ethical issues* (Agrichemicals company)

In the craft chain, participants made much of the fact that they gave out a lot of information, and were 'transparent' because transparency was part of their ethos, and also a selling point:

> *My labels amounted to small, local essays. They included a full declaration of ingredients. I was happy to say as much as I possibly could about what was in the product* (Craft baker and retailer)

For stakeholders, especially in the industrial chain, a major barrier to increased ethical traceability was cost, especially in relation to farmers who might be tipped out of economic viability. This was reinforced by strong doubt as to whether consumers would be prepared to pay more for food with these attributes:

> *The principal obstacle is pitching a standard of production, a specification, that can be delivered cost-effectively by the producers within a global market. I honestly don't think some of the lobbyists understand what a knife-edge it is. We have welfare standards in the UK that go beyond EU standards and the industry is not now competitive* (Farm assurance organization)

Another barrier, mentioned by several interviewees, was the major retailers' dominance of the food supply chain. Interviewees expressed resentment and even fear in the face of what they saw as the retailers' lack of understanding, arrogance and double standards, which they felt not only penalized their own business or sector, but also were a force blocking change towards a more equitable or sustainable supply chain:

When you get a letter from a major multiple saying that organic products should henceforth only carry a 15% premium, it shows they don't understand organic production systems (Craft baker)

The supermarkets say they want British produce, but import at the drop of a hat, often stuff produced to lower standards (Organic farmer)
The buyer for [a major retailer] does not understand. It's all to do with the time taken to grow things. Agriculture has long lead times and the supermarkets operate on a just-in-time system – incompatible (Conventional farmer)

The supermarkets are a big problem. I don't think they want to get into a complex debate with their customers, they are too busy fighting for floor space with their rivals. We went to [a major retailer]. We said we can produce wheat in a more safe way. We could organize a chain where we could guarantee that the wheat was of the highest safety, in terms of chemicals. We already do this for baby food contracts. They said, all food is safe, we don't want to get involved in debates that some food is more safe than other food. They live in the fluffy world of consumerism. Facts are less important than feeling (Merchant)

A third important problem for stakeholders was the difficulty of reaching agreement about which ethical concerns to build into food production standards or traceability systems, and the practical difficulties of turning subjective or ephemeral issues into attributes which could be traced.

For consumers, the overriding factor in bread choice was convenience, underpinned by a complex pattern of idiosyncratic needs and habits. The common denominator was taste, variously defined as tastiness, texture, freshness, crustiness, heaviness (in weight), fragrance, sweetness, substance, ability to 'fill you up', toughness, moistness, light texture and absence of additives. Beyond this, family preference and suitability for specific purposes, such as toast-making, were decisive. Some interviewees bought organic bread, but only one cited environmental as distinct from health reasons.

When asked whether they had ever bought or avoided products on ethical grounds, five interviewees answered with a strong denial, though when they were subsequently shown the list of ethical concerns, they did not reject them as being irrelevant to their bread buying. The other interviewees had either chosen or avoided certain categories of goods for what they considered to be 'ethical' reasons. The items avoided, and the reasons given, included battery chickens (because battery production was felt to be cruel), Nestle products ('They are trying to stop women in the third world breast feeding' and 'They are not paying a fair amount of money for the coffee beans') and sausages (the interviewee had seen an off-putting TV documentary on how cheap sausages are made). Items chosen for ethical reasons, and the reasons given, included products from small producers ('It's healthy for the market to have a diverse supply' and 'They may care more about quality'), organic products (for health reasons and because these systems are felt to be kinder to animals) and fair trade products (to 'try to make things a little bit better'). These responses show that many of the consumers interviewed did consider the ethical implications of their purchases, and on the whole their reasons were rational, though overall their sources of information seemed accidental – such as TV programmes or other media reports. There are also inconsistencies – for example a buyer of free-range eggs, on the grounds of animal welfare and human health, made no mention of choosing chicken meat in the same way. And, as one interviewee pointed out, avoiding something for reasons of personal health, or in order to avoid being cheated, may not be an ethical reason.

One interviewee's description of his purchasing decisions showed a high level of awareness of currently debated ethical dimensions of the food supply. He said he felt he was '*constantly buying food that in some way or another isn't ethical*', and explained the trade-offs and compromises that have to be made between competing ethical considerations, as well as personal, practical considerations. This sense of the bewildering rage of factors to be considered, and the difficulty of balancing them against each other, was echoed by other interviewees, both stakeholders and consumers:

> *It's quite complicated, and one's position changes. Sometimes you might have a weekend when you have more time to think about it, and [then] one would definitely be making ethical decisions ... there's a number of conflicting things to think about, there's environmental, food miles, the farmer in developing countries, then, on the other hand, is it jetted in from Kenya? So you're bombarded all the time with different and potentially conflicting issues. Would you buy an organic apple from New Zealand or a conventional apple from the UK? ... It is actually almost impossible to be completely ethical. So one makes decisions but one nearly always sins* (Consumer)

Two interviewees said they were deterred from making more ethical selections by price. One said she would 'ideally' like to buy organic foods; the other said he justified his current purchases, based mainly on cheapness, '*by saying when I'm in a position for [cost] not to be such a strong factor, then my choices will change*' (Consumer).

Some Conclusions for Ethical Traceability

One of the early findings of this case study was that the wheat-flour-bread chain presents particular challenges both to traceability and ethical traceability. Wheat is a globalized commodity (and was one of the first commodities to be internationally traded in bulk). It has become habitual for millers to blend wheat from many sources (which might mean different farms or different continents), both to achieve economies in handling and transportation and to manipulate the nature of the end product. With some exceptions, such as the Identity Preserved, kosher and organic systems described in this chapter, wheat is tested when it reaches the mill to determine its inherent characteristics – such as the amount of protein it contains, or the presence of certain enzymes – and is then processed on the basis of these characteristics, not on the basis of where or how it was grown, or any other 'ethical' considerations. Most stakeholder interviewees felt that consumers did not care about the provenance of bread or wheat as long as it met their needs and did not cost too much. Consumers, unused to buying flour or bread on the basis of wheat variety or place of origin, did not seem to notice or care where it came from, although some did notice the production system (choosing organic or avoiding 'mass produced' bread) or made what they considered to be healthy choices (choosing wholewheat rather than white bread). So wheat and flour were found to be hard to trace; and on the whole there seemed to be little consumer demand for greater ethical traceability.

But in spite of these difficulties, some examples were found where wheat, flour or baked goods were traceable back to region, farm or even field, though not always for ethical reasons. Merchants commented that in the UK, where most grain is stored on farm until needed by the mill (and therefore unmixed with the product of other farms), offering deeper traceability would allow farmers to differentiate and add value to their product. Supporting this, all the farmers interviewed regretted that their wheat became part of an undifferentiated supply stream once it left their farms, and would have liked to be able to trace it forward into the end product. The main defenders of blending (and the consequent loss of traceable identity) are the millers – one said, '*blending is what millers do*'. And it must be acknowledged that a move to a new paradigm, in which wheat was traded and processed on the basis of traceable ethical attributes, would impede the practice of blending, which has been developed to improve other aspects of quality, such as consistency. It would also require a major restructuring of the physical apparatus involved in storing, transporting and milling wheat, which has been designed to make it efficient to handle large, commingled lots.

Consensus and Ambiguity

Another striking finding was the paradox that although many interviewees doubted the possibility of reaching consensus over which ethical concerns could or should be valorized and traced – because ethics are diverse, subjective and difficult to quantify – there was nevertheless strong agreement about the concerns most relevant to this supply chain (namely human health, methods of production and, for stakeholders, terms of trade). Interviewees were also very familiar with schemes which have in fact 'captured' and quantified ethical concerns, such as fair trade goods. Some concerns, such as working conditions, were felt to be valid but not problematic in this supply chain because they had already been effectively covered by regulation and law. And other concerns, notably voice, were poorly understood. From the responses, it was possible to hypothesize some sort of progression, in which ethical concerns move from the margins of awareness, to a situation where they are recognized as issues that warrant action, to the point where they are – at least temporarily – perceived as being settled or dealt with. Clearly, if this is an accurate representation, the process itself is of great interest and deserves further investigation.

Ethical traceability, though perhaps not so named, is clearly already present in this chain, and all of the stakeholder interviewees felt it was increasingly relevant to their work. But ethical concerns were found to be subjectively (and often pragmatically) interpreted and prioritized, depending on the interviewee's place in the chain – in other words, interviewees had different 'fields of ethical vision'. This raises fundamentally important questions about which attributes should be valued and traced, and who would make this choice.

At present, the attributes being ethically traced are defined and constrained by commercial realities and regulatory rules, and the processes by which decisions are made are largely hidden from the public. If a more ethically traceable chain were envisaged, on whose 'field of ethical vision' would it be based? Would it be wide enough to encompass the ethics of using energy-intensive agricultural methods to produce a crop which is mainly used for animal feed? Or to assess the implications of a UK government study of the cost of food miles, which found that if the 'social costs' of wheat production (such as water pollution, air pollution, loss of biodiversity and impact on health) are calculated, these are lower for organic wheat imported from the USA by ship than for wheat conventionally grown in the UK? (Defra, 2005b:68–73)? Or to address one of the most striking omissions from the current system, which is the impact of the system itself on the nutritional value of the product? Nutritional value is not one of the criteria used by the committee that selects new wheat varieties for cultivation (though it is a consideration when developing crop varieties for animal feed). Nor has the effect of the industrial bread-making processes on the nutritional value of bread been subject to much scrutiny. One interviewee asked, '*If a healthy population was taken as the goal – what would an ethical bread supply chain look like?*' Would there be one dominating model, or several? Would it be practicable to achieve it, even if there were agreement about what it should look like? Ethical traceability as a 'bolt-on' to existing production systems is likely to be of limited value, but in practical terms, given the scale and entrenchment of this supply chain, incremental change is much more likely than fundamental restructuring.

In any case, the difficulty of reaching agreement is not an insurmountable barrier. No system could capture and track all the ethical concerns people had about that system, and in fact differences of ethical interpretation already exist in the food system. People might choose to buy the same thing, such as organic fruit, for different ethical reasons – health or environment. Conversely, an 'ethical' designation, such as the organic label on garlic from China (signalling a low-chemical production system but a lot of food miles), might attract some consumers but repel others, depending on their ethical priorities. One consumer interviewee spoke eloquently of the ethical trade-offs that already have to be made by food shoppers. Ethical ambiguity is probably unavoidable, and ethical traceability systems, if sufficiently flexible and subtle, can help clarify the realities of these choices for consumers.

At present, there is a tendency to 'ghettoize' ethics as optional extras – one retailer interviewee described having two production standards, 'compliance', which met the regulatory minima, and 'compliance plus', which met higher ethical standards and cost more. The readiness to exploit consumers' willingness to pay for selected ethical attributes (mainly, it must be said, those associated with personal health, which may not qualify as an ethical motive for purchase) contradicts the stakeholders' view that consumers would be unwilling to pay for higher ethical standards or the systems to trace them. The fear that adding costs in this way might tip enterprises out of viability was also widespread, and this was in itself seen as an ethical concern. However, the assumption that 'ethical' production is necessarily more expensive than 'unethical' production is misleading, and is sometimes used to obscure more complicated arguments.

This raises another important issue, which is that when some products are designated as more ethical, others, by implication, may be seen (perhaps for the first time) as less ethical. A result is a further form of product differentiation in the marketplace. This unsettles what one interviewee described as the 'fluffy world of consumerism' and is a revelation many stakeholders wish to prevent. It would be naïve to deny that some sectors in the chain would resist moves towards greater transparency for this reason. But producers who believe their own production values are more ethical than the industry standard (notably smaller producers in the craft chain) already explicitly promote this fact, providing a critique of the rest of the chain in the process.

Uneven Information flows

Traceability is about communicating information. The sequence of actions and decisions that converts part of the UK's annual wheat harvest into the millions of loaves that are delivered almost daily to food outlets all over the country is inevitably labyrinthine. It relies on a high level of co-ordination within and between sectors, and this co-ordination depends on a variety of mechanisms for transmitting information, and for ensuring that relevant information is communicated in a prescribed way. Examples include the voluminous records necessary for compliance with assurance schemes or organic production systems, and the detailed product specifications used by the major retailers to communicate their needs to industrial bakers. One way of looking at these systems is as a substitute for trust, one of the ethical concerns that gave rise to most discussion among interviewees. The need for traceability itself, as noted elsewhere in this book, arises when consumer and producer are separated (and by extension when different stakeholders within a chain are separated). The more complex the chain between producer and consumer, the greater the need for trustworthy information to flow along the chain.

But another important finding of this research was that although there is a great deal of communication within the chain, the flow is uneven, with different actors making different uses of information. Some (notably in the industrial chain) felt it was sometimes necessary to block or filter information to consumers, to prevent anxiety or confusion or to avoid discouraging them from buying products. Other actors, mainly in the craft chain, who often had explicitly ethical reasons for being in business, saw the communication of full and accurate information to consumers as part of their ethos – not just a selling point, but an inherently ethical issue. But both these groups of stakeholders manipulated information for their own ends, just as consumers suspected.

One of the most striking findings was the disparity between stakeholders and consumers in relation to information. Whereas stakeholders felt that they had access to plentiful, high-quality information, and had developed mechanisms to control its flow, consumers felt they were recipients of information that was rationed, random, hard to find, biased and unreliable. However, they did not make the most of the information available to them – for example none habitually read labels or checked web sites. They looked for 'proxies' (such as brands) for the qualities they were looking for, rather than seeking out detailed information, and tended to bundle desired attributes together (e.g. organic certification was seen to imply fairer trade and fewer food miles as well as the absence of synthetic pesticides specified by EU law). As stakeholders suspected, consumers had a poor understanding of how bread was produced, and the stakeholders in turn felt aggrieved that consumers had such a jaundiced or romanticized image of the food system. There was general agreement about the need for more 'education' about food production, but little clarity about what this would achieve, because significant shifts in buying habits (if that were the aim) might benefit some sectors of the chain but penalize others. Given that the wheat-bread chain is generally seen as a stable, safe and successful system, it is perhaps surprising to find these levels of mistrust and mutual suspicion, and suggests that communication along the chain is less effective than it could be.

The model of communication used in this chain is still largely 'top-down' – experts (the stakeholders) see themselves as custodians and managers of information, with the public playing a passive, receptive role. The need seems clear for a more sophisticated method of exchanging information, which recognizes that 'consumers' are more than passive recipients of information and that stakeholders are also both consumers and citizens, with ethical values that require room for expression (see Chapter 13 for a discussion of different models of communication). More participatory and inclusive models would allow meaningful consultation between actors along the chain, to negotiate acceptable standards in relation both to production and to the transmission of information – part of the process described by Michiel Korthals in Chapter 11 as 'ethical room for manoeuvre'. This dialogue would enable consumers and other actors to collectively shape food choice.

The Issue is Power

Of the ten ethical concerns addressed in this research, the one that attracted least comment was voice – the extent to which interviewees felt they had a voice in the chain. Many interviewees were uncomprehending when asked about this, and of those who answered, most said they did not feel they had a voice in the system. This does not sit well with a current trend to 'personalize' issues (such as public health) by stressing individual responsibility, especially through an emphasis on choice. On one hand there is a delegation of responsibility; on the other a sense of voicelessness or powerlessness. Ethical traceability may itself prove to be a tool by which the consumers' sense of voice and participation can be increased – especially if systems for implementing traceability can be developed which do not consist of 'top-down' information flows, but which can enlist consumers' and actors' views along the chain. Voice was described by one participant as being '*halfway between loyalty and exit*', suggesting that there might be a tipping point at which a sense of passive powerlessness crystallized into active abandonment of the product or system in question.

Further research could examine the process by which 'ethical consensus' emerges about aspects of food production; how ethical concerns can be translated into acceptable and measurable standards; and how this disputatious process of negotiation can involve consumer/citizens at all stages. Research could also usefully refine the ethical concerns relevant to this and other chains, because some of the project's ten concerns were persuasively contested by interviewees. For example, several stakeholders argued that human health should be broken down into nutritional value and food safety, and this would probably be a useful distinction, given that the majority of health-related measures currently in place focus on hygiene; leaving the nutritional value of the product oddly neglected. Finally, it is also important to note that ethical concerns are not 'owned' by traceability systems (e.g. they may be expressed through ethical policies, Corporate Social Responsibility policies or codes of practice) and interviewees did not always see traceability as a necessary or obvious vehicle for communicating ethical practice. (It is also true that traceability systems can track practices which are not ethical.) Consequently, questions focussing on traceability may not have fully explored how actors in this chain approach and communicate ethical practice.

Ultimately, traceability systems seem to be inextricably bound up with power. The crucial questions – who decides what is to be traced, and who decides what information will be communicated and to whom – are questions about power. Why does the system look the way it does? Whose interests are being served? What is the ideological framework? What determines the direction of travel? Currently, the politics of traceability in the UK wheat-to-bread supply chain is dominated by commercial interests and due diligence (due diligence being a requirement that was introduced by the state, but which is now being managed by commercial interests). But there is a potentially different politics, and the question then is: which interests might start engaging with and altering the balance of power that shapes what is meant and delivered by traceability systems?

Acknowledgement The authors are very grateful to the stakeholder and consumer interviewees who contributed to this research.

References

ABARE (2007) Australian Bureau of Agricultural and Resource Economics website, www.abareconomics.com/interactive/ac_dec07/pdf/wheat.pdf, accessed 10.01.08

ABIM (2006) Association of Baking Ingredients Manufacturers website www.abim.org.uk, accessed 03.03.06

AFS (2006) Assured Food Standards website www.redtractor.org.uk, accessed 05.03.06
AIC (2006) Agricultural Industries Confederation website www.agindustries.org.uk, accessed 04.03.06
Barling, D., H. de Vriend, J.A. Cornelese, B. Ekstrand; E.F.F. Hecker, J. Howlett, J.M. Jensen, T. Lang, S. Mayer, K.B. Staer and R. Top. (1999) 'The social aspects of food biotechnology: A European view'. *Environmental Toxicology and Pharmacology*, 7: 85–92
BSPB (2006) British Society of Plant Breeders website www.bspb.co.uk, accessed 03.03 06
Burnett, J. (1968) *Plenty and Want: A Social History of Diet in England from 1815 to the Present Day*. Harmondsworth: Pelican (First published 1966)
Cauvain, S. and L. Young (2006) *The Chorleywood Bread Process*. Cambridge: Woodhead
Cereform (2005) Cereform company website www.cereform.com, accessed 14.04.05
CMi (2007) Checkmate International website www.cmi-plc.com, accessed 13 May 2007
CPA (2006) Crop Protection Association website www.cropprotection.org.uk, accessed 1.11.06
CSL (2005) *Pesticide Usage Survey Report 202 (Arable Crops in Great Britain 2004)* York: Central Science Laboratory
Curtis, B.C. (2002) 'Wheat in the World' in B.C. Curtis, S. Rajaram and H. Gomez Macpherson (eds) *Bread Wheat Improvement and Production*. Rome: FAO www.fao.org//docrep/006/y4011e/y4011e04.htm, accessed 1.11.06
Curtis, B.C., S. Rajaram and H. Gomez Macpherson (eds) (2002). *Bread Wheat Improvement and Production*. Rome: FAO www.fao.org//docrep/006/y4011e/y4011e04.htm, accessed 1.11.06
David, E. (1979) *English Bread and Yeast Cookery*. Harmondsworth: Penguin (First published 1977)
Defra (2005a) *Better Organic Bread: Integrating Raw Material and Process Requirements for Organic Bread Production: Description of Defra Research Project LK0960*, www2.defra.gov.uk/research/project_data/Morc.asp?I = LK0960, accessed 1.11.06
Defra (2005b) *The Validity of Food Miles as an Indicator of Sustainable Development*. London: Defra
Defra (2007a) *Agriculture in the UK 2006*. London: Defra. Accessed January 2008 via website http:statistics.defra.gov.uk/esg/publications/auk/2006
Defra (2007b) http://statistics.defra.gov.uk/esg/datasets/hstflour.xls
Dove's Farm (2006a) Dove's farm foods website www.dovesfarm-organic.co.uk/bread-baking.htm accessed 15.06.06
Dove's Farm (2006b) information from flour wrapper, bought London, February 2006.
EC 178/2002 *Regulation (EC) No 178/2002*, EU General Food Law http://europa.eu.int/eur-lex/pri/en/oj/dat/2002/l_031/l_03120020201en00010024.pdf
EFSIS-FABBL (2006) website of EFSIS-FABBL Farm Assurance Scheme www.fabbl.co.uk, accessed 04.03.06
ETI (2006) Ethical Trade Initiative website www.ethicaltrade.org, accessed 01.03.06
FAB (2006) Flour Advisory Bureau website www.fabflour.co.uk, accessed 03.03.06
Finch, J. (2006) 'Speculators buy up drought-hit wheat crops to earn their daily bread' in *The Guardian*, 28.10.06, p41
FoB (2006) Federation of Bakers website www.bakersfederation.org.uk, accessed 10.03.06
FSA (2006) Food Standard Agency website, www.food.gov.uk, accessed 21.6.06
FT (2006) 'The Cost of Grain', various authors, *Financial Times* 17.10.06, p10
Genesis QA (2006) Genesis QA website www.genesisqa.com, accessed 04.03.06
HGCA (2006) Home Grown Cereals Authority website, www.hgca.com, accessed 21.6.06
IGD (2003) *Working Group Report: Farm Assurance Schemes in UK: Summary Review and Issues for Development*. Leatherhead: Institute of Grocery Distribution, www.igd.com/cir.asp?cirid = 725&search = 1, accessed 1.11.06
Lepard, D. (2006): 'Give us this day…' in *The Guardian Good Food Guide*, special supplement, 14.10.06
Mintel (2005) *Bread UK 2005*. London: Mintel International Group Ltd
Mintel (2006) *Ethical Attitudes to Food*. London: Mintel International Group Ltd

Montague, D. (2000) *Farming, Food and Politics: The Merchant's Tale*. Dublin: Irish Wholesale Society
Morgan, D. (1979) *Merchants of Grain*. London: Weidenfeld & Nicholson
Nabim (2004) *The UK Flour Milling Industry 2004*. London: National Association of British and Irish Millers
Nabim (2006a) National Association of British and Irish Millers website www.nabim.org.uk, accessed 03.03.06
Nabim (2006b) *UK Flour Milling Industry 2006*. London: National Association of British and Irish Millers
Pena, R.J. (2002) 'Wheat for bread and other foods' in Curtis, B.C., S. Rajaram and H. Gomez Macpherson (eds) *Bread Wheat Improvement and Production*. Rome: FAO http://www.fao.org//docrep/006/y4011e/y4011e04.htm, accessed 1.11.06
Puratos (2005) Puratos company website, www.puratos.co.uk/products_solutions/bakery/bread_improvers/default.aspx, accessed 7.11.2005
PRC (2006) Pesticides Residues Committee website, www.pesticides.gov.uk, accessed 4.03.06
Sainsbury (2007) J Sainsbury plc website, www.j-sainsbury.com/index.asp?PageID = 424&Year = 2007&NewsID = 994, accessed 10.01.08
Sedex (2006) Suppliers Ethical Data Exchange website, www.sedex.org.uk, accessed 22.06.06
Soil Association (2004) *The Organic Food and Farming Report*. Bristol: Soil Association
Soil Association (2006) Soil Association website, www.soilassociation.org/ethicaltrade, accessed 1.11.06
SI 1998/141: *Statutory Instrument 1998 No 141 The Bread and Flour Regulations 1998* www.opsi.gov.uk/si/si1998/19980141.htm, accessed 1.11.06
TCMG (2006) The Traditional Cornmillers Guild website, www.tcmg.org.uk, accessed 03.03.06
USDA (2004) Golan, E.; B. Krissof, F. Kuchler, L. Calvin, K. Nelson, G. Price, *Traceability in the US Food Supply: Economic Theory and Industry Studies*. Washington: USDA Economic Research Service, Agricultural Economic Report number 830 http://www.ers.usda.gov/publications/aer830/aer830.pdf, accessed 1.11.06
USDA (2006) *Quarterly International Trade Report*, February 2006
Whitley, A. (2006) *Bread Matters: The State of Modern Bread and a Definitive Guide to Making Your Own*. London: Fourth Estate

Chapter 7
Traceability and Ethical Traceability in the Greek Olive Oil Chain

Agapi Vassiliou, Emmanouil Kabourakis, and Dimitris Papadopoulos

Introduction

This chapter examines traceability and related ethical dimensions in the Greek olive oil chain, and presents the findings of research on how ethical traceability is perceived by stakeholders in the chain and by olive oil consumers in Greece. Conventional and alternative (organic) olive oil chains are examined and provide the basis for exploring ethical issues related to the olive oil chain. Attention is paid to the knowledge, information and responsibility of actors in the olive oil chain, and to the trust in the chain which is necessary to establish ethical traceability systems. Qualitative interviews were carried out with producers, processors, retailers and consumers, in both the organic and conventional olive oil chains, and the data gathered has been coded and analyzed.

The Olive Tree and Olive Oil

The olive tree (*Olea europea* var. sativa), symbol of the Mediterranean basin, represents one of the most important agricultural crops. Along with vine and cereal growing, it represents the most traditional agricultural activity of the Mediterranean and the most striking feature of its agricultural landscape (Angles, 1999; Castro et al., 1997; Papageorgiou, 1994). The olive tree is one of the oldest known cultivated trees in the world (Fiorino and Griffi, 1992).

Although oliviculture has expanded beyond the Mediterranean basin, the olive remains the constant point of reference for the cultures around the Mediterranean, a deep-rooted link with the past. The sacredness with which the tree is associated indicates how old olive cultivation is. Historians consider the olive tree to be a cultural marker and a compass to explore the development of cultures and civilizations (Ghazau-Gillig, 1994; Standish, 1960).

The International Olive Oil Council (IOOC, 2006a) estimates that 81% of the world's olive oil and 69% of the world's table olives are produced in the Mediterranean zone. As much as 92% and 58% of Mediterranean olive oil and table

C. Coff et al. (eds.) *Ethical Traceability and Communicating Food*,

olives respectively are produced in the EU, reflecting the role that olive cultivation and consumption play in these countries.

Olive farming accounts for significant land use in Mediterranean regions, covering over 5 million hectares (ha) in the EU. The main countries producing olive oil are Spain (2.4 million hectares), followed by Italy (1.4 million hectares), Greece (1.22 million hectares) and Portugal (0.5 million hectares). France is the smallest producer, with 40,000 ha (IOOC, 2006a). The percentage of national utilized agricultural area (UAA) occupied by olives is 7% and 9% in Spain and Italy respectively, but around 20% in Greece. Whereas 6.5% of all farms in Spain have olives as part of their production, this rises to 15% in Greece and 37% in Italy.

Plantations that produce table olives cover a far smaller area than those producing olive oil. In Spain, less than 6% of the total area is devoted to table-olive production, and the figure in Italy is less than 3%. The EU currently dominates the global olive oil market, with over 70% of the world's production. Tunisia, Turkey and Syria account for over 20% of world production. The Greeks are the world's major olive oil consumers, at 26.1 l per person per year (IOOC, 2006b).

The cultivation of the olive tree has been of great importance for the Mediterranean throughout history. The most ancient civilizations in the history of man have left clear evidence that olive cultivation and oil production developed with the rise of the civilizations themselves. The archaeological findings of the Knossos palace in Crete, Greece, are an example. In ancient times the olive tree symbolized friendship and peace among nations, and the champions of the Olympic Games were crowned with a wreath of olive branches. The olive tree also symbolized the arrival of cultivated life, knowledge and divine enlightenment. Greek mythology and the Old Testament provide abundant information on the role of the olive tree in religion and everyday life.

Olive groves determine the character of the landscape and of rural production in the Mediterranean and signify the conquest by the inhabitants of their natural environment (Braudel, 1973). The connection of olive oil to religions in the region, from antiquity up to today, shows the importance the cultivation of the olive has had in Mediterranean life. Olive oil has been the most prestigious edible oil since antiquity. Even today, despite the growth of a number of substitutes, olive oil remains *the* oil for the Mediterranean, a cultural element, and it is closely connected with the nutrition of the inhabitants of this part of the planet (Jacotot and Hagege, 1993).

Olive oil is an essential component of the Mediterranean diet and significantly contributes to its health attributes and the longevity of the people living in the region (Keys et al., 1984; Keys et al., 1986; Trichopoulou et al., 2003; Trichopoulou and Critselis, 2004). Experts have shown that olive oil is associated with optimum health because it provides a high intake of oleic acid, the mono-unsaturated fat that is found in the purest olive oil. Tests by doctors and scientists at the EC have shown that replacing part of the animal fat content of a diet with olive oil can radically reduce levels of cholesterol, which can clog the blood stream (Weinbrenner et al., 2004). Although olive oil is 100% fat and contains the same calories as any other oil, oleic acid is poorly stored in the human body and is used more quickly as an energy source, making it less likely to be converted to body fat.

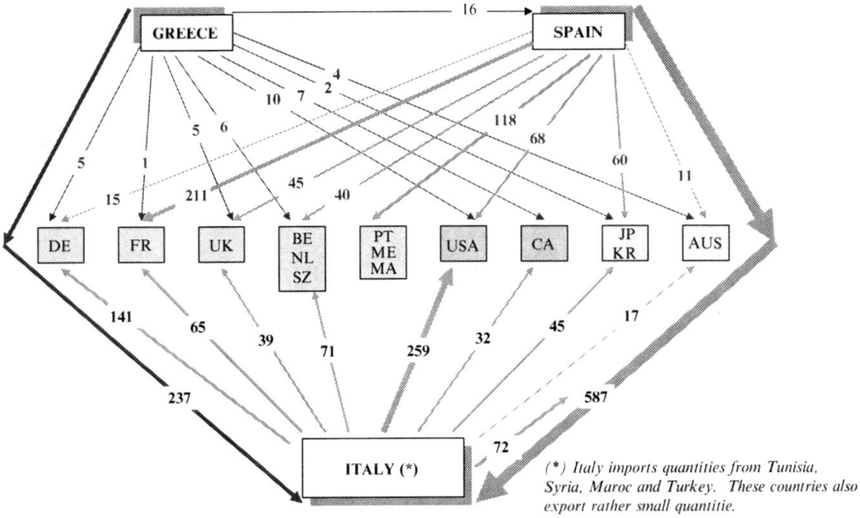

Fig. 7.1 The world olive oil trade, in 2005 (in million of Euros) (Petkanopoulos and Raptis, 2005)

But olive oil is also a commodity product that is marketed globally (Fig. 7.1), although most of it is produced and consumed in the Mediterranean.

Furthermore, it is the most expensive product in the category of edible oils and fats. Today, olives are also cultivated in the new world (California, Australia, Argentina, etc.). The economic importance of olive oil varies from one country to the next but is crucial in the Mediterranean countries. In Spain, for instance, over half a million farmers are involved in olive growing, and the sector annually provides over 46 million working days. In Tunisia more than one tenth of the population earns all or part of its income from olive farming. The economic and resulting social importance of the olive oil sector in Mediterranean countries is clear from its ranking in terms of gross agricultural domestic product and from the employment and wealth it generates. In Spain the olive oil sector contributes around 10%, in Italy 5% and in Greece 15% of the gross agricultural domestic product.

Olive oil comes in several different grades. Extra virgin olive oil, which is considered a gourmet oil, and virgin olive oil are the most highly prized grades. The International Olive Oil Council has developed definitions and standards for olive oil and standards also exist in the framework of Codex Alimentarius (IOOC, 1993; Codex Alimentarius, 2004). These are in general agreement with those of the EC (EC, 1991). Thus, olive oil is fairly highly regulated regarding its chemical composition and quality in the EU and in Mediterranean countries. There are several quality assurance schemes in the industry, and increasingly physical traceability schemes are to be found in the olive oil chain. However, up till now only the organic olive oil chain is fully traceable from the inputs used at farm level to the shops selling organic olive oil. Furthermore, ethical issues are not addressed in existing quality assurance schemes and there are currently no ethical traceability schemes.

The olive chain comprises farm production, which uses external inputs and produces the olive fruits; the olive mills that produce olive oil (the pure juice of the olive fruit); refineries that refine and improve defective oil (which may, for example, lack aroma or have high oleic acid or peroxide values); the packing and bottling houses that usually store and bottle the olive oil, and/or blend it in situations where pure and refined oil are mixed and marketed as olive oil; and the wholesalers and retailers that market the packed olive oil to the food industry or to final consumers.

Given the agro-ecological, socio-economic and cultural importance of olive oil for the Mediterranean, it seemed an appropriate food chain in which to study traceability and related ethical concerns. Greece, the site of our study, is a typical Mediterranean olive-oil producing country, and we examined the ethical concerns of both stakeholders and consumers. Our aim was to investigate the extent to which traceability matters to stakeholders and consumers, and especially how ethical traceability relates to their values and culture. Before presenting the findings, we next give a brief overview of the olive oil chain in Greece.

The Greek Olive Oil Chain

The olive oil chain in Greece involves mainly low-input, extensive agriculture and small-scale production at farm level. Thus, the olive oil industry up to the farm gate is mainly a craft industry with small-scale olive growers taking care of the production of the oil. Farmers produce their own, often extra virgin, olive oil, which they sell to wholesalers (companies or co-operatives) or directly to consumers though their personal contacts and networks. This is a rather short, simple chain. Beyond the farm gate, the chain is complex, industrial and includes big enterprises that collect, store, refine, blend, bottle and market the olive oil (Fig. 7.2). Few of these enterprises are co-operative. Overall, the production sector for olive oil is made up of a large number of companies whose size and activity vary, since the production process – from the reception of the raw material to the disposal of the end product – includes various phases. In addition, several agricultural co-operatives or co-operative unions also operate in this sector, whose main activity is to collect the produce of their members and then take responsibility for processing, standardizing and marketing it.

Farming and Inputs Used

The main management systems are conventional, to a large extent using a system close to integrated production, while organic production accounts for about 5% of the total acreage. Certified integrated farming is not widespread; indeed, until 2006 it was almost non-existent, though it is now booming because it is heavily promoted

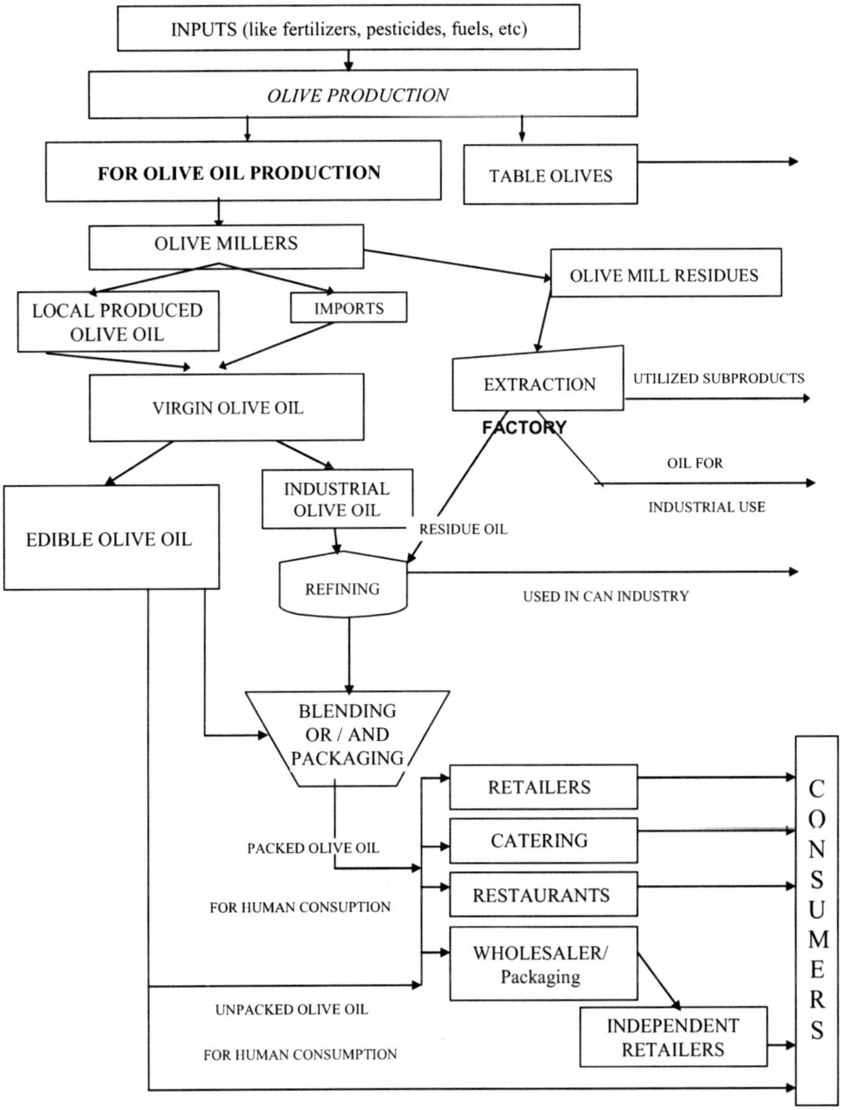

Fig. 7.2 Generic olive oil chain in Greece

by the state and the industry, and farmers find it attractive from a subsidy point of view. The organic production system is fairly similar to the traditional olive production practiced since antiquity in Greece, but uses modern technology, methods and inputs (Kabourakis, 1999). Farms are small in size (about 4 ha on average) and have many scattered olive orchards, typically about 4–6 on each farm. Thus the size of olive orchards is small, often less than 0.5 ha.

Olive trees, mainly traditional old olive oil varieties, are planted in orchards and bear fruit after three or four years. Thus, for establishing new orchards the initial investment is quite high. There are agrochemical shops supplying inputs in every large village and town. The main inputs in conventional growing, which is the predominant system, are synthetic fertilizers (usually combined fertilizers) that are used in rather small quantities compared with annual crops, around 0.5–1 kg per tree per year. Synthetic pesticides are mainly used for the control of the olive fly (*Bactocera oleae*). The pesticides are mainly used for bait sprays as part of a state control programme that covers all conventional olive production, about 3–5 bait sprays per year. Synthetic herbicides are not used extensively, as ploughing is the main weed control method, but this depends on the area. Fossil fuels are used mainly for soil cultivation and transportation of the olives to the olive mills. Irrigation water is used in the small acreage of irrigated orchards, with applications and amounts varying depending on the area and the water availability.

The main inputs in organic olive orchards are composts and/or animal manures for fertilization. Cover crop seeds are also used for cover cropping, mainly leguminous crops. Insect traps are used for the control of olive fly, and fuels for minimum soil tillage and olive transportation to the olive mills.

Olive Oil Production

Oil in olives is found in the mesocarp (fresh shell of the olive fruit) and not in the kernel of the pit (very small amount). The oil should be extracted in a way that allows it to separate from, rather than remain mixed with, the solid material. This is done in processing units called olive mills or olive oil mills.

All the procedures involved in olive oil production aim to increase the efficiency of extraction. The traditional procedure has not been changed a lot down the ages. It is known as the pressing method, and it is still quite popular in many cases and regions. In the 1970s, the centrifugation method of extracting oil from the olive was invented and since 1975 has been widely introduced to Greece. This method is fast, the machinery is compact and efficient, and little labour is needed. However, it requires a large initial investment and trained labour, and it has a rather high energy demand. The pomace produced has high moisture content, with more oil lost to the water component. However, because of its efficiency it has become the most popular method. Various other techniques were also developed, in order to utilize the advantages and eliminate the disadvantages of both methods, e.g. percolation-pressing, percolation-centrifugation and two-phase centrifugation.

Nowadays, certified organic olive oil is extracted at room temperature – a 'cold' process by comparison with the warm water used in conventional processing – by either pressing or certifying or needling (Kabourakis, 1999). Apart from differences in temperature and some other variables, the technique for extracting oil from olives that come from certified organic olive groves is the same as for conventional olives. The choice of technique used depends in practice on the machinery available, the preferences and skills of producers, and market demand.

Olive oil may be extracted from the stone of the olive fruit, which contains less than 1% of olive oil, using solvents such as other seed oils. The stones form part of the solid wastes of the olive oil mills (olive pomace). Refineries also produce edible olive oil by refining defective olive oil extracted from the olive mill residues. Most of the oil extracted from these residues in Greece is exported unrefined to foreign markets where it is refined and used. In Greece there is no organic olive oil blended or mixed with refined olive oil.

Some olive oil is stored at the oil mills, but most goes to commercial or co-operative storage facilities that usually pack it and/or trade it in bulk. There are several packing houses in Greece and a small-scale industry exists to supply packers with packing items such as bottles, cartons, transportation, etc.

Olive Oil Marketing

Olive oil is a globally traded commodity, and the price fluctuates, as determined by the market. Cyclical periods are characteristic of the production of olive oil. Greek production exceeds domestic demand, with large quantities intended for export. Imports are limited, since domestic production more than satisfies demand, but exports fluctuate from one year to another, depending on the amount produced (cyclical pattern of production) and on the demand by importing countries. During the 5-year period 2001–2005, exports in olive oil portrayed an annual average variation rate of approximately -13% (ICAP, 2006).

The sector's growth in Greece is influenced to a large extent by the way olive oil is sold. According to the EU Regulation currently in effect, which determines specifications for the retail of olive oil, it is obligatory to sell olive oil to the final consumer in containers of a maximum capacity of 5 l (in order to discourage bulk trading of the product). However, the proportion of bulk olive oil in the total domestic market (and in exports) remains high. As regards exports, it is important to note that the largest portion consists of cargoes of product traded in bulk, which is usually mixed and standardized abroad, where it is finally sold into foreign markets under foreign commercial brands (Fig. 7.1).

The olive oil chain does not display a high degree of concentration in all links of the chain. The exception is the retailing of bottled olive oil, which is not a major part of consumption in Greece (ICAP, 2005, 2006; Papadopoulos, 2003). Retailing of bottled olive oil is controlled by a handful of companies. Among these, two transnationals, Unilever (Elais, 25% market share) and Paterson Zochonis (Minerva, 24.9% market share) control around 50% of the Greek bottled olive oil market (FFT, 2006). The Greek refined olive oil market, which is rather small, is also very concentrated and is controlled by two companies: Minerva (49.5% market share) and Plias (23.5% market share) (ICAP, 2005).

Organic oil, although a very small fraction of the total production, is mainly exported, either in bulk (30–40%) or bottled. Exported organic olive accounts for 75% of organic production. The 25% consumed in Greece is packed in bottles, boxes or tins of various sizes (from 0.5 to 5 l) and sold mainly in organic stores and supermarkets. Most of the co-operative or private companies involved in organic olive oil have their own small packing houses.

The three leaders of the total packed olive oil sector determine its markets and prices because of their predominant position. Thus, small packing houses with few resources for marketing and advertising try to differentiate in terms of quality, and have a small share of the market.

Over recent decades the food market has dramatically changed in Greece, with the introduction and spread of supermarket chains that squeeze wholesalers and small retailers. This represents a market barrier for small packers, as the supermarkets demand payment for placing a new olive oil in their stores. Furthermore, the shelf space for other edible oils in supermarkets is larger than for olive oil because of the larger profit margins of other seed oils (the margin is estimated to be 12–15% for olive oil and about 30% for seed oils).

Consumption

Olive oil is consumed by more than 80% of the Greek population. Greeks have the highest per capita consumption of olive oil in the world, at 26.1 l each every year (or 0.5 l per person per week) (IOOC, 2006). Consumption is higher in the olive-oil producing areas. All classes of consumers, irrespective of income, education and occupation, consume olive oil.

Olive oil is used as a dressing and is also extensively used in cooking and pastry. Other edible oils are used mainly for frying or in pastry, as Greek consumers consider olive oil to be the healthiest among the edible oils. Greeks purchase only Greek olive oil and preferably olive oil produced locally. Greek olive oil consumers do not take pay attention to information provided by advertising and the mass media, as olive oil is an essential component of Greek culture and food culture. Greek food culture is determined to a large extent by the food ethics of consumers. Prices do not greatly affect olive oil consumption, as olive oil is considered essential for human nutrition.

A high proportion of Greek olive oil production (35%) is consumed by the large number of olive producers and their families ('self-consumption'). The remaining 65% is marketed as bulk (38% of the production) or is sold packed (27% of production). The Greek olive oil market is characterized by a large degree of direct marketing. Large volumes of olive oil are sold directly from producers to consumers in the urban centres, usually in tins containing 17 l of olive oil (well above the legal limit of 5 l). This direct marketing is estimated to exceed 50% of the olive oil consumed in urban centres. Thus, Greeks consume mainly olive oil bought in bulk (about 73%). Even among consumers who buy bottled olive oil, about 23% buy either bulk or bottled olive oil. Most consumers do not ask for extra virgin olive oil and suppose that all olive oil is virgin, especially when bottled, which is often not the case. Overall, the domestic market for olive oil is increasing, as presented in Fig. 7.3.

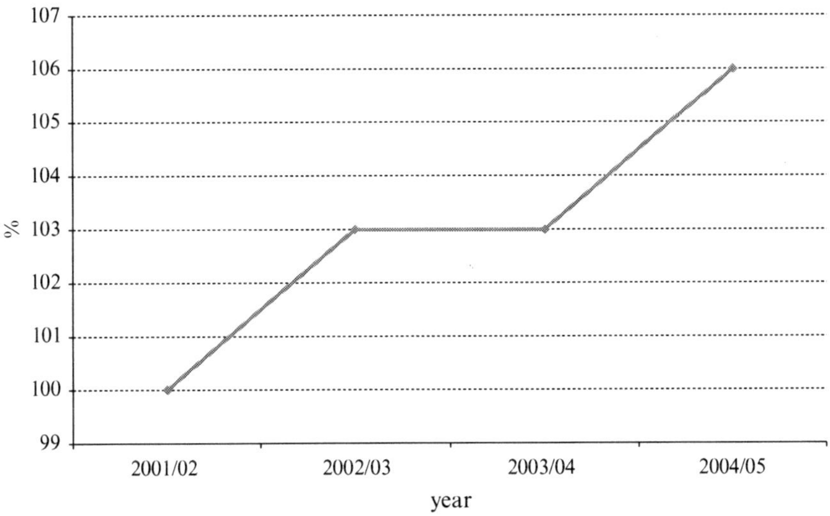

Fig. 7.3 Evolution of the domestic market of olive oil, over time (2001/02 = 100) (ICAP, 2006)

Organic olive oil consumption follows similar trends to conventional consumption, with the exception that it is basically sold packed. The internal market for organic olive has an annual growth rate of about 18%.

Traceability and Ethical Traceability in the Olive Oil Chain

Limited traceability systems exist in the olive oil chain. These systems cover the basic requirements of the relevant EU regulation (Article 18 of Reg. (EC) 178/2002). There are rarely systems in place covering the whole chain, although there are systems for stock control and product recall, especially in large packaging and distribution houses. Existing traceability systems have primarily commercial objectives, such as stock control and legal compliance – in other words, they are concerned with physical traceability, rather than the social, cultural and ethical aspects of the olive oil chain which were the main objects of this research (see Chapter 1 and Table 1.3 for a discussion and list of the ethical concerns used as a framework).

Traceability systems are better established in the organic olive oil chain. They mainly cover the physical attributes of the olive oil, as organic standards and regulations (which require traceability) mainly refer to the inputs used and to the methods of production and processing. Recently, however, there has been an increase in quality assurance schemes in the sector, and these also use traceability systems. These schemes are introduced primarily for commercial reasons and relate to codes of good agricultural practice at orchard level and industry practice in the rest of the chain. These codes are translated into private standards, and these standards are optional. They are not backed by legal requirements but mostly by the commercial interests of the large retail chains. They aim at obtaining minimum quality standards and satisfying consumer demand, primarily in the area of food safety related to health concerns, and secondarily in relation to environmental impacts and the use of natural resources.

On the other hand, it should be noted that the olive oil chain was considered famous for its lack of transparency and traceability. This was especially true in northern Europe and even more in North America, where virtually all olive oil is imported. This was due to the market control exerted by a handful of companies, which purchase olive oil from all production areas, blend it, and then export and distribute it (either in bulk or bottled). Traceability in the various sectors of the chain is described in the following paragraphs.

Farm Inputs, Processing and Transport

Farm inputs used in olive production can be traced by chemical analysis for their residues. These inputs, mainly fertilizers and pesticides, are used extensively and often irrationally in modern olive growing. The inputs used in conventional olive production, which is currently the predominant production method, are mainly synthetic ones. Depending on the way they are used, synthetic inputs may have detrimental health, environmental and economic effects.

However, the amounts of synthetic input residues found in olive oil are relatively small in comparison to other food products and in annual crops. Currently there are only general guidelines and norms (e.g. Codex Alimentarius) but for conventional olive oil in most cases chemical residue analyses are not required and therefore not performed. Only in some countries and/or companies are restrictions set on the permissible levels of certain chemical residues. Input traceability at farm level has a high cost, especially in the olive chain because of the structural characteristics of the olive farms.

The production of oil from olives is an industrial process. The origin of the olive oil produced in an oil mill can be traced at the end of the process, as can its chemical, physical and biological qualities. Furthermore, chemical substances contaminating olive oil during its production, or a deterioration in the quality of the oil caused by using inappropriate processing practices, can also be traced. Although they are desirable, not all of the olive oil mills apply quality assurance schemes. Usually basic quality control is done by the producers (especially the organic ones), who take care of the process and are present in the mill during the processing of their olives in order to supervise the oil extraction.

There is usually appropriate administration in this link of the chain, and various quality assurance systems apply. This is easier in countries such as Greece, where each farmer produces and sells his own olive oil. In other countries, such as Spain and Italy, the majority of the farmers sell the olive fruits to the mill. This is mainly because of the structure of the farms, the size of the mills and the organization of the olive oil market.

Transporting olive oil from the mill to the store or from the bottling plant to the market may affect the quality of the product negatively when it is done in an inappropriate manner, e.g. exposed to the sun or in high temperatures. Such effects can easily be traced when a proper administration and quality assurance system exists. There is usually appropriate administration and different quality assurance systems apply in this link of the chain.

Storage and Bottling

Olive oil is usually stored in containers that hold large amounts of oil of different origin. The origins of the olive oil are thus lost, as oils from different producers, areas and countries are mixed. Besides, different qualities of olive oil are often mixed (blended) during storage. Thus, tracing olive oil during storage is difficult in practice, though technically possible.

Furthermore, during storage the oil may be mixed with other edible oils, which affects its quality. The farms' small size means the storage facilities have to mix the oils. This practice applies especially to conventional olive oil that will later be marketed as a commodity product. On the other hand, organic olive oil in Greece is stored with appropriate documentation and quality assurance schemes apply. It should be noted here that farmers in areas that produce Protected Designation of Origin (PDO) olive oils store and market their olive oil in accordance with the PDO regulations, but the volumes produced in this way are small.

Tracing the olive oil during bottling is fairly easy, since lot numbers are used by the packers in case of defects. If the packer is the olive mill, then there will be some information about the producer and the method of production; if not, it will be known where the oil comes from. There is usually appropriate administration and various quality assurance systems apply in this link of the chain. A few organic olive oil wholesalers and bottling operations already have codes of ethics for their operations, although this is not required by the organic certification systems. Such codes refer mostly to the origin of the olive oil and rarely to the working conditions.

Marketing

In Greece most olive oil is marketed in bulk to Italian wholesalers and packers (Fig. 7.1). It is thus very difficult to trace its origins, and in the case of defects it is difficult to trace a problem, as oil produced from different olive oil mills in a given area is mixed in the storage houses of wholesalers and then exported in bulk or sold to packers. Most of the companies apply quality assurance schemes that examine the chemical and organoleptic characteristics of the product as required by the law. On the other hand, in the niche organic olive oil market the products are traced and controlled, and certification processes take place before marketing the organic olive oil.

Table 7.1 represents existing traceability systems in the olive oil chain, showing the type of information that it is traced as well as the ethical concerns currently addressed. However, it should be stressed that at the moment traceability systems exist and are functional mainly in the organic chain and in low-input systems, which apply codes of good agricultural practice (such as integrated olive production). As presented in Table 7.1, traceability and the information traced currently relate to topics that are demanded by legislation and refer mainly to record-keeping for food safety and economic purposes. Ethical concerns, by and large, are not addressed directly by existing traceability systems in the olive oil chain. Ethical concerns are mainly addressed indirectly, as a result of legislation developed for food safety and consumer protection. Furthermore, although ethical concerns are addressed, there is currently not much traceability (ethical traceability) of these concerns. (The terms in brackets in the final column of the table link to the specific ethical concerns used as a basis for this research, listed in Table 1.3.)

Table 7.1 Traceability and ethical concerns in the olive oil chain

Level in chain	Traceability	Information traced	Ethical concerns addressed
Inputs: Olive trees in new plantations	Farmers buying olive trees for new plantations must purchase olive trees from certified nurseries	Quality of the olive trees and lack of phytosanitary problems. Purity and suitability of olive varieties grown	To ensure that farmers choose varieties that meet agronomic, economic and technological requirements (Methods of Production, Quality)
Agrochemicals	Pesticide registration processes cover human exposure and environmental impacts Fertilizers are checked for composition and unwanted components like heavy metals	Information about: active ingredients; toxicity effects; safe handling, storage and transportation procedures; usage and dosage instructions; environmental effects; hazard warning	To safeguard health of users and consumers, to minimize environmental exposure to toxic substance and to ensure stable production
	Organic certification systems regulate and trace inputs used	Auditable records must be kept, demonstrating compliance with certifier's standards and EU organic regulations	(Human Health, Methods of Production, Quality)

(continued)

Table 7.1 (continued)

Level in chain	Traceability	Information traced	Ethical concerns addressed
Olive production	Assurance schemes/ codes of good agricultural practice cover various aspects of olive production in the orchards and on the farm	Auditable records on use of inputs and the production of olives	Environmental impacts of olive production (e.g. pesticide pollution, soil erosion, etc.); farmer's health hazards; consumer food safety (e.g. pesticide contamination) trust (e.g. ensure that there is no fraud)
	Organic certification systems regulate and trace farming methods	Auditable records must be kept, demonstrating compliance with certifier's standards and EU organic regulations	(Methods of Production, Human Health, Working Conditions, Trust)
Processing – olive oil production	Farmers produce their olive oil, which is administrated	Unique numbers for its farmer and its production	Quality, energy use in the olive oil mills, waste management, pollution and degradation of nature
	Olive oil is tested directly for oleic acid content in order to determine its quality	Specific quality parameters for oleic acid content	(Methods of Production, Quality, Origin and Place)
	Organic certification systems regulate and trace processing methods	Auditable records must be kept, demonstrating compliance with certifier's standards and EU organic regulations	
Transport	Documentation of quantities, quality, etc.	Information about the origin of the load, its quality and the amounts of olive oil transported	Energy use for transportation (Quality, Origin and Place, Transparency)
Storage	Legal documents, lot numbers, etc.	Information on the quantities and the quality of the olive oil stored	Food safety (Human Health, Methods of Production, Terms of Trade, Origin and Place, Transparency)
	Organic certification systems regulate and trace stored olive oil	Auditable records must be kept, demonstrating compliance with certifier's standards and EU organic regulations	

(continued)

Table 7.1 (continued)

Level in chain	Traceabilit	Information traced	Ethical concerns addressed
Bottling	Administration and different quality assurance systems Organic certification systems regulate and trace bottling of olive oil	Lot numbers, quantities, origin and quality of the olive oil bottled Auditable records must be kept, demonstrating compliance with certifier's standards and EU organic regulations	Food safety of the product, labelling and accuracy of information provided (Human Health, Methods of Production, Terms of Trade, Origin and Place, Transparency)
Retailing	The name of the bottling operation and the quality of the product must appear in the label	Quality characteristics of the olive oil (e.g. acidity). Name of manufacturer. Use-by information	Food safety (Human Health, Methods of Production, Quality)

Ethical Concerns Along the Olive Oil Chain

Using the Greek olive oil chain as a case study of food ethical traceability, we conducted research in Crete, Athens and Thessalonica. We focused on Crete, and more specifically on the Heraklion prefecture, as it is the main producing, processing and trading area of olive oil in Crete and a major centre in Greece. Because of the importance of oliviculture in Crete, consumers were interviewed mainly in Thessalonica (north Greece) and in Athens, the largest urban centres of Greece. Stakeholders were interviewed in Crete (mainly in the Heraklion department), as well as in Athens and Thessalonica.

Qualitative data were collected by means of in-depth interviews. A semi-open questionnaire was used, based on a list of 'core questions' shared by all three case studies and a list of ethical concerns (Table 1.3) (see Preface to Part II; Kabourakis, et al., 2006). Stakeholders were visited at the site of their operations, covering all the links in the olive oil chain, from farm inputs to final sale to consumers. Consumers were visited by appointment at their houses or any other convenient place. Often interviewees did not fall clearly into a single category (e.g. some olive oil millers also act as bottling operators and wholesalers). All interviews were conducted on the basis of anonymity and confidentiality. The research took place between December 2004 and January 2006. The research involved 40 elite interviews with stakeholders from the olive oil chain and 32 interviews with olive oil consumers in Greece (5 in Crete, 19 in Thessalonica, 8 in Athens). The analysis of the data was done by coding the interview data in according to actors and themes. In the following paragraphs, the ethical concerns of the actors in the Greek olive oil chain are presented.

Consumers and stakeholders appeared to have different concerns, reflecting the fact that they had different levels of information about and interest in olive oil. It should again be noted here that olive oil is an important element of Greek culture and thus for Greek farmers and consumers. Farmers want to produce the best quality olive oil; they are emotionally linked with olive oil production and they are proud of what they produce. Furthermore, the majority of production is small-scale and artisanal. The existence of so much small-scale production is not justified in purely economic terms, as it does not provide satisfactory returns and economic profits to the farmers. This is demonstrated by (a) the large number of producers; (b) the fact that olive producers do not sell their olive fruit production as raw material to the millers, but produce their own olive oil as well; and (c) by the fact that they are producing the best quality of olive oil (e.g. in Crete more than 90% of the olive oil produced belongs to the top category of extra virgin olive oil).

Consumer interviewees valued olive oil as an important edible fat that not only has high nutritional value but is also an integral part of Greek cuisine and food culture: Greeks have a consuming passion for olive oil. As discussed earlier, olive oil has also been used since antiquity for religious purposes and is an important symbol in the orthodox Christian church. But it has to be noted that Greek society is changing. Modern society is affected by globalization, the global capitalistic economic model and the shift to a post-modern society. New societal trends tend to be towards selfishness, as well as towards purely monetary, market and profit-oriented views of life, which replace family and collective views. It is an open question how such changes will affect Greek culture, relating to olive oil, in coming generations. However, it is notable that our research did not reveal significant differences between the different age groups of consumers interviewed. Nevertheless, younger consumers tended to be more sensitive to concerns related to the environment and to methods of production that affect it.

Overall, the ethical concerns more relevant to the Greek olive oil chain were quality and human health, in spite of the diversity of views and their subjective nature. Quality relates to the oil's chemical composition and physical properties, but also involves abstract aspects such as taste, which are subjective and culturally determined. These concerns are also covered to some extent by quality assurance systems and are not controversial. They are also important for the commercialization of the olive oil. They are related to price levels and the continuation of olive production and consumption in Greece.

Human Health

Stakeholders valued this highly (it was an important ethical concern for 75% of the stakeholders interviewed) and considered olive oil to be the healthiest edible oil, supportive of human health. They often mentioned that olive oil does not pose any health risks (microbiological etc.) and they considered it a safe fat and oil, which does not contain any synthetic or other additives. They mentioned that 'olive oil is safe when it is processed and bottled in units that comply with certain standards' (implying that the processors and especially bottlers should not use substances that contaminate it or practices that diminish its quality). Stakeholders considered that the legislation regarding quality aspects of olive oil (such as oxidization, which relates to freshness) is very clear and strict. They stated that the amount of inputs used in olive oil production (pesticides and fertilizers) is very small compared with other fruit crops or vegetables, and do not pose any risk to consumers. Stakeholders raised the issue of transparency in relation to quality and human health, because of blending:

> *When there is blending there is no transparency. Blending with other oils destroys the quality and if other substances are used then there is also a safety risk for consumers* (Stakeholder)

Human health was valued as an important ethical concern by 97% of the consumers interviewed. It was considered to be important because most of the consumers had their own health and the health of their families in mind when they consumed olive oil. Consumers believed in the healthiness of the olive oil, and did not question it. They attributed positive health properties to extra virgin olive oil, as is widely accepted today by society (scientists, policy-makers, consumers, etc.). On the other hand, it was found that many consumers did not really know what extra virgin means, even though they believed it had quality characteristics that contributed to their health. Consumers were concerned about blending virgin olive oil with refined olive oil or other oils, and the potential impact on their health. Pesticide residues and synthetic substances used in refining were a concern for few consumers. This also appeared to be an important factor for organic olive oil consumers.

Methods of Production and Processing and their Impacts

This was valued as an important ethical concern by 27% of the stakeholders interviewed. Most stakeholders were not concerned about the impacts of primary production as they view it as extensive production which does not pose serious problems either to the environment or to the quality and safety of the products. Some stakeholders even considered olive production to have a positive impact on the environment and landscape. They commented that:

> *Intensive agriculture has negative effects on the environment, but olive production in Greece is not intensive* (Stakeholder)

They were concerned about the olive mill wastes and their negative environmental effects, and commented:

> *Unavoidably, the technology used results in environmental impacts* (Stakeholder)

Stakeholders involved in organic olive production argued:

> *Organic oliviculture [which is a combination of new technology with traditional knowledge] contributes to biodiversity and an improvement of the natural environment and landscape* (Organic stakeholder)

Methods of production and processing and their impacts were valued as an important ethical concern by 59% of the consumers interviewed, although they did not have a clear idea of the magnitude of the impacts. They felt that the methods by which olives are produced make an important contribution to the environment and landscape. Consumers had an idealistic picture of small farms, especially the traditional method of production and were not aware of the changes that have occurred (e.g. the use of synthetic fertilizers, herbicides and pesticides), or of the refining process and practice of the blending different categories of olive oil.

Terms of Trade

Terms of trade were valued as an important ethical concern by 25% of the stakeholders interviewed. The farmers considered it important that they should be paid fairly for their job, and they complained about the wholesalers' high margins. They considered the prices paid to producers to be unsatisfactory. Producers commented:

> *Prices are determined by the trade and/or traders. The producer cannot intervene to affect price formation* (Producer)

The producers complained that there is a very big gap between the price gained by the producer and the price the consumer pays. Stakeholders in the chain also complained that the terms of trade are not fair because consumers abroad do not check the origin of the oil, and allow blending and mixing of different qualities and origins of olive oil. However, one wholesaler said that:

> *The farmers earn enough for the work they do, and farmers do not appreciate the difficulties wholesalers face in the global market for olive oil*

Terms of trade were valued as an important ethical concern by only 19% of the consumers interviewed. In general, terms of trade were not felt to be relevant to the olive oil chain and were not something that consumers were looking for when buying olive oil.

Consumers' main concern was (a) the price they paid and (b) the price level in relation to the producer and the quality of the product that direct marketing offers them. Although they were concerned about farmers' income and the prices paid (one of the reasons for buying directly from olive growers), they complained about the prices charged to consumers and the profits of the retailers, which in their view were high. Consumers, especially in the urban centres, said that because they believed in the product and the direct link with the producers, they were willing to pay the same price they would pay if they bought in a store (the retail price) to a producer whom they knew, trusted and had personal contact with.

Working Conditions

Working conditions were valued as an important ethical concern by only 5% of the stakeholders interviewed. Stakeholders were not concerned about the whole chain and working conditions in its different levels. They said that the regulations covering working conditions are adequate and that working conditions are good, especially at the primary production level and compared with other sectors of the economy. Some stakeholders, especially producers, were concerned about the dependence on migrant labour due to a shortage of domestic workers. Producers mentioned that the following factors: (a) seasonality, (b) bad weather conditions (the olive harvest occurs during winter), and (c) the lack of big farms and fields (because of farm fragmentation) were barriers preventing farmers from having a permanent workforce. This leads to very good working conditions for farm workers but also to lack of seasonal labour, high wages and low productivity.

Working conditions were valued as an important ethical concern by very few (3%) of the consumers interviewed. In general, consumers felt there were no problems in this area, and that workers are paid a fair wage and have very good working conditions. Consumer interviewees commented that farm workers work in a nice environment, unlike city workers who work in stressful conditions.

Quality, Composition and Taste

Quality, composition and taste were valued as an important ethical concern by 87% of the stakeholders interviewed. They said that Greek olive oil is a quality product, especially extra virgin olive oil, with a superb taste. Stakeholders commented that blending and mixing of the olive oil are detrimental to quality. They said that correct/accurate agricultural practices give an excellent product, which can contribute to the good health of consumers. Stakeholders commented:

> *The quality, which is the result of a number of parameters, is excellent. It can be improved further, especially if the farmers will be trained*

Quality, composition and taste were valued as an important ethical concern by 59% of the consumers interviewed. Most of the consumers seemed not to be aware of the different quality categories, and trusted the olive oil blindly, without reading labels or asking about its composition. Many consumers related quality to taste and said they bought olive oil from specific geographic areas because the taste appealed to them and they had become used to that taste. Consumers considered olive oil a quality product and fat.

Origin and Place

Origin and place were valued as an important ethical concern by 65% of the stakeholders interviewed. Stakeholders in the chain complained that overseas consumers do not check the origin of the olive oil, and allow blending and mixing of different qualities and origins of olive oil. They complained that the absence of labelling relating to the country of origin of the olive oil undermines its quality. Stakeholders commented that country of origin labelling should be compulsory. Furthermore, farmers in particular stressed that global trade and bulk exports to Italy threaten their income and the quality of the product. Farmers mentioned that the direct marketing of olive oil, or the sale of labelled oil from co-operatives, guarantees the oil's origin, and guarantees quality for consumers.

Origin and place were valued as an important ethical concern by only 25% of the consumers interviewed. They tended to assume that the oil they bought was of Greek origin, and when they talked about origin, they only mentioned olive oils from different geographic regions of Greece. Many consumers wanted to know more about from where olive oil came from.

Trust

Trust was valued as an important ethical concern by 71% of the stakeholders interviewed. They considered that consumers had trust in the product, and that it was important this should not be lost. Stakeholders commented:

> *There is trust in olive oil because there is knowledge about the way it is produced and it is believed to be very safe in comparison to other products* (Stakeholder)

Furthermore, stakeholders were concerned about the blending of olive oil and its potential effects on consumer trust. Personal contact and trust among the various links of the chain were important factors in maintaining trust in the product. Although chemical analysis is extensively used to determine the quality of the oil, even large companies which purchase oil from different sources value personal contact.

Trust was valued as an important ethical concern by 38% of the consumers interviewed. They trusted unquestioningly that olive oil is a quality product, beneficial to health and without any health risks. Furthermore, consumers appeared to trust farmers, which is why many of them purchase olive oil directly from farmers. Some consumers were sceptical about the practices of large corporations and their profit-driven blending of olive oil. Some consumers were also sceptical about the claims of corporations related to private labels, and questioned the value claims made by such labels (such as 'produced in mountain olive orchards' etc.).

Voice and Participation

Voice and participation were valued as important ethical concerns by 14% of the stakeholders interviewed. Most of the actors in the chain did not consider concern important. Farmers were interested in having their voice heard in the rest of the chain, especially by the exporters/wholesalers, but they felt that they currently have very little voice. Some farmers said that due to their direct links with consumers, they were sensitive to consumers' voices, and consumers could hear their voice and concerns.

Voice and participation were valued as important ethical concerns by 19% of the consumers interviewed. This concern was not easily understood, and often had to be explained to interviewees. Consumers generally felt passive and felt that they did not have a chance to participate or to be heard in the various production stages in the chain. They felt that legislators and large corporations made decisions without any real consultation with consumers. The lack of importance attached to voice and participation probably reflected the lack of understanding of the meaning of this concern expressed by some interviewees.

Transparency

Transparency was valued as an important ethical concern by 55% of the stakeholders interviewed. Farmers and processors felt that transparency is essential for the product and will improve olive oil and its image among consumers. Some reported that at the moment there is a lack of transparency due to the blending of olive oil during the various stages of trading and bottling. An organic olive grower commented that:

> *Transparency and information are only provided in the organic olive oil chain* (Organic grower)

Transparency was valued as a very important ethical concern by 53% of consumers. Consumers felt that there is little transparency in the chain. Some considered that buying direct from producers they know and trust leads to transparency.

Figure 7.4 presents the ethical concerns of stakeholders and consumers in the olive oil chain in Greece. Quality, trust, human health and origin and place were important for all stakeholders. They were less concerned about methods of production and terms of trade. Working conditions and voice were less often mentioned by the stakeholders. Human health, quality and methods of production were important for all consumers. They were less concerned about origin and place and trust. Working conditions and terms of trade were less often mentioned by consumers. Furthermore, consumers are concerned about the transparency of the olive oil chain. Overall, both stakeholders and consumers attached most value and relevance to the concerns of human health, methods of production and quality, with trust also important for stakeholders. Stakeholder and consumer interviewees said that the concern that was least relevant or important for them was voice-participation. Transparency and origin and place were talked about extensively. Working conditions and voice were the least relevant concerns for both stakeholders and consumers.

Traceability and Information Flows in the Olive Oil Chain

Greek stakeholders were eager for information on traceability. The smaller enterprises, especially, wished this information to be more easily available to them, at less cost. No information flow was found between the different links of the chain,

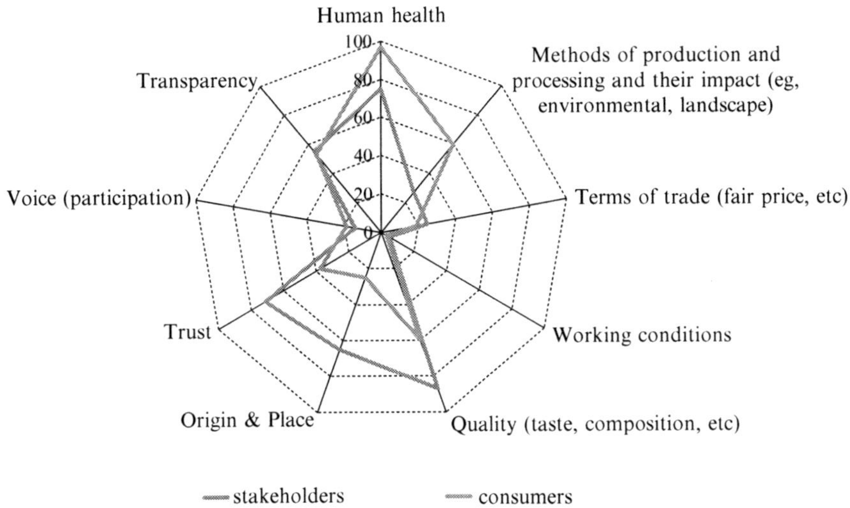

Fig. 7.4 Percentage of stakeholders and consumers that value as important each ethical concern in Greece

and especially from the marketing and retailing end of the chain to the production end. Overall, farmers, processors, and small packing and storage units feel that they are exploited and they have difficulty in obtaining information.

Stakeholders in Greece were found to be generally in favour of traceability, including ethical traceability. None of the Greek stakeholders considered the project's ethical concerns, traceability and their potential links as being irrelevant to the olive oil chain. They did not question the possibility of operationalizing ethical traceability, either in itself or as part of quality assurance schemes. Greek stakeholders did not seem to worry about the costs of administering ethical traceability, and many even said that they were willing to pay such costs. They were sceptical about the related bureaucracy and the way ethical traceability might be applied by administrative bodies. Stakeholders were also sceptical about the transfer of power, as well as about concerns driven by bureaucracy and the certification industry.

Greek consumers do not pay much attention to written information. They prefer and are more likely to trust personal information provided by family, friends, etc. This relates to the large volume of unpacked olive oil sold in Greece. But informal, personal information could also be untrustworthy, as it cannot be validated and need not meet any objective standards. On the other hand, consumers were very sceptical about information presented directly by commercial mass media. They felt that information presented by the commercial actors in the chain was manipulated and untrustworthy. But it was also found that consumers did not read the information provided on the labels of packed olive oil. Consumers often did not pay attention to labelling information on the different grades of olive oil, or concerning blended or refined olive oil. Consumers falsely assumed that all olive oil on the market was virgin olive oil and they did not scrutinize packed olive oil labels.

Conclusions on Traceability and on Ethical Traceability in the Olive Oil Chain

We have found that the olive oil industry has adopted some ethical standards, especially through assurance schemes, but only at some parts of the olive oil chain. The concerns are mainly those that have to do with working conditions and quality (related to taste), in response to food safety legislation and other regulations.

Traceability and potentially ethical traceability in the olive oil chain are limited by the practice of blending by packers/wholesalers, who usually bottle the oil. They blend it in order to manipulate chemical quality and cost. This situation makes it very difficult to trace the origin of the olive oil, or its chemical, physical, ethical and cultural characteristics. Thus, a good way to trace olive oil is by sensory and not by chemical analysis. Mixing and blending are encouraged by the fact that olive oil has a high value compared to other edible oils, and also by the fact that consumers in the non-producing countries are to a large extent uneducated about olive oil. In the organic olive oil chain, a more comprehensive traceability system is applied, but this system represents only a small fraction of the olive oil sector, and does not in any case cover many ethical concerns, such as working conditions, transparency, etc. The use of ethical traceability in the olive oil chain could be instrumental in preventing the communication of misleading and deceptive information to consumers, especially those outside the Mediterranean. This is particularly important, as in those countries a handful of transnational companies control the market. These companies purchase, mix, bottle and trade the olive oil, with the result that the olive oil chain in such countries is complex. Trustworthy information in these complex and long olive oil chains requires ethical traceability schemes.

The ethical concerns discussed were said by interviewees in Greece to be relevant to the production and consumption of food in general, and also of olive oil. When the concerns were explained to them, stakeholders were positive and showed some knowledge of the issues that were discussed. Consumers had basic knowledge of the chain and were interested in the ethical concerns, but their views on the concerns were not consistent. Interviewees prioritized and interpreted concerns differently depending on their place in the chain, and on their knowledge, education, culture and food traditions. Ethical concerns were more relevant to the smaller-scale and the organic stakeholders, who are more likely to value ethical concerns, place of origin, the authenticity of the oil, and food culture.

It is difficult to define 'ethical', while there is more uniform agreement about food safety. Ethical concerns have different meanings in different countries, and concerns such as working conditions have an economic value as well as an ethical dimension. Furthermore, our ethics are our values, which are culturally differentiated, and talking about ethical traceability requires a robust social dialogue. We consider that education and learning on the part of both stakeholders and consumers in the olive oil chain are important for improving traceability and ethical traceability in the chain (Kabourakis, 2000).

We also found that traceability systems, such as the ones applied in organic farming, require extensive record keeping and substantially increase bureaucracy in the olive oil chain. Thus, the introduction of ethical traceability in the olive oil chain, and its institutionalization, involves risks that should be taken into consideration. Such risks include the marginalization of small-scale enterprises (such as the family olive farms that are still the predominant mode of production in Greece), due to their inability to handle demands associated with traceability and ethical traceability systems; a concomitant risk is the reinforcement of existing, powerful interests in the olive oil chain. Ethical traceability systems in the olive oil chain may also increase costs and olive oil prices, because of additional administrative burdens to the chain. On the other hand, the development and application of ethical traceability systems tailored to small enterprises will assist them to survive on the global olive oil market.

Finally, it is questionable whether the provision of more information related to ethical concerns would be of benefit to consumers, who are already overloaded with information. The dissemination of such information also carries the risk of abuse by companies who could use it for marketing or other purposes. Furthermore, small-scale stakeholders as well as consumers seem to be powerless and voiceless in relation to existing traceability systems, which in theory serve them. Thus, there is the risk that institutionalization of ethical traceability may transfer more power to already over-powerful transnational companies, such as the ones that dominate the Greek bottled olive oil market, which are driven merely by commercial interests.

Stakeholders are aware of concerns related to oil production but they did not relate them to ethical traceability. They saw ethical traceability as a tool for improving quality and as an additional instrument for communication about the oil's properties. Traceability, including ethical traceability, was also conceived by stakeholders as a way to differentiate and to add value to their product. Farmers and small-scale stakeholders view traceability and ethical traceability as means to resist globalization, to help them survive in the face of fierce economic competition from large translational companies, to satisfy increasing consumer demands, and to preserve their identity and quality. Furthermore, ethical traceability was seen as a way to end the mass selling of bulk olive oil for export and the defrauding of consumers in non-producing countries who buy blended oil thinking it is pure. Farmers and small-scale stakeholders did not discuss in detail the restructuring of the chain (e.g. different ways of storing olive oil and the related costs) that would be required if blending was to be reduced or eliminated.

Greek consumers in general trust olive oil and are not worried about it. They are interested in improving its quality and are eager to purchase the best quality. Ecological olive oil produced by low-input systems, such as the organic system, was considered the best option by consumers. Thus, ethical traceability was received positively; and although not clearly expressed, there was consumer demand for greater ethical traceability, something that it is indicated by the high levels of self-consumption (where some of the olive oil produced by farmers is given directly to the members of their families) and direct purchasing of olive oil from the farmers.

Consumers seemed willing to pay the extra costs associated with guarantees about the properties and quality of the olive oil they wanted to purchase. As the rest of the stakeholders in the chain also appeared willing to pay extra costs, this is a point worth noting. The feasibility and costs of establishing an ethical traceability system in the olive oil chain should be investigated. However, taking into consideration the saturation of consumers with often irrelevant information, and the reluctance of stakeholders to take on more administrative burdens, the practicalities of ethical traceability schemes should also be investigated.

We concluded that at present there are not many ethical concerns in the olive oil chain that need to be addressed urgently. One major concern that should be addressed, however, is the blending of olive oil of different origins and the mixing of virgin olive oil with refined olive oil and/or other edible oils. This has an ethical dimension, as the consumer assumes he/she is purchasing a product that is natural and authentic, when in fact it is not. Besides, the consumer may pay a high price for a product that does not correspond to his/her wishes. Ethical traceability, although not named as such, was present in the olive oil chain. Similarly, ethical concerns, though not discussed and found to be subjective, were present in the chain. All stakeholders felt that transparency, ethical concerns and traceability in the olive oil chain were increasingly relevant.

Ethical traceability is a rather new concept in the Greek food sector in general. It might prove a particularly useful tool, especially in the case of the olive oil chain, along with future developments regarding the improvement of production and marketing standards dictated by changes in consumers' behaviour and the policy framework.

Acknowledgements We thank the many stakeholders and consumers who provided information and stimulating discussion. The research was supported by the project 'Ethical traceability and informed choice in food ethical issues' funded by the European Union Sixth Framework Program.

References

Angles, S. (1999) 'The changes in the olive-growing geography of Andalusia', *Olivae*, 78: 12–22.

Braudel, F. (1973) *La Mediterranée et le monde mediterranéen a l'époque de Philippe II, tome premier, tome deuxieme*, 4th edn. Paris: Librairie Armand Colin [Greek translation, 1997].

Castro, C., M. Guerreiro, F. Caldeira and P. Pinto (1997) 'The Olive Oil Sector in Portugal: General Aspects, *Olivae*, 66: 12–22.

Codex Alimentarius (2004) *Codex Standard for Olive Oils and Pomace Oils*. Codex Stan 33–1981 (rev. 2–2003).

EC (The Commission of the European Communities) (1991) Regulation No. 2568/91, On the characteristics of olive oil and on the relevant methods of analysis. *Official Journal of the European Communities*. No. L 248.

FFT (2006) *Olive Oil Market in Greece*. Switzerland.

Fiorino, P. and F. N. Griffi (1992) 'The spread of the olive farming'. *Olivae*, 44: 9–15.

Jacotot, B. and A. Hagege (1993) *Olive the Fruit of Life*. Brussels: EC.

Ghazau-Gillig, S. (1994) 'The civilization of the olive tree and cereals'. *Olivae*, 53: 14–19.

ICAP (2005) *Olive Oil. The Olive Oil Sector in Greece*. Athens: ICAP (in Greek with English summary).

ICAP (2006) *Olive Oil. The Olive Oil Sector in Greece*. Athens: ICAP (English summary).

IOOC, 1993 *69th Session, CT/R 461. No. 1*. Madrid, Spain.

IOOC, 2006a World olive oil figures. www.internationaloliveoil.org, accessed 13.10.2006.

IOOC, 2006b EC olive oil figures. www.internationaloliveoil.org, accessed 13.10.2006.

Kabourakis, E. (1999). 'Code of practices for ecological olive production systems in Crete'. *Olivae*, 77: 46–55.

Kabourakis, E. (2000) 'Learning processes in designing and disseminating ecological olive production systems in Crete', *in* M. Cerf, D. Gibbon, B. Hubert, J. Jiggins, M. Paine, J. Proost, N. Rolling (eds.) *Cow Up a Tree. Knowing and Learning for Change in Agriculture. Case Studies from Industrialized Countries*. Paris: INRA Editions.

Kabourakis, E., A. Vassiliou, D. Papadopoulos (2006) *Ethical Traceability in the Olive Oil Chain*. Unpublished Project Report to the EU.

Keys, A, A. Menotti, C. Aravanis (1984) 'The Seven Countries Study: 2289 death in 15 years'. *Prev. Med.*, 13: 141–154.

Keys, A., A. Menotti, M.J. Karvonen, H. Blackburn, R. Buzina, B.S. Djordjevic, A.S.A. Dontas, F. Fidanza, M. Keys (1986) 'The diet and the 15-year death rate in the Seven Countries Study'. *Am. J. Epidemiol.*, 124: 903–915.

OLIAREA Survey (1999) http://agrifish.jrc.it/marspac/olivine/oliarea.htm, accessed 13.10.2006.

Papageorgiou, C. (1994) 'The role of the olive tree in Greece'. *Olivae*, 19: 7–11.

Papadopoulos, D. (2003). *Statistical Analysis and Documentation of the Olive Oil Sector*. Athens: INAGROP, NAGREF (in Greek).

Petkanopoulos, T., T. Raptis (2005) *The Unutilized Greek Extra Virgin Olive Oil*. Thesaloniki: SEVE, (in Greek).

Standish, R. (1960) *The First of Trees: The Story of the Olive*. London: Phoenix House.

Trichopoulou, A., E. Critselis (2004) 'Mediterranean diet and longevity', *Eur. J. Cancer Prevent.*, 13(5): 453–456.

Trichopoulou A., T. Costacou, C. Bamia, D. Trichopoulos (2003) 'Adherence to a Mediterranean diet and survival in a Greek population'. *N. Engl. J. Med.*, 348: 2599–2608.

Weinbrenner, T., M. Fito, R. de la Torre, T.G. Saez, P. Rijken, C. Tornos, S. Coolen, M.F. Albaladejo, S. Abanades, H. Schroder, J. Marrugat, M. Covas (2004) 'Olive oils high in phenolic compounds modulate oxidative/antioxidative status in men'. *J. Nutr.*, 134(9): 2314–2321.

Part III
Ethical Traceability and its Philosophical Implications for Civil Society, Market, State and Democracy

In this third part of the book, the main philosophical (ethical) aspects of ethical traceability are analyzed and evaluated. Drawing on Part I and the sociologists' work presented in Part II, five issues are tackled in separate chapters. The first issue comprises the challenges that ethical traceability poses for existing distinctions between the private and the public, and the corresponding concept of civil society. In Chapter 8, Christian Coff shows that consumer concerns about production practices in the food sector partly change the meanings of these distinctions. Implementing ethical traceability and informed food choice entails creating new kinds of public spheres and a new kind of civil society: the ensuing debate and deliberation by communicative measures could renew democratic and participatory structures (see Chapter 13 for a further development of this idea). The openness of the public sphere is a basic principle of democracy; it follows therefore that the theory of the private character of the market and food production practices needs to be reconsidered if we want more democratic institutions in the food sector. Market actors could even profit from considering how to take the ethical and political concerns of consumers seriously in terms of ethical traceability and informed food choice.

In Chapter 1 we distinguished between different uses of traceability, as a communication tool (between market parties) and as a government tool. In Chapter 9, Liesbeth Schipper elaborates upon this distinction by asking how far concerns about for example animal welfare, the environment, fair trade and biodiversity should be regulated by market initiatives or be the object of laws and governmental regulations. Are these concerns of consumers/citizens to be regarded as mere market preferences (*consumer* concerns in the strict sense) or as public concerns? In answering these questions Schipper makes use of the well-known distinction of liberal political philosophy between conceptions of the good and conceptions of the right. A conception of the good refers to the moral, ideological and religious ideals that people pursue in their lives. In a liberal society, all people have an equal chance to live on the basis of their own conception of the good, as long as other people also have the chance to do this. A conception of the right serves as a general framework that belongs to the responsibility of politics and governments and within which people have this freedom. It is prior to a conception of the good; this priority means that people are not allowed to act contrary to this conception of the right. Schipper makes it clear that when concerns about animal welfare are based on a conception of the good, the proper way to address them is on the market, with labelling the best option. When, on the other hand, consumer concerns are based on a conception of the right, the government should uphold the principles that are behind such concerns. In examining this issue, she sheds light on the questions of whether and to what extent traceability of animal welfare concerns should be considered as a communication tool or a government tool, or whether there is some intermediate solution.

Working within the same liberal political philosophy perspective, in Chapter 10, Volkert Beekman pursues the issue of what market and government respectively should do about consumer concerns by arguing that two distinct versions of ethical traceability can be justified. The first version requires governments to ensure that all consumers are provided with foods that respect some threshold level of, e.g. animal welfare, supported by an overlapping consensus. The second version requires food producers to provide sufficient information about products to allow consumers to make food choices consistent with reasonable, non-superficial values which are not, however, supported by an overlapping consensus. Governments should facilitate this by ensuring that consumers are not provided with misinformation about relevant food characteristics.

In Chapter 11, Michiel Korthals concentrates on the practical issue of how to implement consumer concerns in food chains, in a democratic way, by organizing ethical discussions about conflicting values that include consumers as participants. Consumer concerns are multi-interpretable and dynamic, and there are many different types of consumers, whose choices between conflicting values differ accordingly. It is therefore worthwhile to organize, in a deliberative way, the fundamental ethical decisions in the serial links of food chains in accordance with consumer concerns, and traceability schemes can facilitate this. He argues that instead of considering ethical standards and targets as fixed, as is the case with codes and schemes, it is more fruitful to emphasize the processes by which consumers' ethical concerns are weighed up and shaped. The concept of 'Ethical Room for Manoeuvre' (ERM) is constructed to set out the conditions under which paramount values and their dilemmas can be identified and weighed up. The main aims of the ERM are to make room in all the links of the food chain for regulating and implementing relevant consumer concerns by balancing and negotiating them; to support information systems that are relevant and communicative for various consumer groups; and to organize consumers' involvement in the links of the food chain.

Finally, in Chapter 12, Marco Castagna argues that in the light of traceability's multiple meanings, it is easy to give in to the temptation to solve the ambiguity by curtailing its meaning. Instead, one should recognize its radical ambiguity, based on the belief that a realization of the limitations of an instrument allows one to use it in the best possible ways. Based on this broader interpretation of traceability, he connects the idea of informed choice with the complex notion of consumer/citizenship

Chapter 8
Challenges of Ethical Traceability to the Public-Private Divide

Christian Coff

Introduction: The Challenge of Ethical Traceability

Ethical traceability and informed food choice challenge the way we think of the two main spheres of society: public and private. The public sphere typically includes the state, the media, democratic political deliberation, public opinion, advertisements, books, buildings, plays, public transportation, etc. Public means open and visible. The private, on the other hand, is the sphere of that which cannot be seen or heard. It is the 'un-common' world, characterized by intimacy, individuality, family life, (private) property, private transportation, private businesses, etc.

Ethical traceability and informed food choice challenge this divide because they seem to have a foot in both camps. On the one hand, ethical traceability addresses the common good in society – and more specifically in the food production chain. Ethical traceability is public in the sense that it belongs to the sphere of public goods. Furthermore, it contains elements of the public sphere as public discussion about the content of ethical traceability schemes is paramount for the framing of such schemes (see Chapter 11 for elaboration of this idea).

On the other hand, ethical traceability addresses private consumers who make their choices on the private premises of the market. The roles of 'private' market actors such as producers, retailers and consumers are challenged in that informed food choice enables a political and ethical stance on the part of the market actors. This questions the classical economic idea of an ethically neutral market consisting of non-political consumers and producers. It challenges the idea of food purchase as a purely private matter based on self-interest. The empowerment of consumers through ethical traceability is the extension of the citizen role into the economic sphere of the market.

Taken together, these challenges make up a complex terrain, in which market actors and political actors find it hard to find their feet. This is evident from at least three observations that I would like to point to here.

Firstly, the empirical work on ethical traceability presented in Part II shows that, for some stakeholders in the food supply chains studied, it was hard even to grasp the idea of ethical traceability. Some were sceptical of the idea because revealing certain aspects of production processes or decision-making could raise ethical

C. Coff et al. (eds.) *Ethical Traceability and Communicating Food*,

concerns that might have severe economic consequences. Others doubted that consumers are qualified to take a stand on the very complicated issues of modern 'scientific' food production practices.

The second observation concerns political debates on traceability itself. The most dominant conflict in the debate is between what we in Chapter 2 have called the *safety position* and the *informed choice position*. The *safety position* supports a minimalist traceability approach of 'one step back, one step forward' for food safety reasons only. Hence there is no need to inform consumers. The safety position is based on a traditional liberal approach, where the interference of the state in the private sphere is limited to the protection of the health of citizens/consumers, and the political is excluded from the market (see Korthals, 2004:155 for the liberal approach). The *informed choice position* sees traceability as a means of communicating about food products and hence of informing consumers on, for instance, the ethics of the production history. These two positions have contradictory attitudes to the role of public participation, based on different attitudes towards the public-private divide.

The third observation concerns how the ideas of ethical traceability and informed choice are being perceived in the political sphere. The idea of offering consumers informed food choice based on communication with actors in the food chain seems to fit into neither left- nor right-wing politics. Ethically informed food choice disrupts classical liberal ideas of a non-political market and should for that reason be rejected (see Chapters 9 and 10 for an analysis of a Kantian modern liberal position towards ethical traceability). At the other end of the spectrum, politicians from the left hesitate to support informed choice, because they fear that the self-interest of the market may expand to embrace the sphere of political thinking about the common good. The positions can be summed up as the right-wing's fear of a politicization of the market and the left-wing's rejection of a 'marketization' of the political.

From the perspective of political philosophy, there is definitely a need for clarification of the concepts of ethical traceability and informed food choice and their relationship to the classical public-private divide. How can we think about ethical traceability and informed food choice in the light of the public-private divide?

More practically, we may ask what kind of policy is implied by ethical traceability and informed choice? Is it liberal, is it socialist, is it conservative or is it beyond left and right?

It would, however, be too ambitious to set out to clarify these questions once and for all. Furthermore, it may not make much sense to try to indicate right and wrong ways of thinking about the challenges ethical traceability poses to the public-private divide, as none of them need necessarily be more right or wrong than any other. There is no final key on how to consider ethical traceability and informed food choice in the public-private divide – it is a matter of opinion and conviction. This is not to say that we should not argue about it; dialogue can help develop our thinking about the divide in contemporary society.

Instead, I propose to set out an analysis of how different positions *could* respond to the challenge, considering their ideological and philosophical backgrounds and traditions. This chapter, therefore, in the light of the challenges that ethical traceability and informed food choice pose to us, indicates *a way of searching* for arguments on how to consider the public-private divide.

Informed Food Choice

First, though, it is worth taking a closer look at the idea of informed food choice. The emergence of a massive consumer culture in affluent societies and a strong emphasis on individuality have contributed to a focus on *individual choices* (see for instance Taylor, 1991:37–38). There have been many theories and analyses of choice: rational choice, public choice theory, social choice theory, the politics of choice (Norris, 2005), models of economic choice (e.g. Milton Friedman's book *Free to Choose*, 1979), and so forth. Choices by both citizens and consumers are seen as highly important by liberal policy-makers, for the creation of well-functioning societies, for social responsibility and for the functioning of a free market. This accentuation of choice in rhetoric is not only a sign of a stronger emphasis on the freedom of individuals. It is also a '*mise en discours*' (Foucault, 1976:29) or verbalization of the responsibility of the individual.

Two dominant positions on individual choice can be discerned, one optimistic and one critical. Choice has been a buzz-word in economics and marketing for a long time. Individual choice has been used in liberal, market-oriented rhetoric to describe in positive terms the emancipation of the individual from social ties. In this context, the word choice embodies the freedom to choose one's own individual lifestyle. In affluent, market-oriented societies, consumption choices are seen as a means for the realization of the self and for the creation of personal identity.

There are, however, also critical and academic positions deploring the development of the consumer society. In the book *The Consumer Society* (1970), the French philosopher Jean Baudrillard displays what he calls a 'phenomenology of consumption', pointing to the shallowness and fragility of consumption cultures. The German philosopher Jürgen Habermas, in his book *The Structural Transformation of the Public Sphere* (1962), also regrets the degeneration of the public sphere into a commercial sphere of particular interests. Common to these books is their description of a society increasingly dominated by commerce and of capitalism becoming the prominent ideology. The phenomenology of consumption describes how attention and care turn away from human beings and society and instead focus on commercial objects. Attention and concern are turned towards individual consumption and satisfaction at the expense of care for other human beings and the common good.

For Baudrillard consumption is characterized by self-interest and hence ignorance of the surrounding world. Within the ethical consumption 'movement' this critique is taken seriously, and it is emphasized that consumption should not be seen as a means for purely egoistic satisfaction. Rather, 'the movement' sees in ethical consumption a moral relationship to those living beings (humans, animals and nature) involved in the production history of the product.

Ethical or political consumerism has attracted much attention in recent decades. It exists in at least three forms. The oldest form of political consumerism is the *boycott*. The term was coined in 1880, after the British Captain Charles Cunningham Boycott was 'locked out' by an Irish peasant movement called The Irish National Land League. In protest at poor working conditions on Boycott's farm, the peasants refused to work for Captain Boycott, which led the newspaper the *Daily Mail* to speak of a 'boycott'.

Since then the term has been used in many different fields, but has become especially common in the field of consumption, where it is used to describe the avoidance of certain products for ethical or political reasons. The force of a boycott lies in the explicitness of the 'evil' (e.g. low wages and bad working conditions in third world countries, or environmental pollution) to be avoided. Jonathan Allen formulates it this way:

> We do not know what a perfectly just society would be like to live in, but we do know – some of us know from everyday experience – what it is to live in an unjust, or cruel, or humiliating society… it is often easier to identify evils than it is to recognize and understand goods (Allen, 2001:350).

The idea presented here is that where 'negative morality' or evils such as suffering, envy, corruption, violation, cruelty, betrayal and so forth, are clearly evident, their removal or minimization has priority over the pursuit of goods, and that 'negative morality' plays an important role in formulating ethical judgements. Undoubtedly this can be applied to consumer behaviour, in the form of a desire to escape from foods with 'evil' qualities and production histories. However, this strategy of avoiding 'bad' foods suffers from the problem that as long as we know nothing about alternative foods there is no guarantee that these are any better.

A second form of ethical or political consumerism, the *buycott*, has come to refer to the positive selection of products for their explicit ethical qualities. In buycotting, consumers make use of product information and product labelling for the purpose of selecting products or corporations with specific production practices and qualities that are considered attractive.

More recently, a third kind of ethical consumption has appeared; *discursive political consumerism*. In general, this refers to the discursive expression of opinions about products, brands and corporations. Communication is crucial for discursive political consumption; the idea being to communicate information and opinions about the practices of corporations to other concerned individuals and to the general public. The aim is to 'problematize the politics of products by seeking and relaying information on corporate policy and practice…' (Micheletti and Stolle, 2005:256). So far this strategy has been used to create public dialogue and awareness on products and corporations, in order to initiate negotiations with business to develop ethical business practices and codes of conduct. It has been a successful part of the 'no sweat' movement and its boycott campaigns of, for instance, Nike. Within this kind of political consumerism we may also include communication campaigns by private persons and organizations on specific products and brands. Typically, the

Internet will be used for such activities.[1] Ethical traceability could be considered part of this discursive political consumption strategy in as much as its aim is to communicate recorded information on food products' production histories and their impacts on nature and society.

In addition to these three types of ethical and political consumption, it can be expected that new forms will emerge in the coming years. In particular, the development of new Information and Communications Technology (ICT) and the continuing development of the Internet seem likely to drive ethical consumption in new directions.

Due largely to extensive media coverage, public awareness of the ethical and political aspects of food production is increasing. Examples of ethically sensitive issues include the destruction of the Brazilian rainforest to make way for plantations for animal feed and meat production, and the fatal consequences of industrial shrimp production on local mangroves and farming cultures in some Asian countries.[2] Another example is provided in Chapter 5, which describes the increased long-distance transportation of living animals that results from cost-efficient outsourcing of slaughterhouses to countries with cheap labour in the Danish bacon supply chain. Media coverage of injustices in food production turns food production into a public affair and makes citizens turn to the market with political and ethical concerns. It leads to the social and political mobilization of consumers and reinforces democratic sensibilities. This new awareness of being simultaneously a citizen and a consumer is summed up in the phrase *consumer-citizens* (Scammell, 2003).

Even though ethical consumption and political consumption have been known for decades (see for instance Gabriel and Lang, 1995, *The Unmanageable Consumer*), and seem widely accepted, they are by no means uncontroversial ideas. Ethical consumption, ethical traceability and informed food choice are all concepts that seem to challenge mainstream thinking about societal order, social structures and the functions ascribed to the different spheres of society. Making information on production histories and production practices publicly accessible questions widespread and central assumptions about the privacy of the market, in contrast to the openness of the public sphere. The concepts challenge the binary distinction between public and private.

When consumers address political and ethical questions about societal and environmental issues to a market that mainly functions on the premises of the private and non-political, it necessarily creates tensions and frustrations. Consumers might become dissatisfied with the market's inability to adapt to their concerns,

[1] Examples are www.ethiscore.org, which gives 'ethical scores' to products, and www.natureandmore.com, which reviews products and corporations.

[2] For instance, there have been reports by World Rainforest Movement (www.wrm.org) and Greenpeace (*Shrimp: The Devastating Delicacy*) alleging heavy exploitation, kidnapping and killing of local farmers; cutting down of mangroves that are essential for coastal preservation; devastating aqua-cultural methods based on intense use of chemical fertilizer and pesticides, which all together ruin the land in 3–5 years, after which the corporations move on to new areas leaving more or less a desert behind them.

and corporations may consider consumers' ethical claims to be burdensome and inappropriate to their activities.

Seven Approaches to Public, Private and Civil Society

To analyze how the concepts of ethical traceability and informed choice challenge mainstream thinking about the public-private divide, we should first take a closer look at what can be understood by the notions of public and private.

Louis Michael Seidman (1987:1007) argues that the public-private distinction touches upon three dilemmas:

1. The universal versus the particular
2. Government intervention versus private liberty
3. Openness versus secrecy

All three relate to dilemmas encountered by ethical consumers and to the idea of informed food choice, as the three examples below illustrate:

1. Consumers can pursue their own self-interest or they can pursue universal interests and the common good. The same goes for producers.
2. States and international governmental organizations can intervene to offer greater market transparency and thus make informed choice a more realistic option for consumers. Some, however, would argue that such governmental intervention in the market would limit private freedom.
3. Business and production often rely on privacy and require secrets not to be revealed due to competition with other corporations. This is of course contrary to consumers' claim for openness and information.

The binary distinction between public and private is as old as Western thought itself and still plays a central role in thinking about societal order. The public-private distinction has a descriptive element, as it describes two spheres of society, but it is not merely a descriptive concept. The distinction is not only used to describe reality as it is; it should also be considered an analytical tool, understood as a 'social construction' or conceptual apparatus by which we analyze, interpret, understand and not least organize society. As an organizing category (a notion introduced in this context by Weintraub, 1997:2) the distinction also acquires normative content: it is a way of thinking about society that structures ideas and visions of the common good and the just society.

The Italian political philosopher Norberto Bobbio (1909–2004) considered the binary thinking of public-private so basic to perceptions of society that he called it one of the 'grand dichotomies' of Western thought (Bobbio, 1989). The public-private division emerges as a pattern of social organization: some things are of common interest and thus to be decided upon in public and by the public. Other issues are not of direct interest to others and are thus to be decided upon in privacy. The distinction is, for instance, at the heart of the protection of private individual rights in secular, liberal Western democratic societies and it is deeply incorporated into the structure of much economic theory.

However, the terms private and public are not themselves unambiguous. Public refers in different contexts to such very distinct things as the state, the media, democratic political deliberation, public opinion, advertisements, books, buildings, plays, public transportation or just to a space open to all – a public space (which may, as in the case of cafés, be private property). Often, this multiplicity of meanings seems to confuse the debate on a great many issues for which the public-private distinction is an underlying assumption.

Even though the public-private divide might not be a clear distinction with sharp demarcations, it is common to think of the public as a place where everything can be seen and heard (Arendt, 1958:50–52).[3] In German and Scandinavian languages, with some minor variations, the word for public is *Öffentlichkeit*, which literally means 'openness' (Habermas, 1990:54).[4] *Öffentlich* are, for example, places and buildings that are accessible to everyone. Thus, visibility and audibility are characteristics of the public sphere. It is the common world of human relations and ideally also a realm of active participation.

The private-public dichotomy is traced by Hannah Arendt in her book *The Human Condition* (1958) to the self-governing cities of antiquity, such as Athens and Sparta, as described by Aristotle. Aristotle's famous distinction between private and public is between the household (*oikos*) and the political community in the city state (*polis*).[5] The household is characterized by the 'strictest inequality', as the master of the house dominates and decides over women, children and slaves. The household also comprises economic life. In contrast, the public space of the city (*polis*) is characterized by a relationship between free and equal men – and only men with a property of their own. To participate in the public sphere it was necessary to be free from private inclinations and the daily obligations of a household. One important implication of the household being the realm of obligations was that freedom was located in the political realm. To be free was to be free from obligation and the command of others. Economy, being obligatory, was solely related to the private sphere and had nothing to do with the political (Arendt, 1958:29–31). From this view on the public-private divide it followed that ethics, in the form of the Greek *arête* (virtues), were linked to the public sphere of free men and not to the private sphere of obligation. The private was, so to speak, un-ethical.

[3] I especially make use of Hannah Arendt's *The Human Condition* (1958) and Jürgen Habermas's *Strukturwandel der Öffentlighkeit* (1962) in drawing this general picture of the difference between public and private.

[4] Also the Latin word *publicus* refers to what is shared and open to view (Hannay, 2005:11).

[5] According to McKeon (2005:7) the actual existence of the paradigmatic antithesis of *oikos* and *polis* in atiquity has been questioned by recent research. It has been suggested that Aristotle described an idealistic division of society rather than the actual situation of his time. This again points to the descriptive and the normative character of the distinction. However, for the philosophical investigation undertaken here, the history of this divide is of less importance and shall therefore not be considered in any further detail.

Furthermore, the concept of the public was and still is closely associated with that of democracy. Democracy builds on dialogue between free men and decision-making taking place in open and visible ways – that is, in public (Calhoun, 1997:81).

Arendt herself subscribes to the ancient Greek distinction of private and public, arguing that the public/political sphere degenerates if penetrated by private economic interests. The public sphere should be preserved as a sphere for disinterested (in the sense of free from private interests) discussion and decision-making, with the focus on the best functioning of society and not on individual interests.

Inversely, the private is, as Arendt writes, the sphere of that which cannot be seen and heard. It is the hidden, personal and invisible world, the un-common world, characterized by intimacy, individuality, family life, (private) property, private transportation and private businesses. Shopping preferences would probably also be considered private by many consumers (Coff et al., 2005). It is the sphere of closed doors. Arendt (1958: 58) notes that the privation entailed in the word privacy refers to the absence of others. It is also common to consider the private sphere as composed of two distinct areas: the intimate area of the family and personal relations and the area of economic transactions. Both concern personal ends.

Another important notion that often appears in the debate on the public-private distinction is that of *civil society*. However, as the concept does not exist in other cultures, such as the Confucian, Islamic and Jewish traditions, it is not an organizing concept in every tradition (Post and Rosenblum, 2001:21). Civil society is a modern, post-Renaissance, Western invention and a Western way of conceiving societal order.

McKeon (2005) proposes that we understand civil society as the Enlightenment consolidation of the late-Renaissance category 'society', which appears as the antithesis to the idea of the state in that epoch. Civil society develops along with the rebellious stirrings of the emerging idea of citizenship against feudalism and patriarchal domination. Under feudalism, all activities in society were incorporated into the private household of the feudal lord and 'his' state. As such, creating a civil society can be seen as an opposition to the feudal state and an attempt at redistribution, to create a more 'fair' division of powers in society.

The German philosopher Georg Wilhelm Friedrich Hegel (1770–1831) used the term civil society (in German *bürgerliche Gesellschaft*) more specifically, to mean a social sphere distinct from the state *and* the family. For Hegel, civil society therefore refers to the market as well as to voluntary and social associations. Today, most agree that *associational life* is part of civil society. It is considered a zone for individuals to associate with others for a variety of purposes, free from militarism and violence (Post and Rosenblum, 2001:8). Whether the market should be included in civil society is more controversial. Within some kinds of liberalism the term civil society is used to designate everything that is 'non-state' or 'non-governmental' and it thus includes the market. Within the communitarian and republican traditions, the civil sphere refers instead to a non-governmental sphere, open for public and democratic discussion, and the private sphere to which the market does not belong. The term civil society is thus used differently by different people, which sometimes can lead to confusion and misunderstanding.

In conclusion, it is paramount to note that all three concepts are used normatively as 'organizing' categories structuring the different spheres of society. Because of the strong organizing and normative power of the concepts, they are essential for most political ideologies. Ideas about what public, private and civil society should embrace are fundamental because they define the roles and powers of the different actors and institutions in society. To illustrate this I have chosen seven political and sociological/anthropological positions for which the demarcation between public, private and civil society is especially relevant.[6]

1. **Liberal models.** Demarcation of the public sphere of authority from the private sphere and the market is a major concern within different liberal approaches, from classical liberalism to modern political liberalism. The borderline between public and private is usually drawn between the state and the rest – with the rest conceived as civil society, the market and families. The role of the state is often conceived as that of securing a well-functioning civil society and democracy by controlling armies and different kinds of oppressions that might interfere with the freedom of citizens and the market and ignore democratic rights.[7] There are two ideas within liberalism:

   1) Selfish interests are turned into the common good and general welfare through the functioning of the market (as in the idea of the invisible hand guiding the market to achieve social ends and common wealth). The Dutch physician Bernard de Mandeville (1670–1733) argued that governments should manipulate individualism and egoism for purposes useful to society: 'Private vices by the dextrous management of a skilled politician may be turned into public benefits' (Mandeville, 1714). This was developed into his famous motto: private vices, public benefits.
   2) The state should *not* impose on its citizens a preferred way of life. State intervention in the private (market) can only be justified if it maintains the structure of civil society (including the market) and prevents harm. Gobetti writes that the natural law theorists (such as Thomas Hobbes and John Locke) introduced the idea of:

      [H]arm as the criterion of distinction between private and public... Private jurisdiction extends to all the activities in which an adult engages without harming or endangering others... When harm is done, intrusion on the part of the public authority into the person's private jurisdiction is legitimate (Gobetti, 1997:103–104).

There are, however, disagreements within the different liberal traditions as to where exactly the borderline between private and public should be drawn. Post and

[6] The list does not include all positions, see for example Nielsen (2001:33–37) and Weintraub (1997:7) for other mappings or typologies.

[7] I shall not go further into the relationship and interplay between state and civil society as it is beyond the purpose of this paper, but refer to Post and Rosenblum (2001) for an elaboration on this.

Rosenblum (2001) argue that there are two different positions on this. They call these positions 'congruence' and '*modus vivendi*'. The congruence position argues that the state should actively secure the upholding of public norms in society and that if this is not done society will be fragmented. The *modus vivendi* position favours a lively civil society with few common norms (some overlapping consensus on basic issues like human rights and democracy is enough) and minimum state intervention.[8]

2. **Republicanism.** Republicanism is rooted in the political ideas of ancient Rome (Calhoun, 1997:77). The public and civil society play crucial roles in republican thought because of the importance ascribed to public discourse and action in concert. The political community and public life are opposed to both the market and the administrative state. The republican standpoint assigns sovereignty to the people (rather than to a monarch). The state should seek to cultivate in citizens the qualities of character and virtue necessary to the common good of self-government (Sandel, 1996:25). Public life consists of *active* participation in civil society and deliberation on collective decisions. Citizenship consists, as in the writings by Aristotle, in both ruling the community and being ruled.
3. **Communitarianism.** Civil society is understood as a sphere for public discussion on the development of local communities. The market is often considered a threat to basic entities that bind communities together: the family, tradition and religion. A well-functioning civil society is thus to be based on families and smaller groups in local communities, the 'core institutions' of society. These are generally considered to be the driving force behind the positive social norms on which the good community is founded. Communitarians often believe that the state should secure free education, economic and ecological sustainability, health care and a social safety net.[9]
4. **Critical theory and civil society.** In *The Structural Transformation of the Public Sphere* (1962) Jürgen Habermas considers the public sphere a place for free deliberation and communicative interaction between citizens. In contrast to communitarianism, the focus is not on local communities but on society and civil society. The private spheres of intimacy and economy are considered necessary preconditions for the function of society because it is in the private that the individual is 'formed' and prepared for the public life: public deliberation will only function if the participants are properly prepared for it. However, private and particular interests do not count as valid arguments in the public sphere. Starting from Hegel's tripartite model of modern society consisting of family, civil society and state, Habermas wants to demarcate *civil society* from both the market and the state as a sphere of rational deliberation between free citizens.

[8] There are indeed many liberal writers and traditions. I have for the sake of simplicity chosen to draw a more general picture of liberalism at the cost of a more detailed description of the different positions on the public-private divide. The latter would deserve a paper on its own.

[9] Some well-known communitarian writings are Alasdais MacIntyre's *After Virtue* (1981), Charles Taylor's *Sources of the Self* (1989) and Amitai Etzioni's *The New Golden Rule* (1996).

Civil society is the sphere where private people come together as a public, and it is thus a democratic public domain. This is a sphere where arguments are tested against one another and in the end lead to the better argument's forceless force. It is a realm with an open and pluralistic structure. The role of civil society is to function as a facilitator of communication (as a mediator) between the private and the state.

5. **Public life as sociability.** The social sphere in this understanding is less political and less dependent on rational argument than in the critical theory described above. This is the 'informal' public life, which takes place in urban public spaces like cafés, restaurants, sidewalks, public transportation, public parks and squares, etc., where hegemonic individuals meet and exchange all kind of experiences. The aim is not to express solidarity, but rather to make diversity agreeable. In Europe this kind of sociability is best known in the South, where life to a great extent is (or used to be) lived 'in public'. Philippe Ariès is the author of one of the first academic analysis of this kind of sociability (Ariès, 1960).
6. **Private life as backstage.** The sociologist Erving Goffman (1922–1992) writes in *The Presentation of Self in Everyday Life* (1957) about the threats of too much public exposure and inescapable visibility. Goffman uses the picture of a gas station mechanic who is forced by spatial design to do his service work on cars directly in front of their owners. The surveilled mechanic must constantly be on guard, must watch out that he is *performing* well. Constant openness and monitoring of his work may end up stressing the mechanic and result in unpleasant working conditions. And maybe even poorly repaired cars.

In contrast, the private sphere is unguarded and informal. Privacy, at home for instance, is where one can retreat from the claims of performance and official presentations of the self. For Goffman the private is more authentic than the pretentious public. His analysis of the private sphere in some respects approaches the kind of sociability mentioned above: the private is conceived as a place not only for the isolated individual but also for the private company of friends who are joking, dressed informally, sitting sloppily, mumbling, gossiping, shouting, quarrelling, co-operating, etc. Backstage is also a place to retreat for better performance in public. In a teasing tone, Alan Wolfe writes that privacy is a 'public' good (Wolfe, 1997:186).

7. **Feminist approach.** The dichotomy between public and private is central to much feminist scholarship because women in most traditional cultures have been confined to the private sphere – often under the patron's unrestricted dominion (see for instance Benhabib and Cornell, 1987). The superior public realm, the realm for decision-making on the common good, was reserved the male sex. The feminist critique is that this divide in traditional cultures is gendered and discriminating and therefore should be avoided. The feminist approach is to make the domestic sphere a central focus of academic study to reveal abuse and repression of women.

From the descriptions of these seven different positions on the public-private distinction it is first of all clear that the borderline between the spheres is contested: public and private spheres blur into each other.

Secondly, the roles and the powers assigned to the different spheres are closely linked to political and ideological positions. The normative understandings of the concepts are explicit and prominent. Use of the concepts does not rest on empirical observations of the structures in society; rather, the readings of the concepts are orchestrated by normative undertones of what a good society would be like. In general, the concepts of public, private and civil society invite moral thinking about society, as they touch upon fundamental organizational structures in society, and the role and competencies of different societal actors.

Seven Responses to Ethical Traceability

I now return to the analysis of food purchase to see how the above seven theories on the spheres of the public, private and civil society can be applied as a kind of analytical apparatus to illuminate the discussion on ethical traceability and informed food choice.

From the liberal perspectives described above one of the problems of state intervention is, due to the enormous complexity of the global society, the impossibility of guiding all actors efficiently. As Wolfe (1997:189) notes, under such circumstances society as a whole is much better off allowing private individuals to make private decisions (comparable with the function of the invisible hand). According to the utilitarian liberal position, and also to libertarians, who consider redistributive policies to be a violation of peoples' freedoms, the question of whether or not to introduce ethical traceability and informed choice is to be answered by the market mechanism itself – and not by the state or government. Ethical traceability and informed choice, when imposed by the government, are in conflict with the idea of liberty – at least the liberty of the actors in the food chain (but not necessarily with the liberty of consumers). However, this position need not be sceptical of the concepts themselves, as long as they are a matter of private and not public concern; if consumers demand such information, it is very likely that the market will provide it.

The case looks different from a Kantian liberal position, for example, as represented by John Rawls. Central to this position is individual autonomy (or freedom) and the dignity associated with it. Individual autonomy is, according to Kant, the capacity of a person to be his or her own moral legislator, free from sensuous desires, traditions, personal inclinations, etc., in order to make moral judgements based on rational principles. Together, autonomy and dignity define the human being as an end in itself. The institution of human rights is thus a fundamental way of ensuring that human beings are treated not only as means but always also as ends in themselves. For Kantian liberals the rights of individuals are prior to the good life and considerations about, for instance, the welfare of society. Rights 'constitute a fair framework within which individuals and groups can choose their own values and ends, consistent with a similar liberty of others' (Sandel, 1996:11). From the Kantian liberal position (see Chapters 9 and 10) the case of ethical traceability might seem ambiguous: does the liberty of producers and retailers count for more than the consumers' right to free access to information? Or conversely, is the autonomy and informed choice of consumers a more fundamental right than the producers' and retailers' right to freedom from intervention?

Avoiding *injustice* is also basic to Kantian liberals. In the case of ethical traceability we may possibly speak of two kinds of injustice. One is done to producers and retailers if they are forced (possibly against their will) to provide information about production practices in their corporations. The other injustice is done to consumers when they are prevented from being just to others by a deficit in information about production practices. Again, the dilemma concerns which injustice has a higher claim on being avoided.

The republican view affirms a politics of the common good, in which liberty is understood in terms of self-government, not rights. Sandel (1996:27) writes that:

> The republican begins by asking how citizens can be capable of self-government, and seek the political forms and social conditions that promote its meaningful exercise.

Ethical traceability and informed choice could be a way of pursuing such goals of individual autonomy. The educational aspects and the possibilities of public deliberation associated with informed choice seem to fit well into republican thought. However, from this perspective too informed choice must appear as a dilemma, because informed food choice is so closely associated with the market. The market is not considered a legitimate sphere for public discourse. Much of the same argument would apply for the communitarians, who criticize the corrosive influence of the (global) market on civil society and local communities.

Moreover, the left traditionally has a hostile attitude to the private sphere because private property generates private power out of democratic control. Private interests by nature seek to maximize private utility – often at the cost of solidarity and equality. In this view, informed choice on ethical issues in the food chain will not be a useful political tool to create equality and solidarity, as it is inherent to the market that everybody seeks the highest personal gain and thus disregards care for others. From the perspective of the left, affairs of the private sphere are considered less democratic and less accountable than affairs carried out in public (Wolfe, 1997:189). For these reasons the left tends to reject ethical traceability and informed choice as a route to political participation by consumers.

So much for the traditional political ideologies. Let us turn to something quite different, namely the 'ideal speech situation', as described by Habermas. The ideal speech situation is a model for unrestrained communication, where power differences are as far as possible minimized. In this conception of communication, it is clear that qualified participation in deliberation requires background knowledge of the subject at hand. Ethical traceability, as a way of providing information on food ethics, could be one way of qualifying and preparing consumer citizens for informed choice. This would not be sufficient, however, because the ideal speech situation emphasizes the need for *dialogue* between the actors instead of the one-way transfer of information.

In any case, it is doubtful whether Habermas would consider the market as an appropriate place for political action. The civil society and the public sphere as he described them are sharply separated from the private spheres of intimacy and economic activity. Again we see how the commercial and private interests involved in food consumption and production are considered threats to the unforced communicative structures of the public sphere, which are central to this theory. Instead of acting on the conditions of the market, it would be in accordance with this point of view to arrange civil society forums for rational deliberation on food ethics and ethical traceability. The objective of such forums would be to create a dialogue between civil society and the government on public expectations for food production practices.

So there seems to be a place for ethical traceability in Habermas's version of critical theory, but it is hard to see how informed choice could fit into the theory of communicative action. There is no place for such a concept, as the theory sees the political as part of civil society, which excludes direct deliberation with the economic sphere. On the other hand, if agreement on consumers' informed choice happened to be an outcome of a public and rational deliberation then it would be the task of the state to implement it.

Sociability in the sense used by Ariès, and the gathering of friends in privacy in Goffman's sense, are informal spheres. Such spheres are important for the sharing of knowledge, and it has been shown that much information about food is spread through such informal social and intimate spheres. Ethical traceability as a source of information, for instance via the internet, could be used by one person and subsequently shared by members of local communities and thus function as 'background material' for informal social, private deliberations and in the end informed food choice.

If we are to follow Goffman's fears of losing privacy, the danger is that ethical traceability can lead to inescapable visibility, surveillance and monitoring. The demand for social performance by producers and retailers may then turn into unreasonable burdens and stress. Webcams on farms or in the processing industry are examples that have already been tested in practice. Goffman would probably have argued for the importance of safeguarding private spheres in production, processing and retailing. If these productive activities become public, they increasingly demand official presentations of the self and imply a place with limited possibilities for retreat. Goffman would thus, when debating the use of ethical traceability as a means for informed choice, argue for traceability systems that respect and safeguard privacy. The same line of argument can be applied to consumption. Food purchasing would no longer be restricted to the intimate sphere and to concerns for relatives. It would be extended into a social performance considering the well-being of society. Different concerns might easily be a source of conflict and make food purchasing and consumption an unpleasant activity. Thus, increasing consumers' responsibility towards society may be a threat to the intimate and informal private sphere of food enjoyment.

Feminist movements are traditionally hostile to traditional conceptions of the private sphere, home and family, as they are based on exploitation and repression of women. However, one may ask whether informed choice could be a way for women to gain influence. As most food buying is still done by women, informed choice could be seen as a means for women to participate in and influence matters of ethical and political importance. But, again, some feminists might oppose an approach which assumed that food shopping would remain a female preserve, or argue that informed food choice could make food purchase even more time consuming and demanding, overburdening women.[10]

Situating Informed Food Choice Between Public and Private Spheres

From these seven approaches to ethical traceability and informed food choice, I now turn to a more general discussion of how food purchase and consumption are situated between the public and private spheres.

The marketplace, especially for food consumers dominated by supermarkets, has the character of a public space. It is open to all. Food purchase can be considered as a *private act carried out in a public space.* Weintraub, who supports this idea, remarks:

> If market exchange is considered a 'private' act – on the grounds of being, in principle, self-interested, nongovernmental, and unconcerned with collective outcomes – then it does not cease to be private when it is carried out 'in the public' (Weintraub, 1997:5)

This, according to Weintraub and also to neoclassical economic theory, is the ambivalent character of trading at the market: it is a private act that takes place in public. But can the divide between public and private really be cut that neatly?

Food consumption represents a relationship between human beings and eating together is a fundamental aspect of sociability in all cultures. The analysis of a meal by the German sociologist Georg Simmel (1858–1918) in the essay *Sociology of the Meal* from 1910 draws many parallels to the debate on the public-private divide.[11] For Simmel, the meal is a tension between individuality and community feeling. Eating is individual, as the same food cannot be eaten twice or by two persons; food is ultimately used and transformed during eating and digestion and cannot also be eaten by another person. However, during the shared meal individuality is transgressed and elevated into community and social interaction. Simmel writes that:

> Communal eating and drinking, which can even transform a mortal enemy into a friend for the Arab, unleashes an immense socializing power that allows us to overlook that one is not eating and drinking 'the same thing' at all, but rather totally exclusive portions, and gives rise to the primitive notion that one is thereby creating common flesh and blood (Simmel, 1998:131)

Thus, even though eating is basically egoistic, the common meal has the power to override this egoism, and more than anything else it possesses the ability to institute

---

[10] This can, however, only be a guess as no feminist to my knowledge has written on precisely this subject.

[11] I have written more extensively on Simmel's *Sociology of the Meal* in my book *The Taste for Ethics* (2006).

solidarity and community feeling. Of all the things that human beings have in common, eating is among the most fundamental. Eating is the lowest common denominator, and this is why we find in the meal the best-suited instrument for creating a community: the meal does not demand specific qualifications, and everybody can participate without any conditions or competencies other than being hungry. Individuality and sociability are reconciled in the meal; people eating together enter into a contract or pact. Leslie Gofton touches upon this when he writes that:

> Food not only symbolizes cultural values, but also forms a medium through which social relationships are expressed, from the intimate, face-to-face relations within the family, to the relations between regions and nations (Gofton, 1996:121–122)

These quotations suggest that consuming food is far from being only a private concern. Food consumption breaks with traditional perceptions of public and private.

There are many cases in which public and private flow into each other. Another illuminating example is provided by Nicholas Blomley (2005:284), who lists some of the ways that the private-public division has been challenged. The homeless who are obliged to live their 'private' lives in the public streets and in public view are treated as disorderly, or as the English anthropologist Mary Douglas would put it, as 'matter out of place'. Blomley himself examines how flowers in a bathtub, placed in the public street, are perceived by people passing. The flowering bathtub, surely not a public initiative, was however placed in the public sphere. Hence the bathtub breaks with the traditional and expected categories of public and private. People made very different classifications of it but in general he concludes that peoples' perception of private and public in this case seemed fluid and hybrid. He concludes that the flowering bathtub was an example of 'hybrid ownership' that breaks with the traditional private-public distinction.

There are many examples of the merging of private and public, as in the idea of consumer-citizens described in Chapter 1, so we may ask what is it in our time that has blurred the neat, classical divide between public and private? For Arendt, this blurring is due to the 'invention' of the *social sphere*:

> [T]he rise of the social coincided historically with the transformation of the private care for private property into a public concern (Arendt, 1958:68). The word social is Roman in origin and has no equivalent in Greek language or thought. Yet the Latin usage of the word *societas* also originally had a clear, though limited, political meaning; it indicated an alliance between people for a specific purpose... (Arendt, 1958:23)

In modern times the rise of the social sphere coincides with the emergence of the nation state. Arendt describes the emergence of society as:

> [T]he rise of housekeeping, its activities, problems, and organizational devices – from the shadowy interior of the household into the light of the public sphere... The decisive historical fact is that modern privacy in its most relevant function, to shelter the intimate, was discovered as the opposite not of the political sphere but of the social, to which it is therefore more closely and authentically related (Arendt, 1958:38)

Arendt says here that the opposite of the private in modern societies is perceived as the 'social' and not the political. Does this mean that for Arendt the private has been allowed to become political? Arendt would maybe rather have said that the political has become private, as the political has become dominated by private interests and no longer concerns the common good.

The word privacy in Greek originally meant 'deprivation', but this is not how it is used today, where the private sphere has been enriched by numerous leisure activities and possibilities for self-realization. During the last three centuries, the social sphere has grown out of the private sphere. The decline of the family is paralleled by the rise of the social – at the expense of both the private and the public realms:

> Society is the form in which the fact of mutual dependence for the sake of life and nothing else assumes public significances and where the activities connected with sheer survival are permitted to appear in public (Arendt, 1958:46)

It is, in Arendt's words, the life process itself, which in one form or another has been channelled into the public realm. Necessities of life like running a household are, with the rise of the social sphere, no longer abandoned from the public sphere; instead, they are allowed to appear in public.

Habermas, in *The Structural Transformation of the Public Sphere*, sees the development of the social sphere as an extension of market-oriented relations into the public sphere (Habermas, 1962:225–228). In the social sphere, public authorities and private institutions are merged in such a way that the distinction between private and public becomes obscure or even meaningless (Habermas, 1990:234).

To give an example, Habermas points to traditional liberal theory, which according to him was originally based on the existence of independent, small-scale manufacturers. Early liberal theorists could not have foreseen the political influence and power of large-scale, international corporations. Because of these new centres of power, which emerge as corporations grow and become multinational, there is a growing need for government regulation of the market to secure social equilibrium. Since the end of the 19th century the Western industrialized countries have witnessed increased intervention in the private sphere of the market by governments due to the growth of corporations.

Habermas mentions several cases where it no longer makes sense to use the private-public distinction: workers' rights, social security, tenancy organizations, consumer rights and so forth. Gerald Turkel adds that in cases where private corporations make use of common resources such as fish, water and air, it has become common to introduce public policies aimed at the control of private market activities (Turkel, 1992:220). All these cases are characterized by intervention in the private sphere in the name of common wealth. Today both state intervention in the market and the intervention of private corporations into public affairs are common.

For Arendt, with the rise of the social sphere, private concerns are mixed with the political-public sphere. She seems to regret this, because she believes that only a few people will be able to rise above the conformity inevitably imposed by a life in material production (Calhoun, 1997:82). Habermas agrees with Arendt's analysis of the social sphere and shares her concerns about its emergence. His solution is to revitalize the third sphere of society, namely *civil society*, as a sphere for public and rational deliberation separated from the partial private spheres of intimacy and economic activity.

It is clear that concepts like ethical traceability, informed choice and consumer-citizens do not fit into the classical public-private divide. Neither do they seem to fit entirely with the descriptions of the third sphere of society, such as Habermas's civil society for public deliberation. Civil societies are indeed essential for deliberations on the general desirability of ethical traceability and informed choice, and in the longer run on how they might be applied. But civil society is not capable of capturing the specific act of informed food purchase, which involves both private and public concerns. Food choice is a fairly individualized activity, in which a personal choice is made between one food and another, for whatever reason. Such individual choice cannot necessarily be said to be determined by preceding public agreement arrived at by rational discourse; there might be private concerns that have higher priority. Public discourse may play an indispensable role in enlightening and forming the attitudes of consumers – a parallel can drawn to how media coverage can 'form public opinion' – but it is unlikely that consumer-citizens will reach unanimous agreement. On the contrary, we can expect a variety of attitudes and consumer choices.

At the very moment of food choice, civil society – with its public debate, rational argument and deliberation – is not present. As such, food choices are made in privacy. In the case of ethical consumption, arguments and deliberations may indeed be present, and choices may be informed by discussions in public and civil society. But nothing forces the private consumer-citizen to follow the outcomes of such deliberations.

Arendt feared that the rise of the social sphere would bring private interests into politics and the public sphere. The ideas described here concerning consumer-citizens and informed choice do merge the two spheres into a social sphere, but it is politics and public interests that are brought into the market.

Conclusion

Ethical consumption blurs the divide between public and private. The two spheres are dissolved by the rise of such ideas. Neoclassical economic theory fears that this may disrupt the inherent mechanisms of the free market that optimize production and wealth. But we should not forget that markets are already political, as said earlier. States and international governmental organizations are already interfering in markets with standards, property rules, laws against child labour, securing decent working hours, auditing procedures, etc. Ethical traceability can be seen as a new kind of intervention in the market. At present, this intervention operates at its lowest level and is not responding to current consumer concerns.

Communitarians, on the other hand, seem sceptical of leaving political decisions to the market, as they fear that the economic self-interest of consumers will lead to a decline in solidarity and social coherence. Thus, ethical traceability and informed food choice are at odds with perceptions of how the divide between public and private works. The traditional divides between public, private and civil society are challenged, and we need to reflect on how to deal with this rupture.

But difficulties in handling the challenges posed by ethical traceability and informed food choice should not lead to a rejection of these concepts. It is more fruitful to consider how they could be integrated into political strategies and even expand the political and public spheres.

Intervention in the market should also seek to respect current positive values in the food sector. These are values to be nursed and conserved, and which should not be given up in the name of securing informed food choice for consumers. Among these are the dangers of inescapable visibility, surveillance and monitoring of producers. If the demand for socially responsible performance by producers and retailers places upon them unreasonable burdens and stress, this cannot, in the end, be in the interests of consumers.

Another question which this chapter raises concerns what the consequences would be for consumers if food choice was not a private concern but open to normative judgements about society and the environment? This implies an extension of the responsibilities of the consumer from the close and familiar to more distant and general issues of the common good. Such a broadening of responsibilities might not be received entirely positively as an opportunity to gain more influence, but also seen as yet another heavy burden.

Ethical traceability and informed food choice also raise new challenges for national states and international governmental organizations such as the EC, the WTO and Codex Alimentarius. As has been shown earlier in this book, there are several private initiatives currently seeking to make ethical aspects of food production more visible to consumers. The question, however, is whether this is enough. Governments and international governmental organizations certainly can play a role in implementing ethical traceability and informed food choice. They can, for instance, initiate dialogue about the content of ethical traceability schemes (see Chapter 11), they can play a role in securing the authenticity of information, and they may even have a more powerful role in defining what kind of information should be made accessible to consumers. This could be relevant in cases where certain actors in the food chain, for example producers or processors, wanted to prevent public access to information that could have a negative influence on their public image.

Implementing ethical traceability and informed food choice entails creating new public spheres and a new kind of civil society: debate, reflection and deliberation by communicative measures could renew democratic and participatory structures (see Chapter 13 for a further development of this idea). The public sphere is a basic principle in democracy; consequently, the theory of the private character of the market and food production practices needs to be reconsidered if we want more democratic institutions in the food sector. Market actors might even profit from considering how to take the ethical and political concerns of consumers seriously by means of ethical traceability and informed food choice.

References

Allen, J. (2001) 'The place of negative morality in political theory'. *Political Theory*, 29(3): 337–363.

Arendt, H. (1958) *The Human Condition*. Chicago: The University of Chicago Press.

Ariès, P. (1960) *Centuries of Childhood: A Social History of Family Life*. New York: Vintage.

Baudrillard, J. (1998) *The Consumer Society*. (First published 1970). London: Sage Publications.

Benhabib, S. and D. Cornell (eds.) (1987) *Feminism as Critique*. Minneapolis: University of Minnesota Press.

Blomley, N. (2005) 'Flowers in the bathtub: boundary crossings at the public-private divide'. *Geoforum*, 36: 281–296.

Bobbio, N. (1989) *Democracy and Dictatorship*. Minneapolis: University of Minnesota Press.

Calhoun, C. (1997) 'Nationalism and the Public Sphere', pp 75–102 in Weintraub, Jeff and Krishan Kumar (ed.): *Public and Private in Thought and Practice. Perspectives on a Grand Dichotomy. Chicago*: The University of Chicago Press.

Coff, C., L. W Christiansen and E. Mikkelsen (2005) *Forbrugere, etik og sporbarhed (Consumers, Ethics and Traceability)*. Copenhagen: FDB.

Foucault, M. (1976) *The History of Sexuality*. Vol. 1. New York: Vintage Books. Original title: *Histoire de la sexualité I. La volonté de savoir*, 1976.

Gabriel, Y. and T. Lang (1995) *The Unmanageable Consumer*. London: Sage.

Gobetti, D. (1997) 'Humankind as a system: Private and public agency at the origins of modern liberalism', pp. 103–132 in J. Weintraub and K. Kumar (eds.) *Public and Private in Thought and Practice. Perspectives on a Grand Dichotomy*. Chicago: The University of Chicago Press.

Goffman, E. (1957) *The Presentation of Self in Everyday Life*. New York: Doubleday Anchor.

Gofton, L. (1996) 'Bread to biotechnology; cultural aspects of food ethics', pp. 120–137 in B. Mepham (ed.) *Food Ethics*. London: Routledge.

Habermas, J. (1962) *Strukturwandel der Öffentlichkeit*. Frankfurt am Main: Suhrkamp (1970). English title: *The Structural Transformation of the Public Sphere*.

Hannay, A. (2005) *On the Public*. London: Routledge.

Korthals, M. (2004) *Before Dinner. Philosophy and Ethics of Food*. Dordrecht: Springer Verlag.

Mandeville, B. de (1714) *The Mischiefs that Ought to be Justly Apprehended from a Whig-Government*.

McKeon, M. (2005) *The Secret History of Domesticity. Public, Private, and the Division of Knowledge*. Baltimore, MD: Johns Hopkins University Press.

Micheletti, M. and D. Stolle (2005) 'A case of discursive political consumerism: The nike e-mail exchange', pp. 255–290 in M. Boström, A. Føllesdal, M. Klintman, M. Micheletti and M. P. Sørensen *Political Consumerism: Its Motivations, Power and Conditions in the Nordic Countries and Elsewhere*. Copenhagen: The Nordic Council, TemaNord 2005:517.

Nielsen, H. K. (2001) *Kritisk teori og samtidsanalyse. (Critical Theory and Contemporary Analysis)*. Aarhus: Aarhus Universitetsforlag.

Norris, P. (2005) 'Political activism: New challenges, new opportunities'. in C. Boix and S. Stokes (eds.) *The Oxford Handbook of Comparative Politics*. Oxford: Oxford University Press (forthcoming).

Post, R. C. and N. L. Rosenblum. (2001) 'Introduction', pp. 1–25 in N. L. Rosenblum and R. C. Post (eds.) *Civil Society and Government*. New Jersey: Princeton University Press.

Sandel, M. J. (1996) *Democracy's Discontent*. Cambridge: The Belknap Press of Press Harvard University Press.

Scammell, M. (2003) 'Citizen Consumers: Towards a New Marketing of Politics?', pp. 117–136 in J. Corner and D. Pels (eds.) *Media and the Restyling of Politics*. London: Sage.

Seidman, L. M. (1987) 'Public principle and private choice: The uneasy case for a boundary maintenance theory of constitutional law'. *Yale Law Journal*, 96: 1006–1072.

Simmel, G. (1998) 'Sociology of the meal', pp. 130–135 in D. Frisby and M. Featherstone (eds.) *Simmel on Culture. Selected Writings*. London: Sage. Original title: 'Soziologie der Mahlzeit' (1910).

Taylor, C. (1991) *The Ethics of Authenticity*. Cambridge: Harvard University Press.

Turkel, G. (1992) *Dividing Public and Private: Law, Politics, and Social Theory*. Westport: Praeger Publishers.

Weintraub, J. (1997) 'The theory and politics of the public/private distinction', pp. 1–42 in J.Weintraub and K. Kumar (eds.) *Public and Private in Thought and Practice. Perspectives on a Grand Dichotomy*. Chicago: The University of Chicago Press.

Wolfe, A. (1997) 'Public and private in theory and practice: Some implications of an uncertain boundary', pp. 182–203 in J. Weintraub and K. Kumar (eds.) *Public and Private in Thought and Practice. Perspectives on a Grand Dichotomy*. Chicago: University of Chicago Press.

Chapter 9
Traceability of Animal Welfare: Market or State, Good or Right?

Liesbeth Schipper

Introduction

Are the concerns discussed throughout this book, about e.g. animal welfare, the environment, fair trade, biodiversity, the consumer/citizen, to be regarded as mere market preferences (*consumer* concerns in the strict sense) or as public concerns? Should these concerns be regulated by market initiatives or should they be the object of laws and governmental regulations? If the first is the case, ethical traceability could be used as a communication tool; if the second were the case, ethical traceability should be understood as a government tool (see Chapter 1 for an explanation of these different uses of traceability). However, it could also be the case that the 'either or' does not hold here, and that there are other possibilities. In this chapter I will explore this issue by concentrating on consumer concerns with respect to animal welfare. I will first give an overview of the most salient aspects of this concern and then use the philosophically liberal framework to distinguish between consumer concerns and public concerns and between conceptions of the good and a conception of the right. Then I will look at animal welfare concerns: are they based on a conception of the good or based on a conception of the right, or can some intermediate position be justified? What, finally, are the consequences of the answer to this question?

Concerns About Animal Welfare

The Problem

Animal welfare is of considerable importance for European consumers. Since the 1960s, awareness about the conditions under which farm animals are kept has increased and has given rise to public offence. Animals have always had a special and complex place in our relationship with nature. Meat, and also to a lesser extent other animal products, are among the most common objects of taboo, regulation and avoidance (Twigg, 1983:18). However, the recent rise of public concern about

C. Coff et al. (eds.) *Ethical Traceability and Communicating Food*,

welfare conditions of livestock is of a different character. It is a response to the profound changes that have taken place in animal farming since the 1960s. There has been upgrading of scale (farms with more than 30,000 chickens are becoming normal and so are farms with more than 10,000 pigs), many animals are kept inside, and selective breeding has resulted in highly productive animals. Technologies, such as computer-regulated feeding, have taken workload off the farmer's shoulders. One of the results of this has been that meat and other animal products have become much cheaper. Meat has become affordable for all and sales of meat and other animal products have risen.

These changes have also resulted in welfare problems. Animals suffer from bad health because of the conditions under which they are kept and because of the selective breeding they have been subject to. Broilers for example grow so large in 5–6 weeks that they can hardly walk, or even break legs because of the weight they have accumulated in such a short time. Fattening pigs develop abdominal ulcers. Many animals suffer from abnormal behaviour because of the boredom they experience; for example, they develop cannibalism or mutilate other animals.

Non-governmental organizations (NGOs), animal rights activists and animal scientists have raised public awareness about these and other welfare problems. Despite the fact that a majority of consumers do not have detailed knowledge of farming practices and the specific welfare problems they generate, many consumers are nowadays aware of the existence of welfare problems in industrially farmed animals. The claim that animal welfare is a problem that deserves serious attention is heard more and more often. The *Special Eurobarometer* on 'Attitudes of consumers towards the welfare of farmed animals' showed that a majority of European consumers believe that the welfare of broilers and laying hens is very poor or poor. Forty-four percent of Europeans think that the welfare of pigs is bad or very bad (Special Eurobarometer 229, 2005:10, 18). Furthermore, a majority of EU citizens (55%) state that animal welfare/protection does not receive enough importance in the agricultural policy of their countries. Only a minority (7%) believe it receives too much importance, and 29% believe it receives just about the right level of importance (Special Eurobarometer 229, 2005:64).

The Response

Consumer concerns, lobbying by NGOs and scientific research have led to new European and national legislation for farm animals. Council regulation 2005/1/EC concerns the protection of animals during transport and related operations. A major difference, compared with the earlier directive, is that any member of staff handling livestock during transport should have completed a training course recognized by competent authorities. To the disappointment of many NGOs and political parties, the directive does not give any rules on the maximum distance live animals may be transported.

European Directive 98/58/CE, agreed on 20 July 1998, concerns the protection of production animals. The directive contains very general regulations about the treatment of animals, control and enforcement. Examples of this regulation are: the freedom of the animal may not be restricted in such a way that it causes any harm or injuries; the housing facilities for the animal must be such that harm or injury is prevented as much as possible; animals must have access to sufficient amounts of healthy and appropriate food and water.

This general directive has been elaborated in several more specific directives, concerning respectively the welfare of pigs (91/630/EEC, 2001/88/EC, 2001/93/EC) laying hens (1997/47/EC) and veal calves (9/629/EEC, 97/2/EC). In these directives more concrete requirements are laid down. From 2003 onwards laying hens may not be reared in cages. For owners of battery cages built before 2003 there is a conversion period until 1 January 2012. Enriched cages are allowed, which means that the hens are still caged, but must have access to a nest, must be able to peck and scratch, and they should have appropriate perches. This allowance of enriched cages – a compromise between the producers of cage eggs and the politicians, producers and NGOs who want to abolish cage systems – is felt by many NGOs to be hardly an improvement (Roex and Miele, 2005:117). Beak trimming is allowed in all types of production systems. For pigs there are requirements on the amount of space pigs of a certain weight should have (e.g. pigs of 85–110 kg should have a minimum floor surface of 0.65 m^2), sows must be kept in groups, and sows and fattening pigs should be provided with 'playing material' and a 'reasonable' amount of fibrous fodder. Permanent tethering of sows is, furthermore, forbidden. Concerning cattle, specific standards are set only for veal calves. It is prohibited to keep calves completely in the dark, it is obligatory to feed the calves 'sufficient' amounts of iron, and some minimum size requirements are given (1.5 m^2 free space per calf). Also, calves older than 8 weeks must be kept in groups. Some national governments have passed legislation that goes further than European legislation, e.g. Sweden, Norway, Austria and Switzerland, but most governments have national legislation that is in line with the European legislation.

Uneasiness About the Response

The above animal welfare requirements are seen by many as too minimal (Special Eurobarometer 229, 2005). The amount of space legally required by these three directives is, indeed, not very generous. There are no requirements concerning daylight and access to outdoor space, and no requirement to avoid mutilations such as beak trimming or non-anaesthetized castration. Besides the minimum requirements for calves, laying hens and pigs, specific legislation is lacking for the majority of animals that are kept for farming purposes. There are no rules whatsoever concerning, for example, the amount of space or the type of food that they should be given. The welfare problems mentioned at the beginning of this section are not addressed by the EU legislation. Furthermore, the monitoring and enforcement of the minimum legislation is considered to be far too lax. It is not mandatory to provide information to consumers on production standards and production practices concerning animal welfare (with the exception of egg-labelling provisions that have recently been put into place[1]).

Labelling as the Solution

Sometimes labelling is proposed as *the* public policy that takes consumer concerns seriously and therefore provides an answer to those concerns. However, when one takes a closer look at different types of consumer concerns, it becomes clear that different consumer concerns may call for different types of strategy to deal with them (Brom, 2000:138). Taking into account the degree of acceptance consumer concerns can acquire, it is useful to make a distinction between different sorts of consumer concerns[2] (Brom, 2000:130–131): (1) concerns that matter to all consumers; (2) concerns that matter to specific groups of consumers; (3) concerns that go beyond consumer concerns narrowly defined and acquire the acceptance of consumers as citizens; these concerns are better called public concerns.

A concern that matters equally to everybody in their role as consumer is food safety. Safety regulations are a means to deal with these sorts of consumer concerns. The second group of consumer concerns is of relevance to specific groups of consumers because of the way they want to live their life. Respect for the autonomy of individuals means that they should be accorded a right to live according to their own ideals. Labelling is a good option here. It provides a solution to consumer concerns by giving consumers the opportunity to make choices according to their values. The third group of concerns matters to people in their role as citizens. These concerns are related to ideas about a good society. They are therefore more properly called public concerns than consumer concerns. 'People are concerned about certain products because of the wider impact these products have on their society and the world' (Brom, 2000:131). Because of the fact that these concerns are public concerns, laws and regulations are the best solution.

Right and Good

Now, regarding animal welfare, the question is whether this concern is to be understood as a mere consumer concern or as a public concern? Should it be regulated by market initiatives or should it be the object of laws and regulations? What kind of concerns are consumer concerns anyhow: do they belong to mere consumer preferences, i.e. to conceptions of the good or to conceptions of the right?

---

[1] COM (2002) 626 final: 3.

[2] See also Chapter 10.

The distinction between conceptions of the good and conceptions of the right is, as I will show, linked to the notions of pluralism and tolerance. It is often said that modern Western societies are characterized by pluralism. This means that in these societies there is a plurality of ideas about the good life, and there is profound disagreement about moral, ideological and political issues. For instance people have different religions or none at all. They have diverging political ideas and different ideologies. Politicians, citizens, scientists and NGOs have, for example, discussions about whether global warming is a fact or not and about how to prevent it; they fight over the question of whether the welfare state stifles private initiative; about how the global economy should be regulated; whether nuclear power plants are a solution to the depletion of natural energy sources; if nuclear shields are to be allowed; how to prevent and fight terrorism. Or people quarrel and have deep disagreements over things such as abortion, the rights of homosexuals to marry and immigration policies. The distinction between the right and the good captures the idea that all people should have the equal possibility to live as they please (i.e. in accordance with their conception of the good), within a reasonable and widely accepted legal framework that sets the rules that all people have to abide by (i.e. the concept of right). In this way pluralism is accepted and legitimized, but one is still able to assess which conceptions of the good are legitimate and which are not. So it is not 'anything goes', as is explained below.

Harm and Freedom

When a society is characterized by widely diverging and sometimes opposing worldviews, lifestyles and opinions, it is important to think about how to regulate this pluralism; how to make sure that all these different groups and people live together peacefully despite the sometimes unbridgeable differences between them. One possible answer is that people should be able to live as they please as long as others are able to do so as well. This principle is often, in general terms, referred to as the harm principle. Differences should be tolerated but these differences may not negatively affect others. This is how pluralism and tolerance are intertwined: one reaction to pluralism is to advocate tolerance as a practical and/or moral ideal.

The 19th-century philosopher Mill gives one of the first defences and interpretations of the harm principle. In *On Liberty* he says that he is to

> 'assert one very simple principle, as entitled to govern absolutely the dealings of society with the individual .... That principle is that the sole end for which mankind are warranted, individually or collectively, in interfering with the liberty of action of any of their number is self-protection. That the only purpose for which power can be rightfully exercised over any member of a civilized community, against his will, is to prevent harm to others. (...) The only part of the conduct of anyone for which he is amenable to society is that which concerns others.' (Mill, 1956:13)

From this it follows that:

> 'The only freedom which deserves the name is that of pursuing our own good in our own way, so long as we do not attempt to deprive others of theirs or impede their efforts to obtain it.' (Mill, 1956:16–17)

Mill's definition belongs to the category of negative freedom. Negative freedom is the freedom that is the result of the absence of deliberate external constraints on an individual. It is the domain in which a person, not hindered by deliberate interventions of others, is able to do those things that he is capable of (Van den Brink, 2005:40).

Positive freedom, on the other hand, is related to the notion of autonomy. Positive freedom refers to the capability to independently give direction to one's personal life, to be master over one's own life (Van den Brink, 2005:41). This definition of positive freedom runs the risk of paternalism as people may claim that every person should become autonomous, even when those persons do not value autonomy themselves. It may also lead to the paradox of people claiming to know for others what their autonomy is. This is why it is sometimes claimed that in a liberal society autonomy is not to be enforced but should be illustrated in practice with patience and arguments (Van den Brink, 2005:44). Positive freedom is sometimes described as being 'free to' instead of being 'free from', which is negative freedom. In this sense there is not so much emphasis on the value of autonomy but more on creating the conditions that enable people to make effective use of their negative freedom; the difference between negative and positive freedom is then more about the distinction between what are called *de jure* and *de facto* rights, i.e. formal rights and actual rights respectively.

Justice and Freedom

Rawls, in *A Theory of Justice*, develops a conception of justice in which negative liberties are very important (Rawls, 1999:13). Positive freedom also plays a role, as autonomy is a central notion in his argument. *A Theory of Justice* was one of the most influential works in ethics and political philosophy in the 20th century, as it has set the agenda for political philosophy and ethics since its publication in 1971. His theory dominates contemporary debates, not because everybody accepts it but because alternatives to it are often presented as responses to it (Kymlicka, 1990:9).

A Theory of Justice has a distinctively liberal character.[3] In many liberal theories a distinction is put forward between conceptions of the good and conceptions of the right. Rawls's aim is to develop a theory of justice that pays attention to and structures

[3] Liberal theories are individualist, in the sense that individual rights are considered to be very important. Another distinctive feature of liberal theories is that they include the idea that the ideals of equality and freedom imply economic redistribution. Liberalism is often distinguished from *libertarianism* – of which Nozick's *Anarchy, State and Utopia* (1973) is the most well known example – and *communitarianism*. The largest difference between liberals and libertarians is that the latter claim that economic redistribution infringes on the autonomy of individuals. Sandel's *Liberalism and the Limits of Justice* (1982) and MacIntyre's *After Virtue* (1981) are often regarded as examples of communitarian theories. Communitarianism differs from liberalism in that it is less individualist than liberalism. The values of a group may take precedence over the values of an individual, which may lead to paternalism. For a good introduction to Rawls's theory, and the debate between liberals and communitarians, see *Liberals and Communitarians* (1992), by Mulhall and Swift.

our different intuitions. The principles of justice that he wants to define and explore apply to the basic structure of society. By this he means the arrangement of major social institutions such as the main economic and political institutions.

His theory should be placed in the social contract tradition that goes back to Locke, Rousseau and Kant. In this tradition the content of the conception of justice is seen as a hypothetical contract made in an initial situation – in Rawls's terms, the original position. This is 'a purely hypothetical situation characterized so as to lead to a certain conception of justice' (Rawls, 1999:11). The characterizations are such that, first, no one knows his place in society, his gender, race, class position or social status, nor does anyone know his fortune in the distribution of natural assets and abilities, his intelligence, strength and the like. Furthermore, they do not know their conception of the good. This means that they do not know, for example, what their political ideology is, or their religion. Third, the participants do not know the particular circumstances of their society. For instance, they do not know its economic or political situation (Rawls, 1999:118). These three conditions ensure that the principles of justice are chosen behind 'a veil of ignorance'. This veil of ignorance is meant to make sure that the principles of justice are not biased. It is felt that principles of justice should not be in favour of the more intelligent, healthy or powerful. Justice should also not be such that it is biased towards (groups of) people who have a certain vision of the good life, for example, a religiously inspired one.

Despite the lack of information about the content of their conception of the good (i.e. vision of the good life), the people in the original position *do* know that they will have a certain conception of the good. To be able to live according to this conception, they will want to have as many goods (material as well as immaterial) as possible that will enable them to live the life they will want to live. These goods are valuable independently of the content of a specific conception of the good. Such goods are rights, liberties, opportunities, wealth and income. Rawls calls these goods 'primary goods'.[4]

Under these circumstances,[5] Rawls claims, people in the original position will choose two principles of justice. The first is that each person is to have an equal

[4] Because one does not know what sort of conception of the good one will have, these primary goods are of value to all persons in the original position. If one were to have a conception of the good that did not attach any value to material goods – a conception of the good that monks, for example, could have – one would not be disadvantaged by having access to the largest possible set of primary goods. One could always decide not to make use of the primary goods that are material in character, i.e., wealth and income.

[5] Note that the original position is a purely hypothetical situation. The contract reached in it is thus also hypothetical. The idea of the original position, together with the veil of ignorance, is meant as a thought experiment. If one agrees with the characterization of the original position and with the deductive reasoning that is done on the basis of it, one would also have to agree with the outcome of the agreement. So, 'one or more persons can at any time (…) *simulate* the deliberations of this hypothetical situation, simply by reasoning in accordance with the appropriate restrictions' (Rawls, 1999 (1971):119, my italics). According to Rawls, his interpretation of the initial situation best expresses the conditions that are widely thought reasonable to impose on the choice of principles, and that also lead to a conception of justice that is in accordance with what we find right. To judge the two in relation to each other, until a balance is found, is what Rawls calls a 'reflective equilibrium' (Rawls, 1999 (1971):105–106) and what more continentally inclined philosophers would call a hermeneutic circle.

right to the most extensive total system of equal basic liberties compatible with a similar system of liberty for all. The second is that social and economic inequalities are to be arranged in a way that ensures that they are to the advantage of the least well-off in a society (Rawls, 1999:57, 13, 266). These two principles form the concept of right. The first principle guarantees that all persons in a certain society have an equal right to the greatest possible set of individual rights and liberties. The second principle is about the allocation of economic benefits. The claim that inequality is only allowed when this benefits the worst-off in society expresses the thought that societal cooperation must be to the benefit of all persons contributing to the cooperation. The first principle is prior to the second. This means that individual rights may never be sacrificed for a greater economic good. Individual rights and liberties, thus, play a central role in Rawls's theory.

Right Before Good

The concept of the right, furthermore, is prior to the concept of the good. This means that a conception of the good must be such that it is in accordance with a concept of the right, in the sense that people are not allowed *to act* on a conviction that is in disagreement with the concept of right.

> The principles of right, and so of justice, put limits on which satisfactions have value; they impose restrictions on what are reasonable conceptions of one's good. (…) In justice as fairness the concept of right is prior to that of the good. A just social system defines the scope within which individuals must develop their aims, and it provides a framework of rights and opportunities (…) The priority of justice is accounted for, in part, by holding that the interests requiring the violation of justice have no value (Rawls, 1999:27–28).

This means that persons must act in accordance with the two principles of justice. As the constitution and important laws and regulations would be based on the concept of justice, acts that go against the concept of justice will be punishable by law.

So, a concept of right is independent of and prior to the good.[6] It can also be in contradiction of the good, for example when on the basis of a conception of the good – say a religious one – women are denied certain individual rights. In this case a concept of right takes precedence over the particular conception of the good and so women should not be excluded from rights, even when this is an important element of the religiously inspired conception of the good. A state that acts in accordance with a concept of right, such as Rawls's, is often called a *neutral* state.

---

[6] Theories that distinguish between the (lexically prior) right and the good are *deontological* theories. These are contrasted with *consequentialist* theories. According to latter theories, of which utilitarianism is the best known ethical strand, the right is a maximization of the good. In this case, one should first know what the good is, before one might know what the right consists of.

Communitarians, on the other hand, claim that the concept of the right is not independent of, and prior to, a concept of the good.[7] They contend that this is just another conception of the good and one that presupposes a liberal notion of a person. So, the concept of right is inherently liberal in character, it is claimed, and persons with a conception of the good that is not so liberal, or even is anti-liberal, will not agree on this specific concept of justice.

In Pursuit of an Overlapping Consensus

Rawls has taken this critique seriously and the main aim of his book *Political Liberalism* (1993) is to show that justice as fairness is to be considered as a political conception of justice, and not as a comprehensive conception (in which case it would indeed be just another conception of the good). Rawls claims that since he did not distinguish in *A Theory of Justice* between comprehensive and political theories of justice, readers might have got the impression that one would have to be a full-blooded philosophical liberal (such as Kant or Mill) in order to accept the two principles of justice and the priority of the right over the good. In that case, citizens with a different conception of the good would not endorse justice as fairness. When a state upholds such a conception of justice, it is not neutral.

He should have made it clear, Rawls now claims, that citizens in a pluralist society should affirm the political conception of justice on the basis of their own reasonable[8] conceptions of the good. The problem of political liberalism, Rawls says, is to work out a political conception of justice for a (liberal) constitutional, democratic regime that a plurality of reasonable doctrines, religious and nonreligious, liberal and non liberal, might endorse. Rawls suggests that this is possible on the basis of an *overlapping consensus*.[9] Based on their own comprehensive views, citizens reach an overlapping consensus on the principle of justice as fairness.

---

[7] Communitarianism has been developed mainly as a reaction against liberal political and ethical theories in general and Rawls's theory in particular, criticizing for example the underlying conception of the person, the alleged individualism of liberal theories and the distinction between the right and the good. About the latter, see for example Sandel's *Liberalism and the Limits of Justice* (1982) and MacIntyre's *After Virtue* (1981).

[8] Unreasonable conceptions are conceptions that do not endorse the values of equality and freedom for all. Conceptions of the good are unreasonable if they do not want people with other conceptions of the good than their own to have equal rights to live according to these conceptions of the good given by the state, e.g., because they completely disagree with those theories. Reasonable comprehensive theories thus agree with Rawls's idea that citizens should share equally in ultimate political power, and exercise that power so that each of them can reasonably justify their political decisions to each other (Rawls, 1993:xliv). So, 'it would be unreasonable to use political power to enforce our own comprehensive view' (Rawls, 1993:138).

[9] Rawls emphasizes that an overlapping consensus is not a mere *modus vivendi*; people endorse it not for pragmatic reasons but for moral ones.

The above suggests that there are two sorts of philosophical liberalism: comprehensive liberalism, i.e., the full-blooded version, and political liberalism, and Rawls wants to make it clear in his *Political Liberalism* that his theory belongs to the latter. A preliminary conclusion, then, is that one could give two sorts of arguments for the claim that animal welfare concerns are part of a theory of the right, one based on a comprehensive liberal theory and one based on political liberalism. Within the first line of thought one may claim that animals have rights. The first and most famous argument for this was given by Regan in *The Case for Animal Rights* (1983). One could also show that animals have rights on the basis of the social contract theory, as VanDeVeer does (VanDeVeer, 1983). In both cases animal rights could be part of Rawls's two principles of justice. But, secondly, one could also show that from the position of political liberalism animal rights could be based on an *overlapping consensus* of how animals ought to be treated. Below I give a short outline of these two possible arguments.

Animal Ethics from Comprehensive Liberal Perspectives

Life Preferability and Creation

VanDeVeer proposes a thought experiment that involves, as he calls it, a non-trivial modification of the one put forward by Rawls in *A Theory of Justice* (VanDeVeer, 1983). The most important alteration is that, whereas Rawls supposes that persons know that they are human beings, VanDeVeer on the other hand proposes that individuals entering the original position – VanDeVeer calls this the *pre-original position* – do not know whether they are human beings or not. Individuals entering the pre-original position only know that they are a sentient being.[10]

VanDeVeer claims that individuals in this position would choose, first, the following principle: 'no sentient being should have forced upon it (by deliberate act of a rational being) treatment that makes its life not worth living, i.e. a life such that no life at all is preferable' (VanDeVeer, 1983:155). This is called the 'Life Preferability Requirement'. This principle may be understood as generating a set of negative duties, duties to refrain from treating sentient creatures in certain ways. It does, in itself, not *require* positive intervention. Furthermore, beings in the pre-original position will put constraints on what may be done to existing sentient creatures. Therefore VanDeVeer proposes a related principle: the 'Creation Requirement'. This requirement is that 'no rational being should deliberately cause to exist a sentient creature when it is certain or highly probable that such a creature would have a life not preferable to no life at all'

[10] Rawls is unlikely to accept this modification of the original position, as there are no possibilities for rational discussion between different sorts of animals about a conception of justice (see Nussbaum, 2006).

(VanDeVeer, 1983:156). He thinks that the life of some animals is, at the moment of his writing, not worth living.

He suggests that the life of a calf under intensive rearing conditions is not preferable to no life at all: the conditions in livestock farming are such that a calf had better not to have lived at all than to live under such conditions. He also suggests that in view of further reasonable empirical assumptions, it would seem that the creation of, and consequent radical deprivations imposed on, other animals raised for culinary, experimental and other purposes would be unjustified. VanDeVeer furthermore writes that breeding for slaughter is not unjustified, on the basis of the Life Preferability and Creation Requirements, and that the conditions under which animals are reared diminish the quality of life of some animals but seem to improve the quality of life of others (in comparison with an alternative life in the wild).

Subject of a Life

In *The Case for Animal Rights* (1983), Regan puts forward a theory of animal rights that has larger implications for the way we ought to treat animals. He claims that animals have rights and that humans therefore have the obligation to respect these rights. Regan starts with the assumption that human beings have inherent value. This means that they have a value that is independent of any instrumental value that they have for human beings. This is an extension of the Kantian principle that individuals are a goal in them selves, and should not be regarded as only an instrument to the goal of someone else. Although human beings can be used instrumentally, they should always be regarded as also having a value in themselves.[11]

Regan distinguishes between moral agents and moral patients. Moral agents possess certain highly developed capacities, in particular the capacity to apply moral principles in an unbiased manner and to choose freely to act morally or not. Moral patients, on the other hand, are not to be held responsible for their behaviour as they are not able to understand and formulate moral principles and to act accordingly. To this category belong human babies, young children and human beings who are mentally not well developed. Animals also belong to it. Moral agents as well as moral patients have an inherent value. Furthermore, every person's inherent value is equal.

[11] Kant, though, regards only human beings as autonomous individuals. Therefore, only humans should be regarded as goals in themselves and only they have direct rights. Animals have indirect rights: their rights are derived from the rights of humans. 'The duties one has to animals are not owed *to* the other animals, but rather to ourselves. (…) The duty to ourselves in question is the duty to cultivate feelings that are conducive to morality' (Korsgaard, 2004:15). Kant writes: 'violent and cruel treatment of animals is … intimately opposed to a human being's duty to himself …; for it dulls his shared feeling of their suffering and so weakens and gradually uproots a natural disposition that is very serviceable to morality in one's relations with other people' (Kant, cited in Korsgaard, 2004:15).

Moral agents and moral patients have inherent value because they are not only alive but they are subjects of a life: this is *the subject-of-a-life criterion.*

> Individuals are subjects-of-a-life if they have beliefs and desires; perception, memory and a sense of the future, including their own future; an emotional life together with feelings of pleasure and pain; (...) and an individual welfare in the sense that their experiential life fares well or ill for them, logically independently of their utility for others and logically independent of their being the object of anyone else's interest (Regan, 1983:243).

Animals thus also have an inherent value and therefore they also have rights. It does not matter whether a subject is a moral agent or a moral patient, such as an animal. A subject of a life should never be treated as if it were not a goal in itself, but *only* a means to the goal of another. Everyone who has inherent value has an equal right to be treated with respect.

According to Regan we should accept the principle that we are morally obliged to oppose any practice that appears to violate the rights of individuals unless or until we are shown that it really does not do so. Many forms of present-day-use of animals are not justified, and one may seriously doubt whether they will ever be justified, he claims. It is clear that his theory has far-reaching consequences if enacted in practice.

Suffering

Singer's utilitarian *Animal Liberation* (1975) offers an alternative to Regan's and VanDeVeer's deontological theories and is famous for having put animal ethics on the philosophical agenda. Utilitarianism is a consequentialist moral theory, because the consequences of an act are given first priority. Its basic principle is that an act is right (when compared with alternative acts) if it leads to the greatest possible balance of good consequences or to the least possible balance of bad consequences. The best outcome determines what it is right to do (Beauchamp, 1991).

To be able to assess happiness and pain for a being one has to take into account its interests. Singer claims that one has to take into account the interests of all beings, no matter what sort. This is based on the principle of equality: equal cases require equal consideration. Then, Singer asks, what is the fundamental characteristic that gives a being the right to equal consideration? This is, he says, the capacity for suffering and/or enjoyment or happiness. This capacity is a prerequisite for having any interests at all. This means that equal pain and equal pleasure should be considered in the same manner, whether the pain or pleasure is experienced by a human being or by an animal.

Now, the simple and straightforward principle of equal consideration of pain or pleasure is a sufficient basis for identifying and protesting against all the major abuses of animals that humans practice, he says. Most of the ways in which animals are used should be abolished, such as animal experimentation and rearing them for food. These practices should be abolished, because of the large numbers of animals that suffer and because they are relatively easy to give up. The benefits for humans do not outweigh the burdens for animals. This means that people should stop eating meat, and part of Singer's book shows how to get closer to this goal, for example by giving vegetarian recipes.

Animal Ethics from a Political Liberal Perspective: Towards an Overlapping Consensus on How to Treat Animals

Regan's and VanDeVeer's deontological theories are controversial and so is Singer's. It is not to be expected that a final answer or theory on animal rights, on which all can agree, will see the light. For example, Regan does not recognize the difference between humans being the subject of a life and other animals being the subject of a life. Human animals and non-human animals belong to the same category of subjects of a life. Despite the fact that animals as well as humans could well be subjects of a life, it is reasonable to argue that they are nonetheless different sorts of subjects of a life. Regan does not distinguish between humans and animals and can therefore be criticized for anthropomorphism. Secondly, his justification and foundation of human rights is not accepted by everybody. As for VanDeVeer, his alteration of the original position is not widely accepted. Furthermore, the argument he introduced, i.e. that some ways of living are better not lived because they are not worthwhile, relies too much on a comprehensive conception of the good to be still in line with Rawls's condition that a theory of justice needs to be neutral between different theories of the good (cf. Parfit, 1984). In practice, none of the three theories would find pure adherents, because for the large majority of mankind their implications are too far-reaching. Nor is it to be expected that in the foreseeable future people, other than academic philosophers, will reach agreement on these very different conceptions of the good; people have different religions, belong to different philosophical traditions, they have different political views, and they belong to different cultures.

Now, does this all mean that there is no way to reach some sort of agreement among humans on how we should treat animals? The remarkable thing is that despite the fact that different theories offer different philosophically, ideologically and religiously motivated reasons for animal ethics, they offer quite similar principles and guidelines on how to treat animals at a basic and practical level. These basic principles incorporate common-sense intuitions about the way human beings should treat animals. Take for example Regan and Singer; although these philosophers offer diverging philosophical theories of animal ethics, and despite the fact that they are in a philosophical battle, they largely agree on the practical consequences of their theories. The debate is a philosophical not a practical one. Or think of a protestant quarrelling with an atheist, the first claiming we ought to treat animals well because God has given us certain responsibilities towards our natural environment, whereas the second denies that this is so. At the same time they may very well agree on the idea that we should not needlessly inflict pain on animals. The debate is about the underlying foundations for that belief, not the belief itself.

It is likely that this is the case for the majority of comprehensive theories, whether they are, for example, religiously, philosophically or politically inspired ones. This gives good reason to suppose that an overlapping consensus is possible on animal ethics, because it would mean that on the basis of their own comprehensive theories of the good people can reach an overlapping consensus on the ethics of human–animal relationships. This brings us to the second way to argue that animal welfare concerns are part of a conception of the right.

So then, what would reasonable conceptions of the good agree upon? There is good reason to think that there is an overlapping consensus, based on people's comprehensive conceptions of the good, that animals should be treated respectfully. This principle of respectful treatment can be seen as a basic principle of how to treat animals. The reason that so many, sometimes completely different, conceptions of the good incorporate the basic value that animals ought to be treated respectfully is that this is a very basic, common-sense intuition about the relationship between humans and animals. This means that the overlapping consensus is based on widely accepted premises. It should therefore be appealing to most if not all people.

This is confirmed by various focus group discussions, e.g. those that were held for an EU project about animal welfare, called *Welfare Quality* (Kjaernes et al., 2007). From these discussions it became clear that the majority of the participants thought that if animals are to be used and killed by human beings for their products, they should be treated with respect and live a life without pain and stress as much as possible. In this case, people sometimes referred to a religious outlook on life, but often did not give reasons for this idea; the idea that if you use animals you should treat them well seems to be very much taken for granted.

There also seems to be some shared idea of what norms follow from the basic idea that animals ought to be treated with respect. Results from the Danish investigation into ethical traceability in the pig-pork-bacon supply chain described in Chapter 5, and from the above-mentioned focus group discussions with consumers about consumption and animal welfare, showed that animals should have enough space, that they should have outdoor access, and that the conditions of transporting living animals should be better. These principles are contingent and should be translated into scientifically sound animal welfare parameters. But, despite this, it seems that it is indeed possible for people to agree on the sorts of moral principles and practical living conditions for animals that follow from a basic respect principle.

So, there is good reason, based on empirical investigations, to suppose that there is an overlapping consensus about the way humans should treat animals: when animals are used for their products they should be treated with respect. Consumer concerns about animal welfare are therefore at least partly based on a conception of the right, according to Rawls's political liberalism. As described above, a conception of right should be upheld by a liberal government, which is committed to upholding a distinction between the right and the good. Legal minimum norms regarding the welfare of animals should in that case be upheld by governments, and when these norms are violated a state should take action to rectify this. Of course, this is already the case. European countries already uphold minimum legal norms that in many cases go further than norms upheld in non-EU countries. However, it is very probable that on the basis of what I have called the respect principle, these norms would have to be higher than they are now. It would not be possible to buy products from animals that had not lived in accordance with the respect principle. Norms regarding animal welfare that are currently regarded as being subject to 'market mechanisms' would in that case be the subject of government regulation. Ethical traceability would in that case be used as a government tool (see Chapter 1), that is as tool for governing minimum standards concerning animal welfare.

Conclusion

Animal welfare is an important consumer concern. Recent qualitative as well as quantitative research shows that a large majority of consumers in the EU are concerned about the current level of welfare of animals. NGOs have partly influenced the importance that is attached to animal welfare nowadays. They have influenced the political debate, and media campaigns and media coverage of campaigning by NGOs have been effective in reaching the public. Animal scientists have pointed to many welfare problems in animals kept in intensive livestock systems.

Labelling is sometimes proposed as the public policy that takes consumer concerns, such as animal welfare, seriously. However, it is useful to distinguish between different sorts of consumer concerns, taking into account the degrees of acceptance they can get. These different types of consumer concerns call for different types of answers, not necessarily all consisting of labelling. Three sorts of concerns were distinguished: concerns that matter to all consumers; concerns that matter to specific groups of consumers; and concerns that go beyond consumer concerns and that are better called public concerns. These latter concerns matter to people in their role as citizens. Labelling provides an answer to the second group of consumer concerns. The other concerns should best be answered with laws and regulations. The theory of political liberalism, with its distinction between conceptions of the good and a concept of the right, was applied to the question whether animal welfare should be understood as a mere consumer concern or as a public concern.

I described two theories of philosophical liberalism, comprehensive liberalism and political liberalism, and I have argued that political liberalism offers the best means for basing animal welfare concerns on a conception of the right. On the basis of a diversity of sometimes very different comprehensive theories of the good life, (groups of) people can reach an overlapping consensus on how animals should be treated: animals that are used by humans for their meat and their products should be treated respectfully. I conclude on this basis that the practical norms that follow from this overlapping consensus can best be put into practice by ethical traceability as a *government tool* that applies, monitors and enforces minimum laws based on the overlapping consensus. Animal welfare conditions that go further than this are best answered with labelling. Ethical traceability is then used as a communication tool.

Acknowledgements I would like to thank Volkert Beekman and the editors of this book for their useful comments on this chapter.

References

Beauchamp, T.L. (1991) *Philosophical Ethics: An Introduction to Moral Philosophy*. New York: McGraw-Hill.

Brom, F.W.A. (2000) 'Food, Consumer Concerns and Trust: Food ethics for a globalizing market'. *Journal of Agricultural and Environmental Ethics*, 12: 127–139. Dordrecht, The Netherlands: Kluwer.

CEC (2002) *Communication from the Commission to the Council and the European Parliament on Animal Welfare Legislation on Farmed Animals in Third Countries and the Implications for the EU*. COM (2002) 626 final.

Council Directive 91/629/EC of 19 November 1991, laying down minimum standards for the protection of calves.

Council Directive 91/630/EC of 19 November 1991, laying down minimum standards for the protection of pigs.

Council Directive 97/2/EC of 20 January 1997, amending Directive 91/629/EEC laying down minimum standards for the protection of calves.

Council Directive 98/58/EC of 20 July 1998, concerning the protection of animals kept for farming purposes.

Council Directive 1999/74/EC of 19 July 1999, laying down minimum standards for thc protcction of laying hens.

Council Directive 2001/88/EC of 23 October 2001, amending Directive 91/630/EEC laying down minimum standards for the protection of pigs.

Council Directive 2001/93/EC of 9 November 2001, amending Directive 91/630/EEC laying down minimum standards for the protection of pigs.

Council Regulation (EC) No 1/2005 of 22 December 2004, on the protection of animals during transport and related operations and amending Directives 64/432/EEC and Regulation (EC) No 1255/97.

Kjaernes, U., Miele M., and Roex J. (2007) *Attitudes of Consumers, Retailers and Producers to Farm Animal Welfare*. Cardiff: UK.

Korsgaard, C.M. (2004) 'Fellow Creatures: Kantian ethics and our duties to animals', in G.B. Peterson (ed.) *The Tanner Lectures on Human Values*, vol. 25/26. Salt Lake City: The University of Utah Press.

Kymlicka, W. (1990) *Contemporary Political Philosophy: An Introduction*. New York: Oxford University Press.

MacIntyre, A. (1981) *After Virtue: A Study in Moral Theory*. London: Duckworth.

Mill, J.S. (1956) *On Liberty* (edited by C.V. Shields, originally published in 1859) Indianapolis: The Bobbs-Merrill Company.

Mulhall, M. and Swift A. (1992) *Liberals and Communitarians*. Oxford: Blackwell.

Nussbaum, M.C. (2006) *Frontiers of Justice: Disability, Nationality, Species Membership*. Cambridge: Harvard University Press.

Parfit, D. (1984) *Reasons and Persons*. Oxford: Clarendon Press.

Rawls, J. (1999) *A Theory of Justice*. (First published 1971.) Harvard: Harvard University Press.

Rawls, J. (1992) *Political Liberalism*. New York: Columbia University Press.

Regan, T. (1983) *The Case for Animal Rights*. London: Routledge & Kegan Paul.

Roex, J. and M. Miele (eds.) (2005) *Farm Animal Welfare Concerns: Consumers, Retailers and Producers*. Cardiff: Cardiff University School of City and Regional Planning.

Sandel, M. (1982) *Liberalism and the Limits of Justice*. Cambridge: Cambridge University Press.

Singer, P. (1975) *Animal Liberation: A New Ethics for Our Treatment of Animals*. London: Cape.

Special Eurobarometer 229 (2005), 'Attitudes of consumers towards the welfare of farmed animals'. www.ec.europa.eu/food/animal/welfare/euro_barometer25_en.pdf

Twigg, J. (1983) 'Vegetarianism and the Meanings of Meat', pp 18–30 in A. Murcott (ed.) *The Sociology of Food and Eating: Essays on the Sociological Significance of Food*. Aldershot: Gower.

Van den Brink (2005) 'Vrijheid heeft geen politieke kleur', pp 38–54 in B. Snels (ed.) *Vrijheid als Ideaal*. Amsterdam: Sun.

VanDeVeer, D. (1983) 'Interspecific Justice and Animal Slaughter', in H.B. Miller and W.H. Williams (eds.) *Ethics and Animals*. Clifton, NJ: Humana Press.

Chapter 10
Consumer Rights to Food Ethical Traceability

Volkert Beekman

Introduction

Regulatory and scientific discourse about traceability in food production chains hitherto predominantly focused on the development of traceability schemes as a means towards the end of managing food-borne health risks. The overall aim of developing traceability schemes within this context of risk management in production chains has been to ensure that consumers can trust that their consumption of food products as provided in the market is not risky in terms of health consequences (see Chapter 2 for an overview of conceptual and operational aspects of traceability).[1]

This regulatory and scientific discourse recently witnessed a call to broaden the established notion of traceability to accommodate consumer concerns about product and process characteristics of foods that are not related to health risks. This call envisions the development of something like ethical traceability (see Chapter 1 for an overview of the call for ethical traceability). This notion of ethical traceability thus refers to the traceability of ethically relevant product and process characteristics of foods. It should not be confused either with the idea of ethically responsible food safety traceability or with the idea of a moral hazard analysis and critical control points (HACCP) that looks at 'human failure' as a critical control point (Hirschauer, 2004).

Before engaging in the development of ethical traceability schemes, however, the first question in need of serious scrutiny is the question of why ethical traceability might be a worthwhile objective in the context of the pursuit of a new social contract between food producers, consumers and society at large. In other words, a *justification* discourse about ethical traceability needs to precede an *application* discourse. This chapter presents a contribution to the justification of ethical traceability. The main line or argument is couched in terms of some basic liberal distinctions, since the call for ethical traceability seems to be based first and foremost on intuitions about consumer rights to informed choice on the food market. This chapter is rather strongly embedded in liberal political philosophy as developed – partly

[1] This chapter also introduces the distinction between traceability as a proactive strategy to prevent the occurrence of food safety incidents and trace back (and recall) as a reactive strategy to minimize the negative public health impacts of an emergent food safety incident.

C. Coff et al. (eds.) *Ethical Traceability and Communicating Food*,

in discussion with communitarian and utilitarian political philosophies – after Rawls (1972).[2] Although consumer freedom is high on the agenda of socio-political debate, this chapter explores a relatively unknown territory.[3]

The chapter starts by introducing the basic liberal distinctions to be used subsequently. It then argues that the operational domain for ethical traceability should be positioned somewhere between needs like food safety that entail positive unconditional rights of consumers and positive unconditional duties of producers and regulators with respect to the provision and traceability of food products, and non-reasonable and/or superficial wants with respect to food that do not entail any rights and duties. The basic argument is that ethical traceability is operational within the domain of reasonable and non-superficial wants with respect to food, which entail conditional rights of consumers and conditional duties of producers and regulators. Before concluding, the chapter asks whether calls for the development of ethical traceability schemes are not better – if cynically – understood as veiled attempts to increase corporate power in food production chains.

Some Basic Liberal Distinctions

This section introduces some basic liberal distinctions that inform the main line of argument in this chapter (see Table 10.1). It offers definitions – or rather circumscriptions – of the key terms that are hopefully non-controversial enough to be acceptable for everyone subscribing to the basic values (*liberty, equality and fraternity*) of modern liberal-democratic European societies.

This chapter embraces a pragmatic and modestly perfectionist version of liberalism. It accepts, on the one hand, that in pluralistic democratic societies people have divergent perfectionist visions of the good life but acknowledges, on the other hand, that such societies need to share some minimal basic values. It should be clear that liberalism has quite different connotations in – analytical – political philosophy and in day-to-day European politics. Basically, one could say that 'new labour', green 'realos' and left-wing Christian-democrats and liberals at least implicitly embrace such a pragmatic and modestly perfectionist liberalism, whereas 'old labour', green 'fundis' and right-wing Christian-democrats and liberals accept these minimal basic values but strive after divergent varieties of a fuller perfectionism in public morality. Left- and right-wing populists and fundamentalists, finally, do not even subscribe to one or more of the minimal liberal values.[4]

---

[2] Kymlicka (1990) provides a good introduction to the main discussions within contemporary political philosophy.

[3] Four of the few academic papers critically analyzing autonomy or freedom of choice in a (food) market context are Rippe (2000), Brom (2000), Beekman (2000) and Korthals (2001).

[4] Prominent proponents of a pragmatic and modestly perfectionist liberalism are USA's former President Clinton's advisor Selznick (1992) and UK's Prime Minister Blair's advisor Giddens (1998).

Table 10.1 Some basic liberal distinctions

	Needs	Wants		
Traceability	Traceability	Ethical traceability as a management tool	Ethical traceability as a communication tool	No traceability
Values	Food safety	Reasonable, non-superficial and overlapping food values	Reasonable, non-superficial and non-overlapping food values	Non-reasonable and/or superficial food values
Consumers	Positive unconditional right to be provided with safe foods	Positive conditional right to be provided with foods	Negative conditional right to be provided with information about foods	No rights
Producers	Positive unconditional duty to provide safe foods	Positive conditional duty to provide foods	Negative conditional duty to provide information about foods	No duties
Regulators	Positive unconditional duty to ensure food safety with traceability regulation	Positive conditional duty to regulate ethical traceability	Negative conditional duty to facilitate ethical traceability	No duties

Needs and Wants

This chapter uses the noun ‘need’ to refer to things that are always necessary for survival no matter where or when one lives or what specific values one might adhere to. The noun ‘want’, on the other hand, refers to things that are only necessary in specific spatial or temporal contexts or for the pursuit of certain specific values. Some scholars might question the very existence of universal (i.e. non-contingent) needs. However, such a full-blown scepticism would result in the *reductio ad absurdum* that denies the harsh reality of some ‘primary goods’ (to use a Rawlsian term) as being necessary for human survival.

Rights and Duties

This chapter uses the noun ‘right’ to indicate that someone is entitled to do or not do something, or to have or not have something done to him or her. The noun ‘duty’, on the other hand, indicates that someone has an obligation (not) to provide or do something to someone else. If someone has a right, someone else must have a corresponding duty. If this were not the case, rights would be empty and lose their meaning. It is however not true, as some intuitive accounts would argue, that for someone to have rights assumes that he/she also has duties. One should recognize that the rights/duties vocabulary is one of the strongest in ethics and political philosophy. This is both the power and the weakness of this vocabulary. The rights/duties vocabulary requires strict reasoning and only claims applicability in the domain of the mandatory. The absence of certain rights/duties thus does not imply that it could not be a good idea to arrange certain things in the domain of the voluntary.

Conditional and Unconditional

This chapter uses the adjective 'unconditional' to indicate that some right or duty holds always and everywhere, without further qualification in terms of a *provided that* or *unless* clause. The adjective 'conditional', on the other hand, thus indicates that some right or duty is indeed qualified by such a *provided that* or *unless* clause.

Positive and Negative

This chapter uses the adjective 'negative' to indicate a duty to refrain *from* or a right to be restrained *from* interference with some choice. To have a negative duty thus basically says that you should *not* do something, whereas a negative right says that something should *not* be done to you. The adjective 'positive', on the other hand, indicates that you are entitled *to* (do) something (positive right) or that someone else owes (or must do) something *to* you (positive duty).

Reasonable and Non-Reasonable

This chapter uses the adjective 'reasonable' first and foremost in the rather modest sense that values and consumptive choices will qualify as reasonable, if people arrived at those values and choices under conditions of adequate information or knowledge and on properly functioning markets. The *properly functioning* clause is the market equivalent of the Rawlsian *well-ordered* clause in well-ordered societies. Just as it does not make much sense to talk about justice in a political context that simply does not ensure some basic democratic values, it does not make much sense to talk about reason in a market context that simply does not ensure a level playing field between competitors, i.e. prices need to reflect costs.[5]

---

[5] This is perfectly in line with the idea of a 'negative morality' as defended by Honneth (1995) and Allen (2001).

Furthermore the adjective 'reasonable' is restricted to the (egalitarian) realm of values and choices that do not harm others (*harm principle*) (Feinberg, 1987) and do not entail negative repercussions on distributive justice (*justice principle*) (Rawls, 1972). This restriction is indeed only acceptable for people with a willingness to embrace the basic values of *liberty and justice for all*. The adjective 'non-reasonable', on the other hand, thus indicates that some value or consumptive choice does not meet even a rather limited interpretation of rationality or is at odds with respect for the harm or justice principles.

Superficial and Non-Superficial

This chapter uses the adjective 'non-superficial' to indicate that some consumptive value or preference is an expression of some vision of the right or the good, which implies that not being able to act upon that value has negative repercussions for the constitution of some individual or collective identity. The adjective 'superficial', on the other hand, is reserved for the more shallow preferences without such existential connotations.

Overlapping and Non-Overlapping

This chapter uses the adjective 'overlapping' to indicate that some value is shared and non-controversial within some society that is otherwise pluralistic on moral issues. The adjective 'non-overlapping', on the other hand, indicates that some value is controversial and thus not shared by all people in a certain society.[6]

Providing and Tracing

This chapter uses the verbs 'providing' and 'tracing' to distinguish between the *product* and *information* parts of the product-information combinations that are the subject of divergent traceability schemes. It is, however, important to acknowledge that traceability schemes are always about a 'package deal' of providing products and tracing information.

[6] A first indication of what values are overlapping – and the same holds true for what values are reasonable and non-superficial – in a certain liberal-democratic society is to be found in the relative popularity of the different political parties in the elections for its representative institutions (i.e. the parliament). This knowledge may be deepened by engaging citizens/consumers in participatory processes of opinion-formation (see Beekman et al., 2006) and with qualitative and quantitative social scientific studies using methods like focus groups, in-depth interviews and surveys (see part II of and Schipper's chapter in this book).

Regulating and Facilitating

This chapter uses the verbs 'regulating' and 'facilitating' to distinguish between directive and non-directive strategies of government intervention in food markets. Whereas the former strategies oblige market actors to behave in certain ways, the latter strategies merely entice these actors into certain practices. Regulating is thus a strategy of public management in the mandatory domain, whereas facilitating is a strategy of public–private governance in the voluntary domain.

Consumers, Producers and Regulators

This chapter uses the noun 'consumers' for the people buying, preparing and eating foods that are grown, processed and sold by people referred to as 'producers'. The noun 'regulators' is reserved for people in public, private or public–private bodies that somehow have a responsibility to contribute to a properly functioning market exchange between consumers and producers. This chapter basically tries to argue which rights to informed choice consumers might have and then formulates the corresponding duties of producers and regulators. It is, however, important to recognize that the development of ethical traceability schemes should be based on a balanced account of rights and duties among all affected parties.

Traceability of Food Safety

Rights of Consumers

No doubt seems to exist that each and every consumer needs safe food. This need does not vary with the context of consumption and thus does not require any further qualification. This implies that consumers have a positive unconditional right to be provided with safe food products. This would hold true even if historically – and geographically – the right to safe food has not always – and everywhere – been recognized as such.

Duties of Producers

If consumers have a positive unconditional right to be provided with safe food products, producers will have a positive unconditional duty to provide safe food products. This duty is only qualified by a non-moral clause that what counts as safe is always contingent upon available knowledge. This means that a duty to provide safe food products does not call upon producers to strive after a theoretically and practically unfeasible zero-risk. It does call upon producers to minimize food-borne health risks. Furthermore, in a context of food production chains with more than two links the duty to provide safe food products implies a further duty to trace information about the safety of food products.

Duties of Regulators

If needs inform rights and duties of different groups of actors in society, it will be the duty of regulators in a well-ordered, liberal-democratic society to ensure that rights are respected by corresponding duties. This implies that with respect to consumers' right to safe food products, in a context of food production chains without direct contact between producers and consumers, regulators have positive unconditional duties to secure the safety of food products by developing and enforcing traceability regulations. The whole argument about traceability and food safety is focused on the provision of safe products and the provision of information is only instrumental to that objective.

Non-Reasonable and/or Superficial Food Values

Moving from the domain of needs to the domain of wants, it is not too difficult to see that not all wants or values with respect to food offer sound reasons to argue for rights and duties on the part of different groups of actors. Two borderline cases should suffice to clarify this point. First, nobody will disagree that a cannibal does not have a valid claim to be provided with human meat. Some food values are simply beyond the realm of permissible values within well-ordered, liberal-democratic societies because they impose infringements on the harm principle (this is obvious in the case of the cannibal) or the justice principle. It is of course contingent where the exact demarcation between reasonable and non-reasonable food values is to be drawn.

Second, nobody will disagree that someone's preference for blackberries and distaste for blueberries does not provide sufficient reason to claim a right to be provided with blackberry pudding. Talking about rights and duties would lose any meaning if it were not confined to values that transgress some level of superficiality. Again, the exact demarcation between superficial and non-superficial values is of course a matter of contingent discourse.

Reasonable and Non-Superficial Food Values

It seems that in contemporary European societies the following consumer concerns count as expressions of reasonable and non-superficial food values (for a justification of this reduction of the whole plethora of consumer concerns into six categories see, e.g. Beekman et al., 2006):

- Concerns about impacts on public and personal health (harm)
- Concerns about impacts of genetic modification (harm)

- Concerns about impacts on animal welfare (harm)
- Concerns about impacts on the natural environment (justice)
- Concerns about impacts on international justice (justice)
- Concerns about the preservation of regional foods (justice)

Actually, it might be that concerns about impacts of genetic modification could also be classified as 'justice' or within the domain of the 'good', whereas concerns about impacts on animal welfare and the natural environment and about the preservation of regional foods could also be classified within the domain of the 'good'.

It is important to acknowledge a relevant difference between the first two concerns (health and genetic modification) and the last four concerns. Whereas it is perfectly possible to test products to ensure the accuracy of information about food safety or genetically modified (GM) ingredients, such a proof of the pudding is not available for the latter concerns. In colloquial terms, animal-friendly and animal-unfriendly, environment-friendly and environment-unfriendly, fairly traded and unfairly traded and regional and blended food products are substantially equivalent. These qualities can only be verified with the help of recorded identifications of product (hi)stories through the food chain.

All six concerns might be classified as *substantive* or *end* concerns to distinguish them from *procedural* or *instrumental* concerns about things like transparency, involvement, responsibility, trustworthiness and authenticity. Bluntly speaking, people mostly have procedural concerns because they are substantively concerned about public and personal health, genetic modification, animal welfare, the natural environment, international justice or regional foods. That is why procedural concerns may also be described as instrumental concerns, and that is why it is rather meaningless to discuss these concerns independently of the substantive or end concerns. This would be a confusion of ends and means. Whereas the substantive concerns refer to reasonableness in terms of the harm and justice principles, the procedural concerns refer to reasonableness in terms of adequate information/knowledge and properly functioning markets.

All six concerns share the characteristic that they go beyond merely personal *interests* like availability, convenience, price and taste. They are thus properly termed *moral* concerns.

Rights of Consumers

The fact – or assumption – of value pluralism within contemporary affluent societies implies that more than one set or hierarchy of reasonable and non-superficial food values exists or might exist. However, this does not yet seem enough to argue that consumers have a positive conditional right to be provided with food products that meet their values. It does seem reasonable to argue that consumers have a negative conditional right not to be – deliberately – misinformed about morally relevant product or process characteristics of foods.

However, one should not ignore the possibility of the existence of some overlapping consensus that holds that, e.g. eggs, meat and milk from livestock production systems should adhere to some minimal standard of animal welfare.[7] Then, the focus changes from information to products and it does seem reasonable to argue that such cases call for a positive conditional right to be provided with certain foods. The condition is, of course, that – unlike needs such as food safety – these overlapping values are contingent.

Duties of Producers

If consumers have a negative conditional right not to be – deliberately – misinformed about product and process characteristics that are relevant for some reasonable, non-superficial and non-overlapping food value, it will follow that producers have a corresponding negative conditional duty not to misinform consumers about these characteristics of their food products. Actually, the implications of this duty are significant, since it would imply considerable changes to prevailing marketing practices.

Duties of Regulators

Negative conditional rights and duties, on the one hand, call upon the government to develop regulatory assurances that these rights are respected by corresponding duties. The duty of regulators is, on the other hand, to regulate the development of ethical traceability schemes as a means towards the end of establishing market niches or segments for consumers and producers with similar food values.

Concerns about Impacts of Genetic Modification

It might be wise to provide some further clarification of these points by exploring three possible consumer concerns with respect to the much-debated yet ill-understood case of foods with GM ingredients.[8]

---

[7] The development of an animal welfare monitoring, traceability and labelling scheme is the subject of the European project *Welfare Quality* (see: *www.welfarequality.net*).

[8] See, e.g. Hansen (2004) for an unconvincing argument that with respect to consumer autonomy no morally relevant difference exists between negative and positive labelling of foods with GM ingredients, and Rubel and Streiffer (2005) for a reply with an equally unconvincing argument that respecting consumer autonomy calls for positive labelling of foods with GM ingredients. Much of the confusion in this debate results from the fact that what counts as a positive or a negative label depends on an account of normalcy. When using GM ingredients is the normal situation, the absence of such ingredients is a positive product attribute for people with particular food values. Hence, in this situation that very much resembles real life, a 'non-GM' label would qualify as a positive and not as a negative label.

Health

First, it might be possible to have concerns about negative repercussions of foods with GM ingredients for public or personal health. If these concerns are at least supported by enough scientific uncertainty about such risks to call upon a precautionary approach, the issue will become a case of needs with a negative unconditional right of consumers not to be provided with foods with GM ingredients until their safety has been established beyond reasonable doubt. Producers and regulators will then have the corresponding negative unconditional duties. If these concerns are not supported by some scientific uncertainty, as seems to be the case in real life (see Kuiper et al., 2004), the issue will become a case of non-reasonable concerns with no rights for consumers and no duties for producers and regulators. This first possibility could thus be translated into cases of either needs or non-reasonable wants.

Justice

Second, it might be possible to have concerns about negative repercussions of foods with GM ingredients for either developing countries (an issue of *intra*generational justice) or the natural environment (an issue of *inter* generational justice). This is where 'old-school' Rawlsians (following *A Theory of Justice* [1972]) and 'new-school' Rawlsians (following *Political Liberalism* [1993]) will part. The first group of liberals would argue that it is enough to imagine a hypothetical overlapping consensus on the principles of intra- and intergenerational justice that could be reached, if the power of reasonable argument determined the outcome of socio-political discussions. The second group of liberals would argue that one needs an actual overlapping consensus to support positive conditional rights of consumers to be provided with foods without GM ingredients and to support the corresponding positive conditional duties of producers and regulators.

Naturalness

Third, it might be possible to have concerns about foods with GM ingredients because they interfere with either the order of nature or the creation of God. Such concerns are often voiced in public debate but they cannot count on a supportive overlapping consensus. This is where anti-perfectionist Rawlsian liberals, perfectionist liberals (e.g. Raz, 1986) and perfectionist communitarians part. The difference between 'perfectionists' and 'anti-perfectionists' is that the latter argue that governments should somehow be neutral with respect to different visions of the good life, whereas the former argue that governments should be allowed to promote one or more specific visions of the good life.[9]

Anti-perfectionist liberals cannot see how producers and regulators in liberal-democratic societies could have positive conditional duties to ensure the preservation of the specific visions of the good that inform such concerns. They do, however, acknowledge the existence of negative conditional duties of producers and regulators that should ensure that consumers with these food values are not misinformed on labels that qualify certain foods as non-GM. Perfectionist liberals, on the contrary, do believe that producers and regulators have positive conditional duties to ensure opportunities to choose from a wide range of, e.g. regional foods. Otherwise consumers' freedom of choice would be an empty right. Communitarians even believe that producers and regulators have positive conditional duties to ensure the preservation of specific visions of the good. Hitherto, however, no convincing arguments have been formulated that this latter communitarian position could be in accordance with liberal-democratic governments.

Overlapping and Non-Overlapping Food Values

The above explorations might be summarized in more general terms – transgressing the case of foods with GM ingredients – by distinguishing two subcategories of reasonable and non-superficial food values.

If these values are supported by an overlapping consensus, ethical traceability should be developed as a public management tool to achieve the wanted changes in food *production* processes. It seems that concerns about animal welfare, the natural environment and international justice are based on overlapping, reasonable and non-superficial food values in present-day European societies. These concerns thus warrant regulatory enforcement of certain minimum standards of animal welfare, environment-friendliness and fair trade. Actually, this is not a case of *informed* choice but of *reduced* choice; the objective being to remove animal-unfriendly, environment-unfriendly and unfairly traded food products from the shelves of the supermarket.

If these values, on the other hand, are not supported by an overlapping consensus, ethical traceability should be developed as a public-private governance tool to achieve the wanted changes in food *information* processes. The private side of this governance tool consists of voluntary positive labelling and tracing of food products with specific ethically relevant product and process characteristics (e.g. region of origin or higher than minimum standards of animal welfare, environment-friendliness and fair trade). The public side of this governance tool consists of safeguard-

[9] Wall & Klosko (2003) edited a collection of essays that offers a good introduction to the perfectionism debate in liberal political philosophy.

ing consumers' negative conditional right not be misinformed by corporate marketing. Without this negative right there would be no point in positive labelling, since consumers would then not be able to judge the trustworthiness of the claims on food products. Notice also that a positive conditional information right, which would call for mandatory positive labelling, is almost too absurd to be discussed. Consider, for example, the case of a producer of environment-friendly candy bars, who is for some reason unwilling to label these candy bars as environment-friendly – this is a real life case! Why on earth, and how in practical terms should and could a government enforce positive labelling upon this producer?

A Cynical and Constructive Response to Empowerment[10]

A more cynical response to the whole discourse about consumer rights to make informed choices would argue that this discourse is a veiled – and fraudulent – attempt to shift responsibilities from producers and regulators to consumers. The statement that consumers have a right to make informed choices should then be read as saying that consumers have a duty to make informed choices. This represents a flagrant misunderstanding of an appropriate division of rights and duties in market democracies, and seems only to be introduced by producers and regulators with an unwillingness to act upon their duties.[11]

Current corporate practices of traceability – notwithstanding all 'newspeak' about corporate social responsibility (CSR) – are predominantly framed in terms of control, discipline, power and surveillance. It is therefore pertinent to recognize that food chains are hourglass-shaped, with many relatively powerless primary producers and end-consumers and a few absolutely powerful trading, processing and – particularly – retailing companies (see Bracke et al., 2005). It seems that the chances of ethical traceability schemes achieving the aim of empowering primary producers (to get specific product and process characteristics of foods valued in monetary terms) and end-consumers (to make informed choices for foods with specific product and process characteristics) are pretty meagre. A much more likely outcome is that the development of ethical traceability schemes will further strengthen the already fierce power-position of the intermediate actors in food chains. Intentionally or not, corporate and regulatory traceability initiatives have had this effect hitherto, and there does not seem to be much reason why it should be otherwise this time.

---

[10] This section has been written under the influence of three seminal publications with critical reflections on contemporary food politics: Schlosser's *Fast Food Nation* (2001), Nestlé's *Food Politics* (2002) and Lang and Heasman's *Food Wars* (2004).

[11] A more formal argument that leads to the same conclusion would be that since it is terribly unclear what actor holds what corresponding right to a consumer duty to be informed, the justification of such a duty among consumers lacks a normative basis.

In the past – and probably also in the present – food companies have not been overly scrupulous about using (implicit) lies to sell their products. Thus, claims about animal welfare, environment-friendliness and fair trade which appeal more truthfully to consumers' food values do not stand much chance of being recognized among the misleading marketing suggestions being made by food companies. One could, however, try to tackle this problem by asking for government regulation that focused on safeguarding consumers from deliberate misinformation in food marketing. That would contribute much more to the empowerment of end-consumers – and primary producers – than wishful thinking about corporations taking social responsibility and consumers making informed choices.

Conclusion

The whole argument of this chapter suggests that two versions of ethical traceability find justification. The first is ethical traceability as a public management tool, used to ensure that consumers are provided with foods that respect some threshold level of animal welfare, sustainability (natural environment) or fair trade (international justice), as supported by a contingent overlapping consensus of food values (several Eurobarometer studies suggest that such an overlapping consensus on these issues indeed exists). Such ethical traceability schemes are intended to ensure that corporate standards and quality assurances are verified by documented product (hi)stories. Concerns about public and personal health are excluded here, since food safety is not a want but a need and thus already covered by traceability schemes without the adjective 'ethical'. Concerns about genetic modification are excluded from this whole market application domain, since these concerns are not supported by an overlapping food value consensus. Concerns about regional foods are excluded from the whole market application domain, since it is fundamental to the nature of these concerns that they are only applied to specific food products.

The second version is ethical traceability as a public–private governance tool, used to allow certain consumers to be provided with food products, and sufficient information about these products, that are relevant for reasonable, non-superficial values that are not supported by an overlapping consensus. Governments should then facilitate this in the sense that consumers are not provided with misinformation about product and process characteristics of foods that are relevant for reasonable, non-superficial values that are not supported by an overlapping consensus in contemporary European market democracies.

It is important to notice that these justified versions of ethical traceability should both speak about product-information combinations (and not merely about information), and that in both justified versions of ethical traceability improvements in the tracing of information are instrumental to improvements in the provision of products with some substantive added value. The question then arises of what the two ethical traceability regimes – as management and governance tool respectively

Table 10.2 Two versions of ethical traceability

	Ethical traceability as a management tool	Ethical traceability as communication tool
Values	Reasonable, non-superficial and overlapping food values	Reasonable, non-superficial and non-overlapping food values
Application domain	Whole market	Niche markets
Who	All consumers and producers	Some consumers and producers
Concerns	Animal welfare Natural environment International justice	Genetic modification Animal welfare Natural environment International justice Regional foods
Standard	Minimum	Higher
Emphasis	Products	Information
Labelling	No	Yes
Responsibilities: consumers	None	Willingness to pay higher prices for labelled foods with specific product or process characteristics
Responsibilities: regulators	Develop, apply and enforce regulations to guarantee specific product and process characteristics of foods	Develop, apply and enforce regulations about labelling of foods with specific product and process characteristics
Responsibilities: producers	Obey regulations	Truthful labelling

– might look like. Table 10.2 thus spells out the details of the two versions of ethical traceability.

The second version of ethical traceability might be further developed in terms of policy recommendations with respect to the development of ethical traceability schemes for specified niche markets. It is important to rebut the common misunderstanding that niche markets are small. The defining characteristic of a niche market is that it is a meeting place for just one segment of producers and consumers. This segment might be or become a quite large portion of the whole market. Criteria and indicators need to be developed for a (higher than minimum) standard of labelled food products for specified niche markets. Furthermore, the nature of the demand for and provision of information by consumers, retailers, food processing companies, farmers and input companies (e.g. feed and seed) need to be spelled out. Together these criteria, indicators and information flows specify ethical traceability schemes.

Acknowledgement I would like to thank Tassos Michalopoulos for our lively discussions about the morality of the market. His modestly different conclusions from these discussions are voiced in his paper *The Citizen Goes Shopping* (Michalopoulos, 2006).

References

Allen, J. (2001) 'The place of negative morality in political theory'. *Political Theory*, 29/3: 337–363.

Beekman, V. (2000) 'You are what you eat: Meat, novel protein foods, and consumptive freedom'. *Journal of Agricultural and Environmental Ethics*, 12/2: 185–196.

Beekman, V., E. de Bakker, H. Baranzke, O. Baune, M. Deblonde, E.-M. Forsberg, R. de Graaff, H.-W Ingensiep, J. Lassen, B. Mepham, A. Porsborg Nielsen, S. Tomkins, E. Thorstensen, K. Millar, B. Skorupinski, F. Brom, M. Kaiser, and P. Sandoe (2006) *Ethical Bio-technology Assessment Tools for Agriculture and Food Production. Final Report Ethical Bio-TA Tools (QLG6-CT-2002–02594)*. The Hague: LEI.

Bracke, M.B.M., K. H. de Greef, and H. Hopster (2005) 'Qualitative stakeholder analysis for the development of sustainable monitoring systems for farm animal welfare'. *Journal of Agricultural and Environmental Ethics*, 18/1: 27–56.

Brom, F.W.A. (2000) 'Food, consumer concerns, and trust: Food ethics for a globalizing market'. *Journal of Agricultural and Environmental Ethics*, 12/2: 127–139.

Feinberg, J. (1987) *Harm to Others*. Oxford: Oxford University Press.

Giddens, A. (1998) *The Third Way. The Renewal of Social Democracy*. Cambridge: Polity Press.

Hansen, K. (2004) 'Does autonomy count in favor of labelling genetically modified food?'. *Journal of Agricultural and Environmental Ethics*, 17/1: 67–76.

Hirschauer, N. (2004) 'A model-based approach to moral hazard in food chains. What contribution do principal-agent-models make to the understanding of food risks induced by opportunistic behaviour'. *Agrarwirtschaft – German Journal of Agricultural Economics*, 53: 192–205.

Honneth, A. (1995) 'Moral consciousness and class domination: Some problems in the analysis of hidden morality' in C.W. Wright (ed.) *The Fragmented World of the Social: Essays in Social and Political Philosophy by Axel Honneth*. New York: State University of New York Press.

Korthals, M. (2001) 'Taking consumers seriously: Two concepts of consumer sovereignty'. *Journal of Agricultural and Environmental Ethics*, 14/2: 201–215.

Kuiper, H.A., G. A. Kleter, A. Konig, W. P. Hammes, and I. Knudsen (eds.) (2004) 'Safety assessment, detection and traceability, and societal aspects of genetically modified foods'. *Food and Chemical Toxicology*, 42/7: 1043–1202.

Kymlicka, W. (1990) *Contemporary Political Philosophy: An Introduction*. Oxford: Oxford University Press.

Lang, T. and M. Heasman (2004) *Food Wars: The Global Battle for Minds, Mouths, and Markets*. London: Earthscan.

Michalopoulos, T. (2006) 'The citizen goes shopping: What do the peers have to say about it?', pp. 483–487 in M. Kaiser and M. E. Lien (eds.) *Ethics and the Politics of Food*. Wageningen: Wageningen Academic Publishers.

Nestlé, M. (2002) *Food Politics: How the Food Industry Influences Nutrition and Health*. Berkeley/Los Angeles: University of California Press.

Rawls, J. (1972) *A Theory of Justice*. Oxford: Oxford University Press.

Rawls, J. (1993) *Political Liberalism*. New York: Columbia University Press.

Raz, J. (1986) *The Morality of Freedom*. Oxford: Clarendon Press

Rippe, K.P. (2000) 'Novel foods and consumer rights. Concerning food policy in a liberal state'. *Journal of Agricultural and Environmental Ethics*, 12/2: 71–80.

Rubel, A. and R. Streiffer (2005) 'Respecting the autonomy of European and American consumers: Defending positive labels on GM foods'. *Journal of Agricultural and Environmental Ethics*, 18/1: 75–84.

Selznick, P. (1992) *The Moral Commonwealth. Social Theory and the Promise of Community*. Berkeley/Los Angeles: University of California Press.

Schlosser, E. (2001) *Fast Food Nation. What the All-American Meal is Doing to the World*. London: Allen Lane/Penguin Press.

Wall, S. and G. Klosko (eds.) (2003) *Perfectionism and Neutrality. Essays in Liberal Theory*. Boston: Rowman & Littlefield Publishers.

Chapter 11
Ethical Traceability and Ethical Room for Manoeuvre

Michiel Korthals

Introduction: The Increase and Dynamics of Consumer Concerns

Food production in Europe is in crisis because of ethical consumer concerns and the continuing emergence of safety and health issues, which have resulted in a steady decline of trust in the sector on the part of governments and consumers. Several alternatives to current production methods are proposed, such as more stringent government control (e.g. Dutch policy from 1982–1998) or better co-operation between farmers and technologists (e.g. Dutch policy from 1998–2006). Most of these alternatives have until now had only mixed success, largely due to not very well explicated ethical assumptions and to social barriers, as is well documented by Pretty (2002). Firstly, the ethical assumptions of these alternatives focussed mostly on one or two values, although farming is a mosaic of values. Secondly, they assumed a stable and non-dynamic view of these values. Thirdly, the social barriers that confront directly involved stakeholders (producers, consumers) prohibit them from formulating value dilemmas and proposing new ethical-technological solutions that are alternatives to existing ones. Another barrier is that a certain moral position with respect, for example, to animal welfare could immediately lead to policy measures that were stricter but were not flexible, and could in the long run hamper new ethical solutions. Consumers also mention barriers such as availability and the lack of trustworthy information.

In this chapter I want to show how and why consumer concerns about food production can and should be incorporated into decision-making in food chains by organizing ethical discussions of conflicting values that include consumers as participants. This is valuable because, in the first place, there are several types of consumer concerns with respect to food and agriculture, which are multi-interpretable and often contradict each other, or at least are difficult to reconcile without considerable loss, as I will show later on; many consumer concerns are inherently dynamic as they change over time. Moreover, there are different types of consumers, and their choices between conflicting values differ accordingly. Different weighing models and various types of information are used for making choices. Both these features of consumer concerns make it worthwhile to organize the fundamental

C. Coff et al. (eds.) *Ethical Traceability and Communicating Food*,

decisions made in the serial links of food chains in accordance with consumer concerns.

Subsequently, I discuss common approaches to consumer concerns, such as codes of conduct, stakeholder analysis and assurance schemes, and I conclude that they can be helpful in addressing consumer concerns. However, in cases of dynamic, pluralistic and uncertain developments, sticking literally to some pre-existing ethical code, evaluation scheme or clear-cut normative hierarchy, such as codes or assurance schemes, can be disastrous in killing new, interesting, ethically desirable initiatives. Instead of considering ethical standards and targets as fixed, as in codes and schemes, it is more fruitful to emphasize the structure of the processes in which the ethical weighing of relevant consumer concerns takes shape, and which I call 'Ethical Room for Manoeuvre' (ERM). This concept is constructed to specify the ethically desirable conditions under which identification and weighing of paramount values and their ethical dilemmas can be processed. Finally, I discuss several types of ERMs in food chains, which can be communicated by means of specific traceability schemes to less involved stakeholders. I will also outline the relevance of ERM for implementing ethical traceability (ET) systems.

Three Types of Consumer Concerns

European food consumers have concerns that differentiate according to at least three levels, which result in three types of concern (Korthals, 2004; Beekman, 2004). Consumers have *substantive* concerns about certain ethically questionable structural traits of the food chain, such as lack of animal welfare. Secondly, they complain about the lack of trustworthy information, or even one-sided or distorted information, and lack of objectivity. Thirdly, they complain about lack of involvement with the food chain and an increasing gap between the food chain and consumers, which treats them as complete outsiders (*procedural concerns*; see Chapter 1).

The most common *substantive* consumer concerns that are mentioned are about seven ethical issues (Table 11.1): safety of the food (e.g. the use of hormones and antibiotics in animal feed); quality of the food; healthiness of the food; issues of animal welfare (with criteria such as the five freedoms, the transportation of animals, slaughtering procedures and import/export of animals and animal products); quality of the landscape; environmental effects of food production; and fair treatment of farmers (i.e. good working conditions in both the developed and the developing world). These values are subject to lots of detailed specifications, depending on the circumstances (Donagan, 1993; Korthals, 2004). All these values can be specified in innumerable items, and the concrete tasks and contexts involved are also innumerable. For example, animal welfare can mean intact horns, no lesions and injuries, good feet and limb conditions, etc. Good working conditions can mean that men and women get equal pay, that men and women have access to childcare, etc.

The second set of concerns covers the reliability of the *information* given, and also the relevance of the information in contributing to balanced ethical decisions

Table 11.1 Ethical issues of olive oil production

1. Ethical issues of olive oil production	Authenticity	Sustainability	Fair treatment
2. Co-producers and stakeholders (inc. consumers)	Processing links, also growers	Processing links (also growers)	The whole chain
3. Dilemmas	Authentic production versus sustainable and fair production	Sustainable production endangers authentic production	Fair treatment endangers sustainable and authentic production
4. Arrangements to cope with dilemmas	New technologies and social networks	*Idem*	*Idem*
5. Communication	Local farmers, and consumers not involved in the process	*Idem*	*Idem*

about food choices by both consumers and producers. This set of concerns covers pluralism: the information should not/cannot necessarily be neutral, but at least it should take into account differences among consumers, e.g. that consumers with a preference for organic meat products look for different information about the food chain and want different advice than consumers with other preferences.

The third type of consumer concerns covers the widespread consumer feeling of *alienation* from the food chain. This third concern is the motivation for some people to try to bridge the gap between producers and consumers. Some consumers simply take this gap for granted and don't worry about it, but others find it troubling and try to find out where their food comes from, very often with disappointing results, because they do not get a satisfying answer to their query or cannot get any information at all. Issues of involvement and participation are connected with these concerns.

Multi-Interpretable, Conflicting ('Dilemmatic') and Dynamic Character of Consumer Concerns

With respect to all these concerns it should be borne in mind that they are differently interpreted, often in conflict with each other and very dynamic, in response to a constantly changing world. The multi-interpretable character of the concerns is due to the fact that consumers differ with respect to their ethical orientations, attitudes and purchasing behaviour. There are different types of consumers, and their choice between conflicting values differs accordingly (Rozin et al., 1999; Lang and Heasman, 2006). Different weighing models and types of information are used for making choices. The same applies to producers: their value orientations and attitudes differ enormously across Europe. Attempts to re-establish trust should at least take into account the pluralism of consumers *vis-à-vis* their different ethical orientations, viewpoints and ways of balancing their preferred values.

Secondly, the concerns very often bring about ethical dilemmas with respect to shopping and consuming foodstuffs, which are well known to the more conscientious consumer-citizen. For example, demands to protect the environment and safeguard human health can lead to conflicts with demands for higher standards of animal welfare. Environmental and health requirements by and for human beings can imply, after all, that animals will have less free space to move around and their output (manure) should be heavily controlled (Wagemans, et al., 2003; Stegeman et al., 2003). One of the five demands proposed by the British *Royal Society for the Prevention of Cruelty to Animals* – and adopted in 2001 by the Dutch government committee for the restructuring of the agricultural sector – states that animals must be able to move around according to their normal behavioural pattern. For many farm animals, being locked up in a confined space without fresh air means a limitation of this freedom. There are many cases where the demands of the environment and of public health are at odds with the demands of animal welfare. Another example of a dilemma is that consumers on the one hand want to sustain fair trade, by buying fair trade food products from developing countries, and on the other hand are confronted with the fact that many farmers in developing countries cannot live up to the hygiene standards required by developed countries, thereby possibly imposing risks on the buyers or nearly unbearable financial burdens on the farmers.

One could react to these dilemmas by arguing that they are based in deep-seated emotions and desires, with which humans have to cope anyway. Bernard Williams (1994) is arguing this when he states that the many dilemmas which involve, in particular, friends and relatives are not ethical dilemmas at all, and they cannot be or even should not be discussed, because our essential commitments are at stake – his point is that one should (and does) care about relatives first. He argues that these dilemmas stand for essentially non-cognitive conflicts, and the only way to deal with them is to educate people about their emotions. However, not all ethical dilemmas are conflicts between desires (or involve friends or relatives). Many dilemmas do not resist a solution through communication and reasoning, because they are conflicts between beliefs, which does indeed make it meaningful to discuss them and to look for reasonable solutions. These solutions sometimes have unwelcome and even tragic results (in the sense that a loss is felt if one side loses; Foot, 2002). Ethical reasoning is therefore very important, in the sense of sharing arguments and being criticized by communication partners; it means learning to live with different arguments and in the end making use of arguments you might never have invented on your own. This type of 'communicative reasoning' can make a difference.

Communicative reasoning does not prevent you from having to accept losses, and does not exempt you from guilt or feelings of moral loss. You might feel guilty in a certain case because you didn't choose the best option, which was in fact the option with the largest losses; in that case, many people would choose the second best option and might still have feelings of guilt. However, feeling guilty is not the same as being guilty. So there is not a direct and causal connection between effecting moral losses, feeling guilty, moral wrongdoing and being blameworthy. Suppose I choose a local farmer's meat products, although I know that she, for some good reason, trims the beaks of her chickens, and that more animal-friendly chicken products can be bought 100 km away. In this case I contribute to some moral losses, but also to some moral benefits, but in my view I am not to be blamed for the losses.

Thirdly, consumer concerns change over time, in response to very different and complex events like technological and social developments. For example, it only recently became clear to many consumers that the distance food travels is a considerable concern, which is at odds, probably even inconsistent with, the choice in favour of, e.g. organic or fair trade food (Pretty et al., 2005). Global sourcing of ingredients by large enterprises, be it organic or not organic (i.e. buying products in certain product sectors abroad on the basis of specific requirements), is also a recent phenomenon and consumers are becoming aware of some new ethical aspects connected with this development. This recent concern is added to the ones already mentioned, and can significantly transform them.

In a recent article on identifying and ranking attributes that determine sustainability in Dutch dairy farming, the authors propose to differentiate this concern into at least 36 attributes: 'only one attribute was selected for economic and internal social sustainability: profitability and working conditions, respectively. The list for external social sustainability contained 19 attributes and the list for ecological sustainability contained 15 attributes' (Calker et al., 2005). Although the authors do not specify the exact character of these attributes, it should be clear that all these attributes change over time, as does their relative weight *vis-à-vis* each other. An example of a recent change is the issue of possible greenhouse gas mitigation by crops, which again could be transformed into a new concern for citizen consumers in favour of carbon sequestration (Lemus and Lal, 2005). Indeed, when the public becomes more aware of the risks of global warming, be it triggered by alarmist movies like *The Day After Tomorrow* (by Roland Emmerich, 2004), or seriously informed by the Intergovernmental Panel on Climate Change then the urge for increasing carbon sequestration in crops may well become a consumer concern. The dynamic character of consumer concerns is something to be reckoned with in a serious and structural way.

The multi-interpretable, conflicting and dynamic character of consumer concerns has until now been only partly addressed, by at least two main ethical and social approaches. The first is called 'principalism' and consists mainly of the application of a mix of deontological and utilitarian approaches to food consumption and production. According to the deontological approach in ethics, in cases of ethical uncertainty one should look for principles, rights and duties. According to the utilitarian approach, the consequences of an action chosen are decisive: if they are ultimately negative for the 'greatest happiness of the greatest number' one should refrain from the action. An example of a principalist framework is the ethical matrix of Mepham (1996), which starts out from the standard *opus* in the field of medical ethics by Beauchamp and Childress

(1994). In this approach, four principles are formulated to resolve the ethical problems of nutrition. The first principle is that of respect for autonomy: respect for the right of choice, which also entails the right to information. The second is that of justice: a fair division of advantages and disadvantages, and of risks. The third principle focuses on non-damage: not inflicting damage on human beings and nature. The fourth relates to doing well: contributing to the well-being of human beings and nature, also by avoiding damage. Mepham (1996) applies these ethical principles via a matrix to food production processes, distinguishing therein a number of participants: consumers, various types of producers, a targeted organism (like animals or crops), and nature in its totality. He also combines the principles of doing well and non-damage into the principle of respect for well-being. Even future living beings may be included in the totality. Mepham applies these principles in various cases, and concedes that they are to be ranked, but he insists upon the neutrality of the matrix as a tool for ethical problem solving for deliberating persons. The targeted user of the matrix is the individual person, who solves an ethical problem entirely on his or her own. He does not give any clues as to what a successful deliberation should look like, or according to what criteria a deliberation can count as ethically successful.

This principalist approach (be it mainly utilitarian or deontological or a mix of these two perspectives) has significant shortcomings (Korthals, 2004). In the first place, the principles are ambivalent: they are compulsory in nature, while they are also ideals that we should aim for but can never achieve. Autonomy is something that we should respect, but it also stands for a desired terminus (an ideal or value) that we can never reach; we can therefore act more or less autonomously. The same applies for respect for well-being and its complementary components, doing well and non-damage: it is an ideal that we will seldom achieve. The principle of doing well is even more questionable, because it is hardly possible to do well on a universal basis, since human preferences and characters differ so greatly. Non-damage can be more properly called a principle, since it can entail abstaining from any particular action. The second shortcoming is that, in concrete situations, Mepham's principles are often in mutual conflict (similar to the consumer concerns!). They do not tell us what priority each should have, so they ultimately do not help us to resolve dilemmas. In the end, the principles only operate heuristically in selecting the really worthwhile ethical issues, rather than as absolute commandments. They help us to look at aspects of specific situations and direct our attention to specific characteristics. But they do not cover the entire sphere of meaningful and fruitful ethical concepts. The ethical issues in nutrition are so complex that the various principles are always at odds with each other or inconsistent.

A second approach to tackle the multi-interpretable, conflicting and dynamic character of consumers' concerns includes the perspectives of 'Stakeholder Analysis' and 'Value Chain'. These more economic and sociological perspectives start with an analysis of the main parties involved in a company or a production chain. They try to draw up an inventory of their main economic interests, and persuade stakeholders that it is to their advantage to take these interests into account (Simmons and Lovegrove, 2005). An advantage is that some stakeholder analyses are issue-oriented, for example, with respect to natural resource management in developing countries, and they formulate practical guidelines for making management more participatory and effective (Grimble and Man-kwun, 1995). Some, moreover, recognize environmental concerns, and try to make clear that it is both economically advantageous and ethically acceptable that companies take these into account (Payne and Raiborn, 2001). However, although they are in favour of educating consumers, they want them to be involved only in buying or not buying: 'Consumers ultimately control the failures or successes of businesses and their products. For example, consumer boycotts were the primary reason for the banning of chlorofluorocarbons in aerosol cans for most uses in Canada, the United States, and most Scandinavian countries [...]. Alternatively, when first introduced, "The Body Shop" products rapidly became consumer "must have" items because they sported the "no animal testing" label' (Payne and Raiborn, 2001:9). These approaches do not allow for the involvement of consumers or for symmetrical communication on ethical dilemmas.

Ethical Room for Manoeuvre

The principalist and the stakeholder analysis/value chain approaches have advantages and disadvantages in addressing the multi-interpretable, conflicting and dynamic character of consumer concerns. The disadvantage of the first is the exclusive focus on the individual and on fixed principles; the advantage of the second is the collective orientation, although it excludes important stakeholders (such as consumers). The advantage of the first is its heuristic value in identifying ethical values and dilemmas, an identification the second is not able to perform. To take into account these advantages and to tackle the special character of consumer concerns, I constructed the concept of 'Ethical Room of Manoeuvre' in the food chain. With this concept I want to take into account the multi-interpretable, conflicting and dynamic character of consumer (and producer) concerns by appealing to the social process and structure in which stakeholders cope with dilemmas, and also take into account the problems of applying ethical norms in the food chain in order to address these concerns. The concept circumscribes with a new term the well-established idea that a 'free space' for *deliberation* and *inquiry* can produce solutions to 'hot' or pressing ethical issues. This 'free space' should be sited at the relevant links of the food chain as and when a burning issue arises, but is not conceived as a kind of window-dressing to disguise the neglect of ethical issues. The 'room for manoeuvre' allows one to take all ethical points of view and perspectives into consideration and to balance them, but not to get rid of the ethics: there would be a rather strict ethical and social regulation of the 'room'. The idea is that if consumer concerns do indeed have these special features, and more general substantive norms can't be simply and straightforwardly applied, then one can at least circumscribe the requirements that make the deliberations on these difficulties ethically acceptable.

The ERM has both a substantive aspect, because it requires that all participants involved in ethical issues should be included in addressing those issues, and a procedural aspect, because it requires that participants should be free to examine the opinions of others and to learn over time. The ERM describes a learning process over time and does not allow for the entrenchment of economic or social interests. The concept intentionally wants to counteract the barriers and fears that restrict the ethical capacities of stakeholders and obstruct ethically acceptable ways of solving continuously emerging value dilemmas. It covers forms of communication and deliberation that take into account continuously changing social, normative and technological situations for farming. Using this concept, participants should not react from a fixed and stable normative framework, but in a transparent and ethically acceptable way, coping with dilemmas communicatively by first identifying them and then solving them.

The concept of ERM has strong links with the ideas of constructive technology assessment (Grin et al., 1997) and of co-production. Callon and Latour (1992) originally introduced the concept of co-production as co-production of nature and society in science and technology. Jasanoff (1996) has since coined the term co-production in a more general sense, by transcending the context of science and replacing science with knowledge: '*co-production is the simultaneous production of knowledge and social order*'. It can be seen that ERM is a process whereby innovations in the field of food production are linked to ethical norm-seeking (for a more thorough discussion of this connection, see Chapter 14).

Finally, by specifying the conditions and the structure of ERM and co-production it becomes clear that ERM should counteract opportunistic strategies that reduce the meaning of ERM and coevolution to mere window-dressing. By adhering to relevant rules one can prevent 'free riders' from taking advantage of the fact that the essentially substantial codes are not appropriate for tackling such a dynamic sector as the food sector.

The main aims of ERM should be:

1. The mobilization of personal and collective inputs on relevant ethical issues
2. The pooling of relevant ethical issues, and of required information
3. The specification of interpretations of norms and values with respect to the analyzed ethical issues
4. Arguing for and against special applications and interpretations
5. Because of the continuing learning process, the generation of rationally motivated trust in the outcome
6. The continuing construction of trust can determine more politically oriented discussions, e.g. political decision-making, and their outcomes

The aim of ERM is not directly to put forward political decisions, but to identify relevant ethical issues, to interpret these and to put forward solutions to them, so that they can have a function in managerial or political decision-making.

ERM and Ethical Traceability

ERM should not be permanently located everywhere in food chains and networks, but only be placed in those food chains and links where ethical issues hurt most. By incorporating this device in traceability schemes, traceability could gain an interesting twist for both producers and consumers. As we have shown earlier in the book, traditional traceability schemes stop before the information streams reach the consumer. In the words of Chapter 2: 'Keeping the information within the food chain hidden and not allowing consumers any access runs the risk of not fulfilling important expectations associated with the concept of traceability. To avoid such a risk, traceability should not be used as a purely administrative tool or as a mere safety system, because it is related to highly contested and sensitive issues such as animal welfare, fair trade, traditions and beliefs, environmental protection and sustainability. Rather; it should represent an instrument for establishing effective and responsive policies and institutions based on involvement via informed food choices by citizens/consumers.'

Traceability needs to be restructured in a way that reconciles informed consumer choice and consumer sovereignty (Korthals, 2001).The features of ERM allow traceability schemes to be implemented and to become ET schemes.

Traceability takes a different shape depending on the interests, values and features of the sector. ERM calls for the involvement of ethically conscious consumers and producers and their differing ethical opinions. A complex interplay between end-users and producers in the chain is necessary. Food chains are probably able to cope with pluralism and with the concept of ERM, which means that it is not necessary to stick zealously and totally to ethical rules, but that ethical considerations have to be balanced and negotiated with co-producers and stakeholders, and that it can be proved to third parties that these considerations have taken place. Because of this procedural feature of ERM, any implementation of ERM has to comply with certain non-negotiable standards.

ET through ERM can:

(a) Assist ethical consumers in making their food choices on the market according to their own ethical beliefs, by making clear that ERM has been consciously implemented
(b) Increase the ethical responsibility of consumers, because they see that it pays off
(c) Increase the share of ethical products on the markets
(d) Assist producers in tackling ethical problems in an acceptable way

One important advantage of ERM is that it enables full compliance with the rather strict rules on consultation which, according to the EU General Food Law, Articles 7, 8, and 9 (EU, 2002), should be organized with consumers on technological innovations in the food chain.

From a governmental point of view, there is a lot to win by implementing ERM. But there are also advantages from a management point of view. Although producers will probably have lots of difficulties with potential negative information that could harm the factory or the farm (such as disease levels), there are also advantages.

As Trienekens and Hvolby (2001) have outlined in several papers, the demand drive is now recognized as an important feature of modern food chains, and it requires networking and evolving partnerships within chains. Total integration of the different links of the chain leads to fixed and less responsive food chains; networking the various ERMs, however, could be a good alternative. Trienekens argues:

> Developing partnerships in the perspective of co-operative action seems to be relevant for food supply chains. Motives for this are, e.g., development of competitive power, need for quality, safety and sustainability of food produce and flexibility to react fast on changing markets. Several forms of risk can be reduced this way, risk of fluctuating prices, risk of quantity/quality features (e.g., transport of pork, scheduling of pork finishing capacity with slaughterhouse and meat products processing capacity), risk of food safety and hygiene. (Trienekens and Hvolby, 2001)

Again, because of the special features of consumer concerns, it is better to develop partnerships which include ethically conscientious consumers.

Three Types of Traceability

Recent consumer concerns such as animal welfare, the use of genetically modified ingredients and bovine spongiform encephalopathy (BSE) have shown that the need to trace and authenticate the contents of food products has never been more urgent for governments and managements (Lees, 2003). There are several ways to approach the concept of traceability and to establish new techniques of food authentication and information applicable to particular sectors, such as dairy or wheat production. Roughly, one can differentiate between traceability systems as management tools, as government tools or as consumer tools. These three different approaches have overlapping similarities (see also Chapter 1).

Traceability as a Government Tool

Governments are mostly oriented towards upholding minimum standards of safety, animal welfare and environmental protection and looking for possible flaws, mistakes or fraud. Governments try to balance the interests of producers (and defend their share in global production) and the interests of consumers, at least in the area of safety. In this way traceability is a means of surveillance, of policing certain standards to ensure they are met, a kind of bio-politics, as Foucault would call it. The government ensures a minimum level of, in particular, safety standards in the supply chain, and establishes a structure that ensures producers provide consumers with sufficient information. First, traceability should maintain food safety and protect consumers from being misled, by supplying consumers and control authorities with relevant information. Secondly, it should enable a simple, fast and targeted recall of products in emergencies.

Traceability as a Management Tool

Firms are interested in brand protection, in showing that they comply with the minimum standards set by governments or themselves, and in easy product recall. The main function of food traceability as a management tool is brand protection and a traceback system for recalls. During the Food Safety Summit in Washington, DC, in 2005, 'Brand protection' through traceability was discussed as a major advantage for food suppliers. Wal-Mart and McDonalds created their own traceability systems in order to anticipate possible recalls. According to experts, traceability schemes are not a costly evil; they have lots of advantages. Food companies create an operational system of record, which enables end-to-end traceability, including forecasting and scheduling, customer and inventory management, expiration date control, label management, specification and recipe management, supplier and purchasing information and 'executive dashboards' to assist executives with traceability This is advantageous for companies, because it enormously reduces their recall time and, as a consequence, lowers their insurance costs by reducing traceback time to store shelf. Moreover, it makes it possible to cut returns of expired products according to an optimized inventory plan.

Some corporations, in particular the socially responsible corporations (SRCs), try to detect common denominators in the values people have with respect to the treatment of animals or fair trade, and produce foodstuffs accordingly. Consequently, they inform consumers about these standards and assure them that production, etc., has indeed been done according to these standards. Here, too, giving consumers information on all concerns is not the final target of the traceability system.

Traceability as a Consumer Tool: Ethical Traceability

However, there is also a third way of looking at ET: ethical consumers are not in the first instance looking for fraud etc., but for goods that have been produced in a way that satisfies their ethical concerns, which are various and also differently interpreted and valued (pluralism). Consumers have different values, different appreciations of food production; some of them are more oriented towards animal welfare, some more towards fair trade, etc. In view of the recent rise of traceability schemes for food safety, it would be a very interesting idea to widen traceability schemes to include more ethical criteria, such as these consumer concerns. However, their often ambivalent, flexible, dilemmatic and dynamic character can raise doubts about the possibility of incorporating them satisfactorily in food chains. ET schemes should therefore not comply strictly with consumer concerns *per se*, but with structural rules (such as the inclusion of and access to relevant information) that guarantee that sufficient attention has been paid to them. The concept of ERM can give some help here for companies that really try to take consumer concerns into account, and to meet the demands of consumers.

Types of Ethical Room for Manoeuvre

Several types of ERMs can take shape in food chains, and their results can be communicated to less involved stakeholders by means of specific traceability schemes, with the possible consequence that they will be motivated to be more involved. ET in the sense of ERM reports on the main features of the processes and results of the ethical deliberations that gave rise to the final agreement on the direction of the production process in the particular link of the chain.

First, one type of ERM can be implemented in food chains that predominantly use craftsman-like skills and small-scale organizations. In this type of food chain, ERMs are partly already institutionalized; if not, they can be organized on a local or provincial basis because of their consumer-driven production base and the preferences of their consumers to keep distances as small as possible. An example of this type could be the one described by Carnes and Karsten (2003) in their article on the diverse community networks for sustainable food systems that are covered by the Pennsylvania Association for Sustainable Agriculture (PASA). Community or urban farming often allows for easy communication between producers and consumers on ethical issues.

Secondly, food chains that are large scale and/or national and make use of intensive processes (such as the UK wheat-flour-bread chain, described in Chapter 6) require ERMs that consist of representatives of the various consumer concerns, such as animal protection movements, environmental organizations and others. It is most fruitful to place these ERMs at the interface between the main links of the chain, where processing and standardizing starts and where the relevant consumer concerns (animal welfare, human health, environment) and producer concerns (profit, labour) are put in the balance. An example, given by Marsden and others (2000), is Llyn Beef Cooperative Producers in Wales, which connects non-farmers with farmers and suppliers and in this way stimulates urban–rural communication (and development; compare Pretty, 2002).

Thirdly, in the case of export-driven food chains (such as the Greek olive oil chain or the Danish pork chain, described in Chapters 5 and 7 respectively), ERMs should be run by representatives of various international organizations, such as non-governmental organizations (NGOs) and consumers' organizations. With respect to the Greek olive oil chain, one of the main consumer concerns is the authenticity of the oil. Because it seems to be impossible to produce enough genuine virgin pressed (or 'cold- pressed') olive oil, with an oleic acidity of less than 1%, there is a strong urge to mix it with refined oil (Visser, 1986). If done properly, the taste losses are not that large; however, the consumer should be informed about it. The ERM should start with a list of provisional ethical problems and on the basis of this list formulate a list of co-producers and stakeholders. So, with olive oil, we have the issues of authenticity, environmental sustainability and fair treatment of farmers. The co-producers (i.e. the producers directly involved in the chain) and the stakeholders concerned with these issues should start their debate with these issues, taking into account the deficits of their own, mostly one-sided interests' positions.

The second step should be to start with ethical dilemmas, and the third step is oriented towards looking for and constructing new ethical and technological arrangements and assessing them. Finally, the response of stakeholders in other links of the chain will be considered, with the aim of increasing the success of the ERM. The whole process should have an iterative character.

In all cases, the ERMs in the chain should communicate their decisions to the broader public, by means of ET schemes. These ethical schemes can then be taken into account by wider circles of other, less involved consumers, who may then be motivated to become either less or more involved.

An additional advantage of the implementation of ERMs in the relevant links of the food chain could be that it engages consumers more in the food chain, which gives companies the opportunity for user-centred innovation. As is already the case in the software business, user-centred innovation pays out twice over, for both firms and end-users (Hippel, 2005). Why not with respect to food production? More and more people are willing to spend time improving the quality and processing of food. Hippel (2005) gives ample examples that innovation in software moves from the laboratory to the kitchen by the introduction of toolkits and platforms or platform products for users. Ethical and technological learning processes can go hand in hand in user-centred innovation.

Finally, a lot of research has to be done with respect to unpacking the idea of ERM: what form of ERM will fit what context and food sector? What competency and behaviour does it require of the different partners? And which factors can enhance the success of the ERM and which barriers endanger the success of the ERM? How exactly can ERM be linked to ET and consumers' informed food choice in the various food chains and networks? These are a few of the questions that need further research.

Conclusion

In this chapter, first three types of consumer concerns, i.e. substantive, information and procedural concerns, are described. A lot of these concerns are conflicting (dilemmatic) and dynamic in character, which means that it is very difficult *a priori* to determine their full range of attributes and their weight in a final decision. In connection with the recent rise of traceability schemes for safety and authentication of food, it would be a very interesting idea to widen the traceability schemes to include more ethical criteria, as expressed in consumer concerns. However, their ambivalent, flexible, dilemmatic and dynamic character seems to be an obstacle in incorporating them in the chain in a satisfactory way. ET schemes therefore should take into account not only the consumer concerns themselves, but (even more) the fact that sufficient attention has been paid to them. The concept of ERM can give some help here for both consumers and producers. With this concept we can formulate socially and ethically acceptable criteria that make it clear that sufficient attention is being paid to these concerns.

However, these rooms cannot be installed in all contexts: in stable and consensual contexts it seems much more apt to use established forms of ET such as certifying and labelling. Another requirement of ERM is that the ethical decisions are taken seriously and implemented in other parts of the chain. So they are not a licence to do what one wants, a kind of 'anything goes' arrangement. Moreover, ET through types of ERM requires special types of communication with consumers not involved in the ERM (see Chapter 14).

References

Beekman, V. (2004) 'Sustainable development and future generations', *Journal of Agricultural and Environmental Ethics*, 17/1: 3–22.

Calker, K. J. v., P. BM. Berentsen, G. W. J. Giesen, and R. B. M. Huirne (2005) 'Identifying and ranking attributes that determine sustainability in Dutch dairy farming', *Agriculture and Human Values*, 22: 53–63.

Callon, M. and B. Latour (1992) 'Don't throw the baby out with the bath school! A reply to Collins and Yearley', pp. 343–68 in A. Pickering (ed.) *Science as Practice and Culture*. Chicago: University of Chicago Press.

Carnes, L. M. and H. D. Karsten (2003) 'Building diverse community networks for sustainable food systems: Guiding philosophies of the Pennsylvania Association for Sustainable Agriculture', *American Journal of Alternative Agriculture*, 18(4): 174–184.

Donagan, A. (1993) 'Moral dilemmas, Genuine and spurious', *Ethics*, 104, 1: 7–21.

European Commission (2000) *White Paper on Food Safety*, 12 January 2000.

Foot, P. (2002) *Moral Dilemmas and Other Topics in Moral Philosophy*. Oxford: Clarendon.

Grimble, R. and C. Man-kwun (1995) 'Stakeholder analysis for natural resource management in developing countries: some practical guidelines for making management more participatory and effective' Natural Resources Forum, 1995, 19(2): 113–124.

Grin, J., H. van de Graaf and R. Hoppe (1997), *Interactive Technology Assessment: a Guide*. The Hague: Rathenau Institute.

Hippel, E. v. (2005) *Democratizing Innovation*. Cambridge: MIT Press.

Jasanoff, S. (1996) 'Beyond epistemology: Relativism and engagement in the politics of science', *Social Studies of Science*, 26/2: 393–418.

Korthals, M. (2001) 'Taking consumers seriously: Two concepts of consumer sovereignty', *Journal of Agricultural and Environmental Ethics*, 14(2): 201–215.

Korthals, M. (2004) *Before Dinner*. Dordrecht, The Netherlands: Springer.

Lang, T., and M. Heasman (2006) *The Unmanageable Consumer*. London: Earthscan.

Lees, M. (ed.) (2003) *Food Authenticity and Traceability*. Cambridge: Woodhead Publishing.

Lemus, R. and R. Lal (2005) 'Bioenergy Crops and Carbon Sequestration', *Critical Reviews in Plant Sciences*, January-February (24)1: 1–21.

Mepham, T. B. (ed.) (1996) *Food Ethics*. London: Routledge.

Marsden T., J. Banks J. and G. Bristow (2000) 'Food Supply Chain Approaches: Exploring their Role in Rural Development', *Sociologia Ruralis*, (40)4: 424–438.

Payne, D.M and C.A. Raiborn (2001), 'Sustainable development: The ethics support the economics', *Journal of Business Ethics*, 32(2): 157–168.

Pretty, J. (2002) *Agri-Culture: Reconnecting People, Land and Nature*. London: Earthscan.

Pretty, J. N., A. S. Ball, T. Lang and J. I. L. Morison (2005) 'Farm costs and food miles: An Assessment of the full cost of the UK weekly food basket', *Food Policy*, 30(1): 1–20.

Rozin, P., C. Fischler, S. Imada, A. Sarubin and A. Wrzesniewski (1999) 'Attitudes to Food and the Role of Food in Life in the U.S.A., Japan, Flemish Belgium and France: Possible Implications for the Diet-Health Debate', *Appetite*, 33:163–180.

Simmons, J. and I. Lovegrove (2005) 'Bridging the conceptual divide: lessons from stakeholder analysis', *Journal of Organizational Change Management*, 18 (5): 495–513.

Stegeman, A., A. Bouma, A.R.W. Elbers, M.C.M. de Jong, G. Nodelijk,. F. Klerk, G. Koch and M. van Boven (2003). 'Avian influenza A virus (H7N7) epidemic in the Netherlands: course of the epidemic and effectiveness of control measures', *Journal of Infectious Diseases*, 190(12): 2088–2095.

Trienekens J. H. and H. H. Hvolby (2001) 'Models for supply chain reengineering', *Production Planning and Control*, 12(3): 254–264.

Visser, M. (1986) *Much Depends on Dinner*. London: MacMillan.

Wagemans, M. J. M., J. M. G. Hol and C. M. Groenestein (2003). '*Onderzoek naar de ammoniak- en geuremissie van stallen LIX: Welzijnsvriendelijke huisvestingssystemen voor dragende zeugen in kleine groepen*', Agrotechnology & Food Innovations, Rapport B740, 53.

Williams, B. (1994) *Ethics and the Limits of Philosophy*. London: Penguin.

Chapter 12
Interpreting Traceability: Improving the Democratic Quality of Traceability

Marco Castagna

As argued in previous chapters, the concept of traceability can be used as a communicative tool for the reconstruction of the 'history' of food products, allowing consumers to take an 'active' stand on the ways foods are produced. Traceability is an *ambiguous concept* (Mattéi, 2003:36). Located at the crossroads of economy, law and communication, the notion plays a fundamental role not only in juridical-administrative dynamics but also in marketing strategies. Given its multiple meanings, it is easy to give in to the temptation to solve the ambiguity by formalistically curtailing its meaning. This paper, instead, tries to take into account this radical ambiguity, based on the belief that a realization of the limitations of an instrument allows one to use it in the best possible ways.

The approach to the problem presented here will allow us to view the ambiguity of the term traceability as a dialectical force illuminating the importance of its interpretive character. Based on this broader interpretation of traceability, the idea of informed choice will allow us to go beyond the limited notion of the consumer towards a more complex notion of *consumer/citizenship*, an essential component of the idea of democracy.

'The Trace', 'To Trace' and 'Traceability'

Notwithstanding its currency, the term traceability dates back only a few decades and is still not documented in most dictionaries. Even the authors of the collection of essays *Traçabilité et responsabilité* (Pedrot, 2003), while specifying the technical and juridical nature of the term and its ties with the concepts of responsibility and control, do not offer an unequivocal definition.[1]

The indeterminacy of the concept reflects, thus, the various ways of perceiving it: that of legislators, who positively associate it with the concepts of *prevention*

[1] For more technical definitions of traceability see Chapter 2.

C. Coff et al. (eds.) *Ethical Traceability and Communicating Food*,

and *responsibility*; that of producers (of commodities or services), who lament its *complexity* and *cost*; and that of consumers, who question its *effectiveness*.[2] Let us try then to re-problematize the philosophical implications of the verb and noun *trace*.

The trace is, first of all, something that refers back to something that happened in the past. On account of this ability to refer back, the trace is, therefore, a sign, and, as such, it is necessarily oriented in two different directions: towards that of the author and that of the interpreter. Hence the word *trace* and the act of tracing are ambiguous: they indicate both the action of the one who leaves a trace and the action of the one who follows it.

The French philosopher Emmanuel Lévinas sharply distinguishes *trace* from *sign*. Only after having explored the reasons behind this distinction, can one grasp the sense in which all signs are also traces (while not all traces are signs). In his essay, *La trace de l'autre (The Trace of the Other)* Lévinas writes that:

> [T]he trace, compared to other signs, has still this aspect which is exceptional: it signifies beyond any intention of producing a sign and beyond any project of which it would be the intention (or the goal: *visée*). (Lévinas, 1967:199)

Therefore, adds Lévinas, when, for example, we pay with a cheque because we intend to 'leave a trace' of the transaction, the notion of trace does *not* show all its meaning. The latter emerges fully in the following observation: 'He who left traces while erasing his traces did not mean to say or do anything with the traces he left' (Lévinas, 1967:200).

The decisive aspect is precisely this excess. Something escapes the control exercised by the signifying action. Herein lies the specificity of the trace. But, Lévinas concludes that:

> [E]ach sign is, in this sense, a trace. In addition to what a sign signifies, it is the passing of he who has left the sign. The significance (*significance*) of the trace doubles the meaning (*signification*) of the sign produced in view of communication. (Lévinas, 1967:200)

This is the point. The use of the term traceability inevitability sets in motion a process that goes beyond intentional communication. Someone, such as a food producer, who intends to create traceability, sets in motion a process in which content is communicated – but this is also a process in which the content escapes the sender's control. There is an element of unpredictability as to the reactions to traceability and to how the information communicated will be interpreted.

As a preliminary conclusion we can say that food itself is a trace of its own production history. Foods, however, for obvious reasons, do not communicate this story explicitly themselves. Food remains an enigma for the consumer. The task and challenge of traceability is to make the history of food explicit.

---

[2] Korthals in Chapter 11 systematically addresses three perspectives on traceability: as government, management and consumer tool.

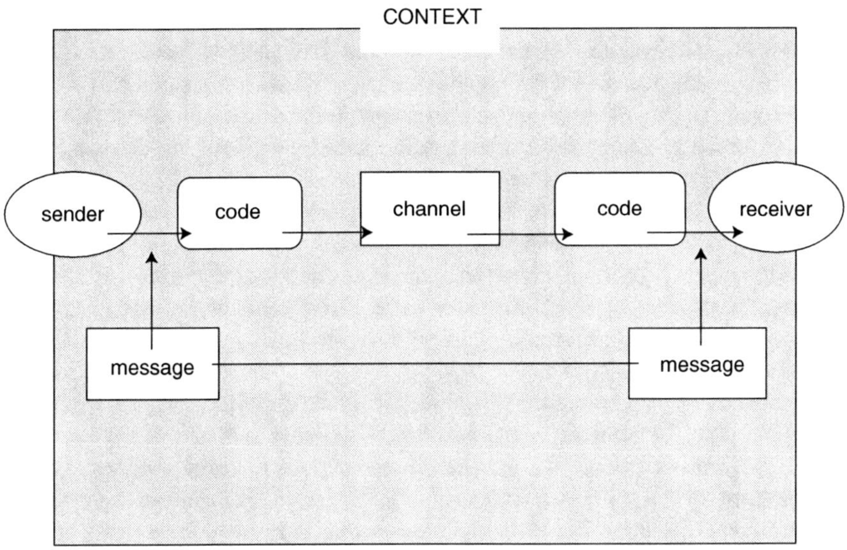

Fig. 12.1 Jakobson communication model (Jakobson, 1960)

Interpreting Traceability

It is necessary to distinguish, at this stage, between different modes of communication evoked by traces. At the end of the 1940s, the American mathematician Claude Shannon proposed a model of the communication process for telephone systems that were being developed at the time (Shannon, 1948).

According to this model, there is communication when a *message* goes from a *sender* to a *receiver*. But, in order for the entire process to take place, it is necessary to have a *context*, in which the message can be grasped; a *code* – regulating the nature of the message – shared by the sender and the receiver; and a physical *channel* that would allow the transmission of the message.

In a 1958 essay, republished in a 1960 collection (Jakobson, 1960), the Russian linguist Roman Jakobson noted that each of the six elements indicated by Shannon performs a specific *function* (Fig. 12.1).

Gradually, an arbitrary reception of Jakobson's scheme had made it an increasingly unsatisfactory model,[3] which assumes an ideal communicative context and can

[3] Stefano Gensini offers the following incisive comment: 'The model of the message is built around a metaphor called "of the channel," or, if you will, around a "postal" vision of communication.... Taken literally the model of the message leads one to assume that signs are a sort of physical container where one can insert whatever content one wishes and then "send it," as one would send a letter to a friend far away: once the envelope is opened, if sender and receiver share the same code, communication succeeds; otherwise it fails" (Gensini, 2002:41, my translation).

be successfully applied only to 'mechanical' communications (as for example an exchange between two computers), where there is no need for the *interpretation* that inevitably characterizes the (verbal and non-verbal) language of human beings.

In other words, the model assumes that messages are constructed according to precise rules and that the relation between expressions and meanings is 'univocal', so that in order to understand a message the receiver has only to associate each expression with the equivalent meaning.

The Italian scholar Umberto Eco elaborated an interpretative semiotics that is arguably the most influential and penetrating analysis of the communicative process. According to Eco: 'Signs are the provisional result of codifying rules that establish transitory correlations in which each element is, so to speak, authorized to combine with another element and form a sign only in given conditions established by the code' (Eco, 1976:73–74).

The novelty of the communicative model introduced by Eco (Fig. 12.2) is that it locates the place of production of meaning in the area between the text and the reader. Differently from the previous model, meaning is not univocally established from the start by the linguistic expression but is rather the result of a process of interpretation, in which various factors have a role, and in which the main actor is the receiver of the message.

In his *The Limits of Interpretation* (Eco, 1994), in re-elaborating the medieval interpretative triad *intentio auctoris/intentio lectoris/intentio operis* Eco defines the intention of the author as *traces*, which are inscribed into the text and prefigure the process of cooperation between author and receiver (*interpretative cooperation*).

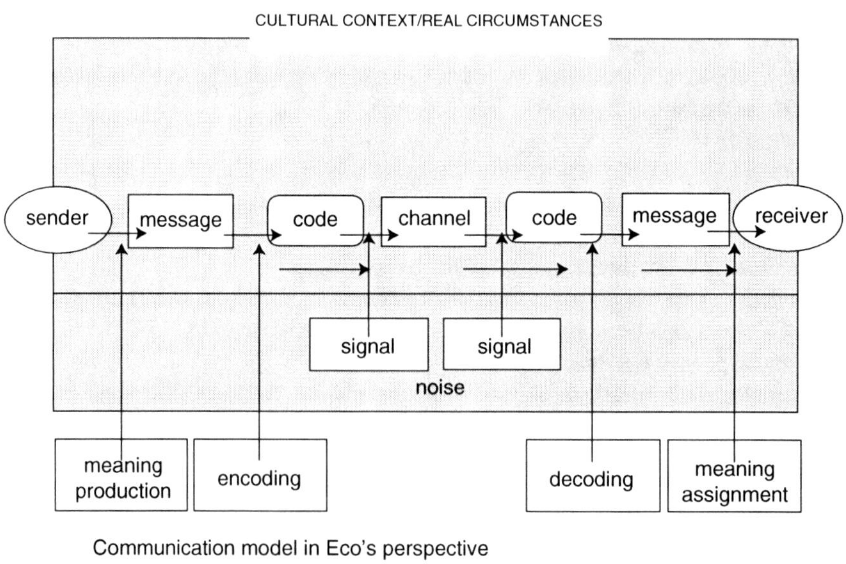

Fig. 12.2 Communication model in Eco's perspective

Any isolated text is therefore incomplete for at least two reasons: because it always requires a grammatical competence on the part of the receiver and, more important, because it is always permeated by a 'not said' (not manifested on the level of expression), which requires active and self-conscious co-operative movements on the part of the reader in order to be actualized on the level of content.

Without the intervention of the reader, the text cannot signify anything: its meaning is not an intrinsic property of the text, being instead located between the text and its possible interpretations. Among the various factors that influence the reader's interpretation of a text, there is one to which Eco attributes special importance: the topic of the text. While reading, the interpreter confers meaning to linguistic expressions and is guided by the goal of finding a consistency within the text.

Without entering into the details of semiotic analysis, it is obvious that Eco's model goes beyond the concept of interpretation as simply the decoding of a message, towards emphasizing the interpretative act. The authority of the reader is such that – in the topical reconfiguration of the narrative world – the reader can choose whether to *interpret* the text (i.e. strive to hypothesize the intentions of the author and respect them) or *use* it (i.e. freely infer meanings).

For example, the presence of vials of blood with names on them leads to very different interpretations depending on whether they are found in a biology lab or in the dungeons of a Transylvanian castle. The 'history' of traces is reconstructed on the basis of 'hypotheses' that are different from direct decoding: nothing can give us the certainty that we are about to meet a disciple of Dracula (or Bela Lugosi) in the second case and not in the first, other than an interpretative process based on cultural as opposed to natural data.

In other words, traceability in its primary meaning does not ensure transparency and unambiguous communication. This emphasizes the importance of the communicative context and of the interpreter.

But what are the elements which, in the practice of communication, could make informed choices possible and effective? And what are the conditions for those elements to be enacted?

The question of communication and choice opens up a new scenario for the figure of the interpreter, which in the case of food consumption is the consumer. The latter is first of all the receiver of a message sent by means of traceability. But he or she is not a passive receiver, who must simply decode a message with a predetermined content. In this case, the communicative act is not aimed at unearthing a univocal meaning (that the sender meant to send to his potential receivers). On the contrary, notwithstanding the goals of the sender, the interpreter must organize his or her knowledge for the purpose of performing an action. Informed choice is not simply an act of knowing, but involves an actual person who interprets, because it orients his or her practical behaviour. Behind informed choices, therefore, lies an interpretative act guided by specific goals.

The most recent studies in consumer psychology tend to go beyond the notion of a purely rational subject who bases his or her choices on purely utilitarian reasons (Siri, 2004). Beliefs, habits, needs – in short, the personality and world view of each consumer is what makes him or her a 'self-aware' subject. Self-awareness is what guides the process of evaluation that precedes purchase.

In this regard, there has been some significant work done in psycholinguistics, which has studied the role in the interpretative-evaluative process of two categories: the representation of the self and the representation of the world.[4] The imagination of consumers also seems 'guided by their view of themselves' (Siri, 2004) and this allows them to identify within the communicative or semiotic aspects of the food chain particular and singular aspects of their individuality (see also Chapter 4 on the four main narrative strategies). It is therefore evident that the fruitful interpretation of a narrative implies much more than organizing textual information in a structure of meaning: it implies a constructive process, a projection of one's meaning into the story. The fruitful interpretation of a narrative implies a review of significant aspects of the self and for this reason can play an important role in the maturation of a personality, allowing it to explore itself and its emotions through affections and by testing the systems of beliefs that give meaning to reality (Levorato, 2000:81).

As one's views of the world and of the self affect the interpretation of traceability, in turn the latter can influence or modify one's views of the world and of the self. When traceability becomes significant it excites the imagination and organizes experience, constructing meanings by providing models of behaviour and of feeling – its interpretation can produce in consumers a change in the perception of reality.

Conclusion

The Italian philosopher Norberto Bobbio, who studied the question of rights extensively, in a now classic work called *The Age of Rigths* (Bobbio, 1995) examines human rights from a socio-historical perspective, stressing the progressive multiplication (not to say proliferation) of rights over time, which occurred in three ways: (a) through an increase in the number of things considered worthy of being protected; (b) through the extension of certain rights to subjects other than humans; (c) by no longer viewing humans as generic entities, or humans in the abstract, but rather considering them in their specific ways of being in society (as infants, old people, sick people).

The generational classification of human rights identifies three generations of rights:

1. *First generation rights or civil and political rights* are pure individual rights, which can be distinguished as *negative rights* (which limit the power of the state with regard to individuals) and *positive* ones (political rights that allow citizens to choose those who govern)

[4] Starting from the early studies of cognitivist psychology up to the more recent ones, psycholinguistics has produced a great deal of work on this topic; among the main studies: Abelson (1979); Bruner (1986); Dunn (1988); Jackendoff (1987).

2. *Second generation rights or social rights* are not, like the former, an answer to threats to individual liberty, but commit the state to removing obstacles to the well-being of individuals and ensuring social protection (fundamental rights such as the *right to work*, the *right to education*)
3. *Third generation rights or the rights of peoples*, which include the right to self-determination, peace, development, the environment

Alongside these three classical rights, there is discussion of a *fourth generation of rights* or *new rights*, which relate to questions of bioethics, the environment and genetic manipulation, and concern the individual in her or his concrete biological individuality. While social rights have mainly a redistributive importance, these new rights concern more immaterial goods (privacy, health, safety and especially information).

Now, if, as Bobbio states, rights are historical, i.e. they develop in response to the new needs of society, what type of social organization of food production is reflected in the rights of consumers to the traceability of food products and informed choice? What type of 'well-being' or 'good life' does this social organization promise citizens?

By obliging producers to be transparent, that is, by forcing them to inform consumers about ingredients and modes of production, ethical traceability prevents producers from competing by economizing on product quality. In other words, traceability, by making information mandatory, shifts competition towards quality of life and therefore towards innovation in quality, which promises an increase in well-being and good life.

But the strategy of transparency must also acknowledge its limits. One cannot delegate to consumers, in the name of informed food choice, the responsibility of choosing in cases where the scientific debate over the safety of certain food products has still to reach a conclusion. It would be immoral to burden individual consumers with the evaluation of risks when scientists have yet to provide an answer. In these cases, the juridical tool for protecting consumer cannot be transparency of information, but only prohibition.

Due to its interpretive character, traceability should be accompanied by reflections on norms, meanings and values underlying the production histories revealed by ethical traceability:

> Democratic society must, so to speak, double itself, and create a symbolical space for meanings, values and norms destined to remain (within certain limits) valid and full of concrete meaning, in the face of possible disconfirmations by actual reality. Saying that it is a *symbolical space*, means acknowledging that it is an impalpable yet decisive reference to something that is not present as an actual fact, that cannot be referred to an actual fact, that does not derive from an actual fact: and it is precisely this movement towards the *elsewhere* or *alterity* of meaning, of value or of the norm, that defines and characterizes the specific nature of democracy. (Ciaramelli, 2003:11)

Traceability becomes more ethical and democratic only if it favours *the informed choices of consumers/citizens*, which do not treat the 'narration of food' as a simple collection of data, but rather as a meeting point for the construction of 'other' stories, capable of narrating, through food, the diversity of people and values.

References

Abelson, R.P. (1979) 'Differences between beliefs and knowledge systems', *Cognitive Science*, 3: 355–366.

Bruner, J. (1986) *Actual Minds, Possible Worlds*. Cambridge: Harvard University Press.

Bobbio, N. (1995) *The Age of Rights*. Cambridge: Polity Press

Ciaramelli, F. (2003) *Lo spazio simbolico della democrazia*. Troina:Città Aperta Edizioni.

Dunn, J. (1988) *The Beginning of Social Understanding*. Cambridge: Harvard University Press.

Eco, U. (1994) *The Limits of Interpretation*. Bloomington: Indiana University Press.

Eco, U. (1976) *A Theory of Semiotics*. Bloomington: Indiana University Press.

Gensini, S. (2002) *Elementi di semiotica*. Roma: Carocci Editore.

Jackendoff, R. (1987) *Consciousness and the Computational Mind*. Cambridge: MIT Press.

Jakobson, R. (1960) 'Linguistics and poetics', pp. 350–77 in T.A. Sebeok (ed.) *Style in language*. Cambridge: MIT Press.

Levorato, M.C. (2000) *Le emozioni della lettura*. Bologna: Il Mulino.

Lévinas, E. (1967) *En découvrant l'existence avec Husserl et Heidegger*. Paris: Vrin.

Mattéi, J.-F. (2003) 'Traçabilité et responsabilité', pp. 35–44 in P. Pedrot (ed.) *Traçabilité et responsabilité*. Paris: Ed.Economica.

Pedrot, P. (ed.) (2003) *Traçabilité et responsabilité*. Paris: Ed.Economica.

Shannon, C.E. (1948) 'A Mathematical Theory of Communication'. *Bell System Technical Journal*, vol. 27, pp. 379–423, 623–656, July, October.

Siri, G. (2004) *Psicologia del consumatore*. Milano: McGraw-Hill.

Part IV
Conclusions and Outlook

This fourth and final part of the book points to ways by which we may take the concept of ethical traceability forward. Crucial to the advancement of ethical traceability is the commitment of all actors along the supply chains, including public policy makers, to maintaining and furthering transparency and to enabling greater public access and so participation in the food system. Participation by the public embraces their identities as both consumers and citizens. Hence, improved communication about ethics is linked to enhanced public participation in the framing of what is to be communicated about food.

The chapter on communicating ethical traceability, which starts this fourth part of the book, reflects on different approaches to the communication of food, and on the degree and nature of public participation incorporated into these different approaches. The authors identify three stages or steps with different degrees of public participation. The first step provides a one-way flow of information from key actors along food supply chains (be they producers, manufacturers, retailers or public regulators). In this fashion information provided about food is carefully controlled and filtered by these gatekeepers along the supply chain before it reaches the public. The second step provides a reciprocal dialogue between food supply chain actors and the public, where the needs of the public are sought and recognized. However, although such consultation does occur in other sectors (such as technology management) such consultation in the food sector occurs infrequently and/or takes place relatively late in the corporate decision-making processes. Consequently, such consultation plays little or no role in the framing of key food production decisions, for example concerning the complexity or nature of a product's ingredients and additives. Instead it is often an essentially reactive exercise by the supply chain actors. Finally, a more imaginative and inclusive approach is based on what are termed 'co-production strategies', whereby the engagement of public concerns is anticipated and integrated at the beginning of the decision-making process on food products, hence incorporating the ethical dimensions that the public might wish to know about the history of the product. It is likely that only a vanguard of interested and active consumers and citizens, or their representatives in civil society organizations, are likely to be engaged in co-production processes. Nonetheless, the authors see these different approaches used in combination as providing for a more open, transparent and accessible food system, with more opportunities for ethical traceability to occur.

The key findings of each of the different streams of research from the book as whole are presented in the concluding chapter, which also includes policy recommendations. Both the ethical or philosophical studies from the project and the three empirical case studies contribute to our understanding of the current workings of traceability systems in food supply chains and the potential and rationale for a greater focus on ethical dimensions in traceability systems. These findings echo the need for a more transparent and open flow of knowledge back and forth along food supply chains. The synopsis of the collective findings from this research project provides a checklist or departure point for the advancement of ethical traceability in the contemporary agri-food sector. The policy recommendations are directed at policy makers in the corporate sector, civil society organizations and public authorities alike. The nature of contemporary food governance is that key decisions and standards-setting reach across corporate, civil society and public sectors – each has a role to play, often with different sectors acting in combination.

Contemporary thinking within the highest levels of the European Commission around the issues of ethics and the traceability of food is provided by Mariann Fischer Boel, European Commissioner for Agriculture and Rural Development, and Margaritis Schinas, Head of Cabinet to Markus Kyprianou, Commissioner for Health and Consumer Protection. The appendix presents transcripts of the speeches they gave at a conference on ethical traceability held at the European Commission in Brussels in September 2006. Both speakers emphasize the EU's actions in promoting the ethical dimensions of food production and choice through different regulatory initiatives and policy instruments. Margaritis Schinas highlights the initiatives by DG SANCO to overhaul and improve food labelling, including nutritional information. He distinguishes traceability and labelling as being different policy instruments with different goals. However, he acknowledges that effective traceability provides the basis for the information that is put on labels, and so is complementary to accurate labelling. Accurate labelling is identified in the speech as a pre-condition 'to enable consumers to make an informed choice'. Similarly, in the area of animal welfare, more access to and exchange of information is a feature of the EU's Action Plan on the Protection and Welfare of Animals.

Commissioner Boel applies food ethics and traceability to key dimensions of food quality in European agricultural production. The recent reforms to the Common Agricultural Policy, in the form of the Single Farm Payment, are characterized as encompassing ethical dimensions of farming, notably through 'high standards of environmentally friendly land management, animal welfare and public health', enforced through the instrument of cross-compliance (cross-compliance is a series of standards that farmers need to meet in order to receive their subsidy payment in full). Regulations around food provenance, notably the use of Geographical Indications and a proposed EU organic production logo, are cited as further examples of ethical traceability being enshrined in European law. However, the Commissioner warns that this degree of regulation around the provenance of food and the processes of its production are opposed by other countries at the level of world trade agreements (a point elaborated in Chapter 3). The institutions of world trade decision-making are one stage further removed from public scrutiny and access. In an era of globalizing food supply, this is a reminder that ethical traceability will have to negotiate its way across multiple levels as well as different sectors of food governance.

Chapter 13
Communicating Ethical Traceability

Volkert Beekman, Christian Coff, Michiel Korthals, and Liesbeth Schipper

Introduction

In the first chapter ethical traceability was defined as '*the ability to trace and map ethical aspects of the food chain by means of recorded identifications*'. The first chapter also emphasized that access to information is a key issue for consumers who are concerned about food production processes. If such consumers are to make informed food choices on the basis of their ethical considerations, it will be necessary to make information on production histories available. This is exactly what can be done by implementing traceability. Ethical traceability therefore needs to entail a communicative endeavour which, in contrast to the legislation currently in force (see Chapter 2), *includes* the consumer. The current chapter addresses how ethical traceability should include strategies for ensuring appropriate information and communication with consumers.

Consumer concerns play a central role in this book's discussions about ethical traceability. Chapter 2 has already indicated that one of the means of addressing these concerns is to be found in better communication about them. It is therefore important to acknowledge that regulatory, corporate and scientific bodies in the agri-food sector have recognized that they should be more accountable and responsive to the public at large, and should involve them in decision-making processes concerning technological applications to food whenever possible (Rowe and Frewer, 2000).

Since broader discussions about public involvement have a direct impact on the issue of communicating ethical traceability, this chapter will start with a short overview of recent discussions about the role of consumer concerns in regulatory and corporate decision-making processes. It begins by discussing the claim that (better) communication with consumers is quintessential for a socially responsible and responsive food production system. Next, three approaches to communication are discussed: firstly one-way information, secondly the use of participatory methods and finally co-production. The final section of the chapter outlines some of the implications for the purpose of communicating ethical traceability.

C. Coff et al. (eds.) *Ethical Traceability and Communicating Food*,

Recent Discussions on Communication Strategies

This section introduces three different types of communication. It starts with a short overview of relevant arguments to emphasize the importance of communicating about food ethical issues with citizens/consumers.

Background: Arguments for Communication

The first argument for improving communication with citizens/consumers about the ethics of food production is purely strategic. Since citizens/consumers will ultimately eat food utilizing or derived from technological innovations, corporations should reveal to consumers the 'true' nature of technological developments and innovations in food production and processing. It is still not uncommon for scientists and corporate representatives to employ this line of argument. The most (in)famous case would of course be Monsanto's communicative approach during the introduction of genetically modified foods in Europe in the 1990s (see, e.g. Chapter 3 for an illustration). Public refusal of this and other food technological innovations would imply that invested resources were utterly wasted. Hence, purely strategic arguments already seem to be enough reason for food companies and scientists alike to pursue early knowledge of public wants, concerns and possible scares. This, however, only warrants a call for a rather limited form of communication: provide citizens/consumers with information that is expected to have a positive impact on public acceptance of a certain food technology and play down any suggestions of dangers and risks.

A second argument for communicating about (the ethical aspects of) food production is a normative one. Since technological developments will always have normative implications after their societal introduction, citizens/consumers as affected parties are entitled to voice their anxieties about these developments. This normative argument builds on the argument that when it comes to matters of ethics and morality, entrepreneurs and scientists are in no position to claim an expert role. Communicative and participatory processes in corporate and regulatory decision-making are thus necessary to ensure that democratic ideals are reflected and acknowledged in the development of food technologies. Since societal values should determine the considerations of corporate and regulatory decision-makers with respect to the impact of their activities on the environment, animal welfare, health and other ethical issues, this normative argument to engage citizens/consumers in deliberations about the priorities in food production chains is compelling (Korthals and Miele, 2004).

A third argument for communicative interaction with citizens/consumers about food production is epistemological. Entrepreneurs and scientific experts tend to be experts in only one field of knowledge, and the integration of knowledge from different fields cannot be left to these experts but should be a matter of public discourse. Citizens/consumers now enter the picture as producers of (integrative) knowledge rather than as sources of moral knowledge. Some scholars even argue that lay people can also be experts on rather specialized issues, such as risk identification and the provision of data about ecological conditions (e.g. Wynne, 1996). However, claims about the epistemological symmetry of lay and expert knowledge should not be exaggerated. To be sure, lay people are quite knowledgeable in certain respects, but they generally lack the methodological rigour of academic science.

A fourth and last argument for communicating about food ethical issues refers to the idea that if consumers are to make autonomous choices on the food market, they should have access to relevant information that enables them to make such choices. This is the informed choice argument.

It should be clear from the preceding chapters that not all of these arguments for communicating about the ethical aspects of food production are relevant to the idea of ethical traceability. The strategic and epistemological arguments will therefore not concern us in the remainder of this chapter, whereas the informed choice argument and the normative argument will be addressed in the one-way information and participation/co-production strategies respectively.

One-Way Information Strategies

One-way information is provided 'top-down'. Three one-way information strategies can be distinguished. The first builds on the idea, popular in risk management of the early 1970s, that citizens/consumers lack expert knowledge and understanding of the issues at hand in food production. This knowledge deficit was considered to be the reason that citizens/consumers arrive at 'wrong' conclusions about food risks. The aim of many studies on risk communication with respect to food in that era was therefore educational. The goal was to develop ways of influencing citizens/consumers to adopt the 'correct' views of scientific experts (Hansen et al., 2003). The reason for informing citizens/consumers about foods is thus purely strategic in this context, namely to create acceptance and avoid criticism (see the first argument in the preceding section).

It is evident that the 'deficit model' of providing one-way information to citizens/consumers adopts strategic arguments to communicate about consumer concerns. The deficit model does not take these consumer concerns very seriously, since consumers/citizens are seen as lacking the capacity to make their own informed decisions. On the contrary, their ignorance calls for education. This strategy hands the provider of information, i.e. companies, regulators or scientists, full authority in deciding what kind of information should be channelled to citizens/consumers. It thus ignores the fact that experts are not a unified community – one expert may frame questions differently from another in a different discipline and so tend to give different information from the other. Hence, the deficit model is not a suitable start for communicating ethical traceability.

Secondly, traceability schemes could be used for a deliberate non-information strategy. This use of traceability currently dominates the agri-food sector. The non-information strategy focuses traceability efforts on food safety and internal supply chain management and control. The verifications of production practices in the food chain that it entails are not intended for communication with citizens/consumers. Rather than giving detailed information about food production processes, this non-information strategy conceals information with symbolic labels and brands. These symbolic signs are not used as an invitation to citizens/consumers to search for information behind the label/brand and thus to trace the (hi)story of food products, but rather as a 'stop sign' (see Chapter 2 for an elaboration of this strategy in the context of the EU's regulation of traceability).

The dominant argument behind this non-informative use of traceability schemes is that traceability does not in and of itself add (ethical) value to food products and production processes. Traceability is thought to be capable only of verifying existing values in the food production chain (USDA, 2004). This argument seems reasonable at first sight. The aim of traceability as a management tool (see Chapters 2 and 11) is indeed to record and verify existing values in the food chain, and as such traceability adds no (ethical) values. This perspective, we can conclude, does not value transparency and communication, which in Chapter 1 were termed procedural concerns around food. Traceability schemes as a non-information strategy are not intended to create transparency and visibility for citizens/consumers: the use of recorded data is purposely restricted to entities within the food chain. This second one-way information strategy is not suitable for ethical traceability as a communicative tool.

Thirdly, one-way information may be provided in order to enable citizens/consumers to make better, i.e. informed, choices (this strategy thus builds on the informed choice argument in the previous section). Examples of this strategy are:

- Labels providing information about ethical issues
- Web pages providing information about ethical issues
- Brochures providing information about ethical issues
- Other information about ethical issues as provided by producers, NGOs, the media, retailers, wholesalers and food authorities

The one-way flow of information in this strategy is shown in Fig. 13.1.

Within this strategy information ultimately originates from actors in the supply chain and third parties such as food authorities. This one-way information strategy is suitable for communicating ethical traceability but entails some pitfalls:

- The provider is in full charge of the information flow
- The information is often but not necessarily used for marketing purposes

The role of traceability schemes in this strategy is to verify and legitimize labels and brands and thus to assure trustworthiness of labels, products and companies. The strategy relies on the assumption that consumer trust will be enhanced by the implementation of traceability schemes by companies. Good reasons, however, exist to question this assumption. First, it can be argued that as long as traceability

Producers, retailers, wholesalers, food authorities
⇓
Information
⇓
Consumers

Fig. 13.1 One-way flow of information in the food supply chain

and verification schemes are controlled by the companies themselves and without independent audit and control, the likeliness of enhancing consumer trust is small. Labels derived from civil society organizations, such as FairTrade, illustrate the potential for trust-based alternatives to corporate brands to find acceptance among consumers. However, fair trade also needs to provide a clear auditing and accreditation process to support its integrity. Communicating a certain amount of information to citizens/consumers about the process of establishing traceability schemes is imperative to enhance consumer trust. Second, trust relations are bolstered by a certain level of openness and transparency.[1] Trust can never be built on a complete renunciation of information. Something must trigger the trust relation. In the case of food ethical traceability, disclosure of some traceability information can be such a factor. Traceability as a verification scheme thus needs to be combined with a policy of openness, transparency and communication to build trust.

A general pitfall for all one-way information strategies is that they communicate about (ethical) values in the food chain as if these were static and unchangeable values indifferent to perceptions among consumers and other stakeholders. Values, however, are hardly ever that static in real life, as was argued in Chapter 11 with the concept of 'Ethical Room for Manoeuvre'.

Participatory Strategies

Apart from one-way information strategies, various other approaches for public engagement exist that entail reciprocal communication processes in corporate and regulatory decision-making. Generating input from citizens/consumers is the main objective of these participatory strategies and also what differentiates them from the previously discussed one-way information strategies (Rowe and Frewer, 2000). Participatory strategies can be divided into three types: feedback, consultation and participation (Folbert et al., 2003). These three different types differ in the stage of influence (early or late) and in the degree of deliberation (debate, discussion).

[1] On the other hand Luhmann (1968) makes the observation that it only makes sense to talk about trust in situations where one does not have all information, since if one had all information no trust would be needed.

Feedback is a type of participation within which citizens/consumers are asked for their opinions without having the opportunity to discuss or to debate. Consultation does give room for discussion and dialogue, not only between the decision-makers and citizens/consumers but also among citizens/consumers themselves. Participation, finally, asks citizens/consumers for their views and for their advice, thus giving them ample room for discussion and input. The difference from consultation is that with participation citizens/consumers can introduce their own definitions of problems and possible solutions (see Beekman et al., 2006 for a more extensive overview of participatory strategies).

Feedback

The first type of participation is feedback, which is as one-directional as the one-way information strategy but in the opposite direction – communication goes from consumers to food chains. Two methods stand out as commonly used techniques to get feedback from citizens/consumers. The first method – focus group discussions – is regularly used in a corporate setting, whereas the second method – referendum – is normally confined to the political domain.

Focus group – A qualitative interview format in which a small group (typically 5–12 persons) is gathered and guided through a structured discussion.

Referendum – A vote on a specific issue, involving all affected citizens in a region or nation.

Consultation

The second type of participation is consultation, which unlike feedback seeks to establish a dialogue between citizens/consumers and food chains, but unlike participation does not do this at an early stage of decision-making. Several methods are available to stimulate this dialogue, some of them more apt to be applied in a corporate setting than others.

Stakeholder dialogue – Interaction between corporations and their stakeholders that can take many forms. Stakeholder workshops are popular. Stakeholder processes are, however, often poorly managed and the expectations are often poorly defined and unrealistic (Grimble and Man-kwun, 1995; Payne and Raiborn, 2001).

Future/scenario workshop – A common format is that participants are recruited representing various stakeholders and discussions focus on a social issue/problem. Participants are guided through a structured debate in three phases. In the first phase participants are allowed to criticize anything related to the issue, without being contradicted. In the second phase visions about the issue in question are formulated, without paying respect to barriers. In the third phase strategies for realizing the visions are discussed. In a scenario workshop, moreover, the participants are presented with different scenarios for the issue at stake.

Public hearing – A widespread and common means of consultation, where the public is invited to participate in the decision-making process, either at public meetings or through a call for (written) comments.

It is important to note that these consultative methods run the risk of being used only in a strategic way. This is an important consideration that should be taken into account when organizing or analyzing consultation processes with the aim of including citizens/consumers in discussions about ethical issues. Besides the general consultative methods already mentioned, specific tools have been developed to facilitate communication about the ethical aspects of food production and consumption. The following tools may be applied to flesh out the processes of consultation:

Delphi – The Delphi method was designed to bring the combined knowledge and abilities of a diverse group of experts to the task of quantifying variables that are either intangible or shrouded in uncertainty. It is structured around the notion of a virtual committee, where the exchange of ideas is conducted remotely through a series of opinion exchanges (in the form of rounds). Anonymity of the participants is central to the process. This feature aims to eliminate external power relations and personal influences that may interfere in the debate of key issues.

Ethical matrix – The ethical matrix is a principle-based methodology that aims to guide decision-making by appealing to principles based in both deontological and consequentialist ethical theories (Kaiser et al., 2007).

Multi-criteria methodologies – The framework for using a multi-criteria methodology in technology assessment was developed as an alternative to consensus-oriented deliberative methods and economic models of valuation and risk/cost-benefit assessments.

Participation

The third and final type of participation is participation in a more narrow sense of the word and characterized by a two-way dialogue in an early stage of decision-making. The most famous participatory method is the consensus conference, albeit mainly employed in a governmental and not in a corporate context.

Consensus conference – The basic idea of a consensus conference is to give lay people a voice in the political process by selecting a panel of lay people (12–15 persons), who are given the power to set the agenda in a pending (often techno-scientific) controversy. At the end of the conference the lay panel produces a document presenting their consensus on the issue at hand.

Co-Production Strategies

The final communication strategy to be addressed in this chapter has been termed 'co-production'. This strategy is based on dialogue and a co-operative style, and characterized by a common policy among different social actors. Examples of co-production are Korthals's concept of 'ethical room for manoeuvre' (see Chapter 11) and the better-known concept of 'co-evolution' that stems from the political sciences (see: Grin et al., 2004).

As discussed in Chapter 11, Callon and Latour (1992) originally introduced the concept of co-production as co-production of nature and society in science and technology. Jasanoff (1996) has later used the term 'co-production' in a more general sense by transcending the context of science and replacing science with knowledge: '*co-production is the simultaneous production of knowledge and social order*'. Co-production can perform several functions, either simultaneously or consecutively. The first is securing the trust of consumers. The second is fact and norm finding. The third is guaranteeing accountability and responsibility of agencies in charge of branding, labelling and assurance schemes (Dryzek, 2000). Habermas's (1984) idea of communicative action lies at the core of co-production, but co-production does not rely on his more abstract and controversial account of justifying moral judgements. Co-production does not aim at justification but at the discursive application of moral judgements to complex and uncertain (knowledge) problems. Co-production is about involving citizens/consumers or their representatives in:

- Labelling and labelling strategies
- Information and communication strategies
- Overall mission, strategy and practices of companies and chains
- Resource decision-making (what sources to use and how)
- Innovatory decision-making (upstream in the innovation process)
- Certification schemes (like fair and ethical trade, sustainable production) with respect to developing these schemes and in monitoring and evaluation

Implications for Communicating Ethical Traceability

This section asks which of the communication strategies discussed above are relevant in the context of ethical traceability. It therefore provides an evaluation of these strategies and looks at the arguments that underpin the various strategies. Table 13.1 summarizes the communication strategies that were discussed in the previous sections.

Table 13.1 Overview of communication strategies (Modified from Folbert et al., 2003)

		Participation			
	Information	Feedback	Consultation	Participation	Co-production
Style of governance	Authoritarian	Consultative	Consultative	Participatory	Co-operative
Role of participants	Receiving information	Providing advice	Providing advice	Providing advice	Being partner
Input of participants	Listening	Voicing ideas and opinions	Voicing ideas and opinions	Voicing ideas and opinions	Voicing and acting
Influence by participants	None	Opinion-formation	Opinion-formation	Opinion-formation	Decision-making
Stage of influence	None	Ex-post	Ex-post	Ex-ante	Ex-ante
Interaction and deliberation	Monologue towards consumers/ citizens	Monologue towards decision-makers	Dialogue between consumers/ citizens and decision-makers	Multilogue between onsumers/ citizens, stakeholders and decision-makers	Multilogue between consumers/ citizens, stakeholders and decision-makers

One-Way Information Strategies

Among the one-way information strategies discussed – the deficit model, the non-information strategy and the informed choice strategy – only the informed choice strategy was found useful for communicating ethical traceability. The underlying reason is – as the name of the strategy already indicates – the informed choice argument.

A pitfall of this communication strategy is that it misses communicative opportunities building on the ability of citizens/consumers to talk back to information-providers and actors in the food supply chain. Furthermore, since providers are in charge of the information flow in one-way information strategies, these strategies do not allow citizens/consumers to express the kind of information-need they have. Consumers in European countries are constantly bombarded by information that does not interest them; lots of information flows do not attract their attention because language, pictures and topics do not connect with their own interests, values and preferences. Consumers with small children, for example, very often have different ethical and cognitive concerns compared to elderly or young people. This strategy also does not facilitate any reciprocal communication between consumers and producers. Finally, one-way information strategies deal with values in the food chain as if these were static and unchangeable.

This means that communicating ethical traceability exclusively through one-way information strategies fails to grasp the procedural and democratic potential of ethical traceability schemes. It also lacks the flexibility and capacity of more participatory strategies to improve the understanding of (regularly contested) ethical issues and values. It does not acknowledge the ability of communicative strategies to enhance understanding between involved actors. This implies that an important – democratic – reason for communication with citizens/consumers cannot be fulfilled by using only one-way information strategies.

Furthermore, when informed choice is used as a strategy, one should be careful not to use it as a pure marketing strategy. This is not as straightforward as it may seem. In practice it is often hard to distinguish between information strategies and marketing strategies. It must be clear that the information is trustworthy, and the origin of the information must also be made clear. Finally, the idea that consumers have a right not to be misinformed should be given due consideration in relation to the idea of informed choice (see also Chapter 9).

Participatory Strategies

Participatory strategies provide the opportunity for citizens/consumers not only to be informed but also to inform other parties in and around the food chain. Such strategies offer citizens/consumers the chance to give feedback, to be consulted in a dialogue and even to participate on an equal standing with other stakeholders. They make it possible to tailor the information provided by producers to suit the special needs and preferences of consumers, because the strategies give consumers the opportunity to communicate with all partners about their own preferences. Participatory strategies make it possible to produce tailor-made information to suit the questions, issues, interests and values of specific, concerned groups.

When communicating ethical traceability is understood in a participatory way, procedural values such as openness and transparency concerning the production history of a food product become pivotal. Ethical traceability may then be used as a means to add values along the food chain or, to put it the other way around, to develop (new) values among citizens/consumers as they come to learn and understand more about the rationale and benefits of specific modern food production practices. Ethical traceability is then not only a means to achieve verification; it becomes an end and value in itself by opening up substantive issues for discussion and deliberation.

Participatory strategies have pitfalls as well. Again, they can be used to legitimize a (new) technology rather than start an open discussion: (1) Feedback may be asked for pure marketing purposes only; (2) consultation and participation may have the hidden agenda of seeking public acceptance for a new technology such as genetic modification (GM). Furthermore, participatory strategies can be too solution-oriented and thus have a tendency to ignore the changing and dilemmatic nature of consumer concerns. Thirdly, participatory strategies require time, energy and activity on the part of the participating consumer/citizen, and not everyone is willing or able to make these investments in the process. Finally, the issue of 'who participates' is crucial; very often the consumers most affected by an ethical decision are not included in the participatory process. One runs the risk that only people who have outspoken ideas but no real involvement participate.

The quality of participatory strategies thus depends (not unlike the one-way information strategies) on (the absence of) a true commitment to the following democratic ideals:

- Transparency
- Openness and inclusiveness
- Treating consumers as autonomous individuals with certain rights
- Deliberation

Co-Production Strategies

The last strategy, co-production, demands a high degree of actor involvement, not least on the part of consumers. Among the benefits is the high degree of understanding of the issues at stake that such a strategy would provide. This would entail a better understanding of the compromises that are necessary in almost all kinds of food production. A compromise has, for instance, to be made between animal welfare and price. Perfect animal welfare (whatever this is!) would probably lead to very high prices and, conversely, very low prices would be the outcome of very poor animal welfare standards. Such understanding would help consumers make deliberate and informed choices on a very high level. The profound understanding of production practices and decisions also implies a high level of embeddedness on the part of actors in the food chain. As the outcome of co-production is a common policy, strong links develop between the parties involved.

The disadvantages of co-production are that it can be time-consuming, merely window dressing and exclusive instead of inclusive. With respect to the first disadvantage, it takes more time to include more people, because there is a higher risk of disagreements. However, this could pay off in the long run, because fundamental objections to a certain type of production will find a way to become public anyway. The second disadvantage is not unique to co-production: it takes time for a company or certified food chain to show that it does indeed take consumer concerns seriously. With respect to the third disadvantage, it can be expected that only dedicated individuals among consumers and producers will be willing to participate in such processes, which means that others are excluded and can feel neglected. Only a careful selection process can help here.

Conclusion

In conclusion, communicating ethical traceability should be anticipated as a three-step process that includes the following three approaches:

1. Providing sound information to citizens/consumers (one-way information strategy on the basis of the informed choice argument)
2. Establishing a reciprocal dialogue with citizens/consumers (participatory strategy on the basis of the normative argument)
3. Establishing a deeper engagement between dedicated citizens/consumers and producers (co-production strategy on the basis of the normative argument)

On its own, the first strategy runs the risks that the information given will fail to interest some or most consumers, because it will not connect with their own ethical and cognitive information preferences. Consumers nowadays are subjected to an overload of information that simply does not tell them anything, because the information totally neglects their own ethical and information preferences. The second strategy looks more promising with respect to communicating issues of ethical traceability, because it takes the two-way information process seriously, and thereby recognizes the specific information needs of different consumers. However, this strategy runs the risk that the communication will take place at a late stage in corporate decision-making processes, on the basis of definitions of problems that have not been subjected to critical scrutiny by citizens/consumers. The third strategy looks more promising with respect to the joint development of (the premises for) ethical traceability schemes. However, this strategy is probably too demanding to count on the engagement of a large number of producers and citizens/consumers. The three approaches together do, in our opinion, offer a new and fascinating strategy for involving consumers in the methods and decisions of the food supply chain.

References

Beekman, V., E. de Bakker, H. Baranzke, O. Baune, M. Deblonde, E-M. Forsberg, R. de Graaff, H-W. Ingensiep, J. Lassen, B. Mepham, A. Porsborg Nielsen, S. Tomkins, E. Thorstensen, K. Millar, B. Skorupinski, F. Brom, M. Kaiser, P. Sandoe (2006) *Ethical bio-Technology Assessment Tools for Agriculture and Food Production. Final Report Ethical Bio-TA Tools (QLG6-CT-2002-02594).* The Hague: LEI.

Callon, M., B. Latour (1992) 'Don't throw the baby out with the bath school! A reply to Collins and Yearley', pp. 343–68 in Pickering (ed.) *Science as Practice and Culture*. Chicago: University of Chicago Press.

Dryzek, J. (2000) *Deliberative Democracy*. Oxford: Oxford University Press.

Folbert, J.P., V. Beekman, J.C. Dagevos (2003) *Praten met het publiek; Ontwikkelingen op het terrein van publieksraadpleging in het levensmiddelenbeleid [Talking with the public; Developments in public consultation in food policy]*. The Hague: LEI.

Food Traceability Report, April 2006, Volume 6, Number 4.

Grimble, R., C. Man-kwun (1995) 'Stakeholder analysis for natural resource management in developing countries: Some practical guidelines for making management more participatory and effective'. *Natural Resources Forum*, 19/2: 113–124.

Grin, J., F. Felix, B. Bos, S. Spielstra (2004) 'Practices for reflexive design: Lessons from a Dutch programme on sustainable agriculture'. *International Journal of Foresight and Innovations Studies*, 1/1–2: 126–149.

Grunwald, A., G. Banse, C. Coenen, L. Hennen (2005) *Internet and democracy. Analysis of Network-Based Communication from Cultural Aspects. TAB Working Report No. 100.* www.tab.fzk.de

Habermas, J. (1984) *The Theory of Communicative Action: Reason and Rationalization of Society*. Cambridge: Polity Press.

Hansen, J., L. Holm, L. Frewer, P. Robinson, P. Sandoe (2003) 'Beyond the knowledge deficit: Recent research into lay and expert attitudes to food risks'. *Appetite*, 41: 111–121.

Hardt, M., A. Negri (2000) *Empire*. New York: Harvard University Press.

Jasanoff, S. (1996) 'Beyond epistemology: Relativism and engagement in the politics of science'. *Social Studies of Science*, 26/2: 393–418.

Kaiser, M., K. Millar, E. Thorstensen, S. Tomkins (2007) 'Developing the ethical matrix as a decision-support framework: GM fish as a case study'. *Journal of Agricultural and Environmental Ethics*, 20/1: 65–80.

Korthals, M., M. Miele (2004) 'Trust as communication on ethical dilemmas: Possibilities and commitments of consumers of animal friendly products'. *EurSafe 2004 Science, Ethics and Society*, 98–100.

Luhmann, N. (1968) *Trust. A Mechanism for the Reduction of Social Complexity*. Chichester: Wiley.

Lukawetz, G. (1999) 'Online-communities als Demokratisierungsagenten und Marketingparadigma'. Deutsche Gesellschaft für Online-Forschung. www.gor.de/gor99/programm.html

Nielsen, H.K. (2006) 'Internettet og den demokratiske offentlighed' [The Internet and democratic publicity]. Working Paper Centre for Digital Æstetik Forskning Aarhus University.

Payne, D.M., C.A. Raiborn (2001) 'Sustainable development: The ethics support the economics'. *Journal of Business Ethics*, 32/2: 157–168.

Rowe, G., L.J. Frewer (2000) 'Public participation methods: A framework for evaluation'. *Science, Technology and Human Values*, 25/1: 3–29.

USDA (2004) *Traceability in the U.S. food supply: Economic theory and industry studies*. United States Department of Agriculture, Agricultural Economic Report Number 830.

Wynne, B. (1996) 'May the sheep safely graze? A reflexive view of the expert-lay knowledge divide', S. Lash et al. (eds.) *Risk, Environment and Modernity: Towards a New Ecology*. London: Sage.

Annex 1: Enabling Consumer Involvement Through Information and Communication Technologies

A new way of informing and communicating on traceability, different from the methods presented above but drawing on some of their elements, is emerging from developments within Information and Communication Technologies (ICT). Emergent ICT is often characterized by its 'many-to-many' communication capabilities and thus seems to enable new forms of 'participatory democracy'. It is often promoted as the new hope for democracy (Lukawetz, 1999). This idea is based on two characteristics of modern ICT: (1) All kinds of information is made accessible to all citizens/consumers. (2) All citizens/consumers can raise their voices by feedback mechanisms. Democracy can thus be radicalized, in the sense that it is based on transparency and open to all for participation.

It is tempting to see in the apparent ease of the new communication technologies, and the many possibilities they offer for consumer information and communication, an opportunity for enlarging and widening participatory strategies around food. However, since these technologies are still in their infancy, there are only a few experiences of how such systems can be used by consumers for making informed food choices.

For example, different kinds of barcodes or Radio Frequency Identification (RFID) on food packages can be used to indicate unique serial identity numbers (numbers that identify food). These can be transmitted to a computer, a cell phone or a personal digital assistant (PDA, a small, mobile computer that can work either online or offline). The computer, cell phone or PDA then uses the unique identity number to link to specific services on the Internet, which provide information about the product to the consumer. Depending on how the systems are set up, information in the system can theoretically be provided by everyone (producers, retailers, NGOs, researchers, public institutions, private persons).

Such systems are still at the experimental stage, but are being developed. For example, in 2006, 25 supermarkets in and around Tokyo made traceability information available to consumers. Tags with large amounts of information were attached to food packages. Consumers accessing information via their mobile phones could learn about what chemical inputs had been used, when the vegetables were harvested, and who brought them to market. Consumers could also talk into the system to give feedback. It turned out that consumers also wanted to receive information about soil composition, organic (versus synthetic) fertilizer content, the use of pesticides and herbicides and even the name of the farm where the product was grown (Food Traceability Report, 2006).

ICT does offer new spaces for information sharing and hence for recognition between consumers and food producers/processors. It is gradually becoming a new room for social practices, providing new opportunities for information, communication and participation. At the same time, this new informative and communicative space is embedded in the three traditional forms of social institutionalization, i.e. market, state and civil society (Nielsen, 2006; see also Chapter 3 for a discussion of these institutions and food governance). There is a need to reflect more critically on the role of ICT in communicating ethical traceability in order to identify the potential risks and benefits of using ICT for informed food choice.

The question is whether ICT will be able to live up to its promises and possibilities, and so offer a many-to-many communication tool providing new opportunities for a liberal democracy to reach all citizens and respect the idea of open deliberation. How will citizens/consumers make use of ICT to relate to society and other human beings? Will it be used to promote self-interest in a new 'communicative capitalism' (Hardt and Negri, 2000) or will it be seen as a means to consider society and the common good in a broader context, recognizing the autonomy and well-being of others?

According to Nielsen (2006) this depends on how and for what purpose ICT is used in the future. If it is dominated by commercial uses, the chances of improving deliberative democracy and bottom-up democracy will be poor. If, however, it is used for non-commercial purposes, the possibilities for developing new ways of deliberation and public discussion will be tremendous. Such developments depend to a large extent on the possibility of using ICT not only as a source of one-way information but also as a means of becoming an active participant with the chance to make a difference concerning the common good.

At present the use of ICT for commercial purposes, such as entertainment, is by far the most widespread, while the potential for developing interactive communication, such as Internet forums for discussion, has not been developed to any great extent and certainly not enough to serve democratic and deliberative purposes (Nielsen, 2006; Grunwald et al., 2005). According to Nielsen the development of interactive and democratic communication structures is not something that is created automatically; it is not something that somebody will offer to Internet users but rather something that must be created by the participants themselves. Thus one should acknowledge that pure user- or consumption-oriented perspectives will not develop communication or the 'C' of ICT.

Even with high involvement by participants, the development and use of ICT face many new challenges. A first challenge concerning the use of ICT for informed food choice relates to the form of knowledge and how it is communicated. ICT can provide intellectual – and we could even say 'virtual' – information to consumers/citizens. We should ask what the consequences are of consumers having no direct contact with the food chain. That is, no direct contact with real life and real people, but only a mediated and virtual contact. Does this separation imply less involvement by consumers and hence less ethical concern? What does the lack of sensuous and face-to-face interaction mean for consumer involvement? A second challenge concerns the issue of wider public deliberation, and its importance for democracy, and the more individualized approach of the dominant ICT strategy. Communication, even if it is two-way and provides opportunities for feedback from consumers to producers, seems to favour more individualized communication and food choice, partly because of the limited possibilities for deliberation. ICT solutions seem to consider responsibility at the individual level and not as an outcome of democratic deliberation and critical public reflection. The final challenge contradicts the previous challenges, in a sense, in that it argues that if ICT lets more people express themselves and make their opinions public, then there will be less time to listen. In short, there will be more 'voice' than 'ear'.

In conclusion it is evident that ICT offers new possibilities for making information public as well as new ways of communication with consumers that seem to fit the idea of communicating ethical traceability. However, many questions concerning ICT remain unsolved. ICT will most likely offer new communication tools that can be used both for the provision of one-way consumer information and for feedback, and perhaps also new kinds of participation.

Chapter 14
Conclusions and Policy Options

Christian Coff, David Barling, and Michiel Korthals

Ethical traceability was defined in the first chapter as 'the ability to trace and map ethical aspects of the food chain by means of recorded identifications'. Traceability is currently mainly used for a range of sometimes overlapping purposes, notably for the management of food safety and public health recall, control and verification, supply chain management and for assurance schemes around food quality and provenance. Only in the latter cases is traceability communicated to consumers; these cases therefore constitute an exception to the general intention of EU Food Law and Codex Alimentarius, which do not consider it appropriate to extend the principle of traceability as far as consumers (as shown in Chapters 2 and 3).

The common idea throughout all the studies presented in this book has been that traceability, as used to map the production history of food products, can also be used to map ethical issues in the food chain that relate to production practices.[1] Equally important from the outset was the question of how ethical traceability could be used for enabling informed food choice. By reconstructing or mapping the production history of foods - as can be done by means of traceability schemes - consumers are empowered to become part of a communicative process concerning food production practices. The analysis in Chapter 4 shows that traceability, i.e. the history of the product, is in fact already used in food advertisements as a commercial communication strategy. In this use of traceability, in advertisements, telling the story of a product ascribes identity to the product, which in turn allows disembedded modern consumers to reconnect with food in a 'non-superficial', cultural and ethical manner.

The users and beneficiaries of this ethical traceability information about food are primarily consumers. The major ethical concerns of consumers are therefore outlined in Chapter 1. Consumers are *substantively concerned* about health, animal welfare, the environment, terms of trade, working conditions, etc. Consumers have *procedural concerns* about the reliability of information, transparency, voice, participation, etc. This analytical distinction is important as it calls our attention to the fact that different measures must be taken in order to address the concerns properly.

[1] Food security and food safety are also ethical by nature but they are not the issue of this book.

C. Coff et al. (eds.) *Ethical Traceability and Communicating Food*,

However, as the sociological case studies show, information on ethical issues relating to production practices could be of equal importance for actors such as farmers, processors and retailers in the food supply chain, as it reveals 'value-laden' qualities that contribute to the identity of the food product. Therefore, ethical traceability is in the interests not only of consumers but also of actors in the food supply chain.

This book presents many different perspectives on ethical traceability and informed food choice. Firstly, these include the perspectives of different actors in the food supply chain (such as farmers, processors, retailers and consumers), revealing their different views and expectations. Secondly, they cover the perspectives of different academic disciplines (such as political science, practical philosophy, sociology, jurisdiction, science and technology studies and even embracing reflections on management and advertising by private corporations). The diversity of perspectives illustrates the complexity of the issue at hand. The multiplicity of views and attitudes shows that many perspectives will need to be considered in framing and forming the future of ethical traceability and informed food choice. The multiplicity of considerations presented here makes it clear that ethical traceability is indeed a controversial term and that there is widespread disagreement on how and for what purpose it could be implemented. However, it is likely that ethical traceability will be negotiated and developed further in the future, given its potential for communicating food and ethics. It is an opportunity to open up the food supply chain to the eyes both of consumers and of actors within the food chain, and hence to increase visibility and transparency.

The question then arises of how all these different perspectives match each other. Is there a thread running through the chapters which in the end ties them together and makes us able to present a common and consistent reflection on ethical traceability that can be used for future deliberation on the issue?

In order to answer these questions, we set out in this last chapter to summarize the main conclusions from the preceding chapters and to consider how the different approaches can be brought together into a (necessarily nuanced and complex) 'blueprint' for ethical traceability. In this sense, this final chapter presents results and conclusions that are the result of a common effort from all the contributors to this book, not just those of the named authors of this chapter. Finally, towards the end of the chapter, we present some policy options on ethical traceability and informed food choice that have emerged during the writing of the book.

Findings from the Food Supply Chain Case Studies

The sociological studies investigated ethical concerns in three important supply chains. The main findings of the three case studies (presented in detail in Part II) are summarized below.

The Danish Bacon Supply Chain

The major ethical concern in the bacon supply chain was found to be animal welfare. Concerns arose from breeding to fork, and included, for example, the transport of living animals, the type of stable systems (fixation) used, and the use of growth promoting substances. In general, the retailers said they found the idea of ethical traceability interesting but expressed scepticism about consumers' willingness to pay for advanced traceability systems. However, many of the consumers' ethical concerns are already addressed by traceability systems, but the information is not used actively or proactively in relation to consumers. Consumers felt that they needed more information about bacon production, especially about some of the invisible attributes, such as origin, use of medicine or animal welfare. From the stakeholders' point of view, quality was the most frequently mentioned concern, and here quality was mostly understood as a dimension of food safety, specifically the absence of disease-causing bacteria.

The Wheat-Bread Supply Chain

Ethical concerns were perceived to be increasingly important in this chain. Although many interviewees said that the ethical concerns were too subjective to be standardized, there was strong agreement about which concerns were most relevant to the chain, namely human health and methods of production. For stakeholders, terms of trade was also important. Transparency and trust were discussed at length, though not so often identified as important. The least important concerns in this chain were animal welfare and voice. Stakeholders said that traceability was limited by the routine practice of blending wheat from different sources for convenient handling and to manipulate quality and cost. Thus, traceability back to a single farm is often impossible. A response to this challenge has been to set quality standards (by means of assurance schemes) for the whole UK supply base, to reduce risk. However, examples were found where traceability back to farm was supplied, at extra cost, by means of 'Identity Preservation', in which consignments with specific characteristics are kept separate throughout the chain. Existing regulations and assurance schemes cover several of the ethical concerns discussed in this book. Producers in the chain felt well supplied with information. Consumers, by contrast, felt that information was withheld and unreliable. Traceability schemes were not necessarily seen to be linked with ethical practice, which might be addressed by, e.g. codes of practice. In general, ethical concerns were more actively communicated by craft-scale rather than industrial-scale operations.

The Olive Oil Supply Chain

The dominant ethical concerns in the olive oil supply chain were trust and transparency. Ethical concerns were more relevant to the smaller-scale and organic stakeholders, who valued different types of olive oil and food culture. Traceability and potentially ethical traceability in the olive oil chain were widely said to be limited by the practice of blending oil by olive mills and packing houses, in order to manipulate quality and cost or for convenience. Education of both stakeholders and consumers in the olive oil chain is important to improve traceability and ethical traceability in the chain.

It is striking that the relative importance of various food-related ethical concerns varies among the chains. It is hardly surprising that animal welfare is not a major concern in the wheat-bread or olive oil production chain. But there may be deeper differences. In general, substantive ethical concerns dominate over procedural concerns (i.e. emphasis is on animal welfare or terms of trade rather than on visibility, transparency and voice). If we look more closely at this finding, we see that voice seems to have particularly low priority for consumers in the UK and Greece, whereas this is not the case in Denmark. Such differences may be due to different political cultures; for instance, it has been shown that some countries emphasize representative democracy while others adhere to a more procedural and deliberative notion of democracy.[2] In the former, voice is implemented by voting; in the latter, other procedures targeted at including citizen opinions are welcomed, such as consensus conferences and the like.

We have summarized the empirical findings in a generalized form in Table 14.1.

Table 14.1 Key findings from the sociological case studies

Key findings of case studies
1. Ethical issues arise in any food chain. Ethical traceability – the ability to trace and map ethical aspects of the food chain by means of recorded identifications – is a response to emerging ethical issues. The form it takes is affected by the mode of production (e.g. artisanal or industrial) and by the length and complexity of the chain.
2. The depth and range of ethical aspects of the '(hi)story' of foods are hidden. However, assurance schemes do exist that tell aspects of those stories. Ethical traceability aims to create fuller visibility along food chains.
3. Some relevant information about ethical issues already exists in the chain but is not being communicated. The potential for opening up this information is considerable.
4. The issues being ethically traced are at present being defined and constrained by commercial realities and regulatory rules, which are subject to public and private pressures.
5. Actors in the food chain prioritize and interpret ethical concerns differently depending on their place in the chain, having varying 'fields of ethical vision'.

[2] See for instance Porsborg, A., J. Lassen, P. Sandøe (2007) 'Democracy at its Best? The Consensus Conference in a Cross-national Perspective', *Journal of Agricultural and Environmental Ethics* 20(1): 13–35.

Conclusions from Philosophical Investigations

The philosophical investigations (detailed in Part III) have as their main aims first, to investigate ethical traceability as a concept and second, to connect this concept with key concepts of ethical and political philosophical literature, such as informed choice, public and private, civil society, government, market and (individual) responsibility. The following summarizes the main conclusions of the philosophical investigations.

Consumers' ethical concerns reflect their quest for ethical information on food production practices. However, informed food choice requires food production processes to be visible and transparent. This can be achieved by reconstructing and mapping the production histories of foods. Traceability schemes are a suitable tool for such purpose. Improving visibility and the possibilities for recognition between actors in the food chain also enables consumers and other actors to restore and develop more meaningful relations with food.

However, implementing ethical traceability and informed food choice is not as straightforward as this implies. Both ethical traceability and informed food choice are concepts that challenge the traditional divide between the private and public sphere – and between consumer and citizen. They both bridge the public–private divide: state, civil society and market structures are all involved and their contours are reshaped. From a traditional point of view, this makes ethical traceability and informed food choice highly contested concepts. However, they do tackle important societal challenges emerging from the dynamics of modern and post-modern food production processes.

Ethical traceability can be instrumental in preventing the communication of misleading and deceptive information to consumers. In complex chains, ethical traceability schemes can be helpful in establishing trustworthy information. In addition, ethical traceability can be used to ensure that consumers are provided with foods that respect some minimum standards of, for instance, animal welfare, sustainability or fair trade, as supported by a contingent overlapping consensus of food values in the population. The issues that fall within the overlapping consensus are suitable for governmental responsibility; but market partners can always do more than respect minimum food standards, such as with animal welfare.

It is argued that some consumers have an interest in traceability schemes which require ethical discussions of values in the food chain and which include them as participants. Ethical traceability schemes therefore have a twofold target: first, to make producers in the chain aware that they are acting upon ethical issues, and second, to communicate that these ethical decisions can be improved by explicitly taking consumer concerns into account, by means of various forms of deliberation and consultation.

Risks of Implementing Ethical Traceability

Translating the ethical issues encountered in food chains into operational and practical terms will be challenging. There are a number of risks, which must be considered carefully; there is even a risk that implementation may cause more harm than good. For example, there is a cost to implementing and administering (ethical) traceability systems, and this may damage consumers (who must bear higher prices) and producers (who must assume increased administrative burdens and also carry higher costs). It can be expected that big corporations, with higher levels of capital, will be in a better position to handle such costs and therefore gain a competitive advantage compared to small-scale producers. This tendency might be exacerbated if traceability systems are based on high-tech and capital intensive information and communication technologies (ICT) solutions. Third World producers are especially vulnerable to such a development.

Other risks are associated with the 'information' provided by traceability schemes. First of all it is essential that ethical traceability schemes are designed to capture relevant and useful information, and that this information is not manipulated. If this is not the case, the information may contribute to the creation of mistrust rather than enlightenment. Secondly, if the information is not structured and edited to suit consumer needs, ethical traceability runs the risk of overloading consumers with the time-consuming tasks of reading and learning about foods. Furthermore, making so much information accessible could lead to too much responsibility being placed on the shoulders of the consumers, instead of on politicians and corporations. Finally, the large amount of information accumulated by traceability schemes -for instance, on individual consumer preferences - could be misused for highly targeted marketing strategies against the will of consumers. The main risks are listed in Table 14.2.

Recommendations and Policy Options for Ethical Traceability

Traceability has been implemented in the agri-food sector but ethical traceability has not. There is much talk about consumers' informed choice and most actors in the food supply chain and elsewhere support the idea in principle. However,

Table 14.2 Risks associated with the implementation of ethical traceability

Risks linked to ethical traceability
• Increasing administrative burdens and control
• Higher food prices
• Possible exclusion of small-scale enterprises due to their inability to handle demands associated with additional traceability
• Reinforcing already powerful interests in food chains as traceability can be used to exclude small-scale corporations
• Not recognizing the full range of potential voices (e.g. excluding smaller producers due to their inability to establish traceability)
• Use of inappropriate indicators, auditing and monitoring that can lead to misleading information, false evaluations and ineffective policies
• Information overload for consumers
• Abuse of information (for instance for marketing purposes)

informed food choice as concerns ethical issues in the agri-food sector is still limited. Only a few food products are provided with information on the ethics of the product's production history.

This book has highlighted the fact that a great deal of information about food ethics exists in the supply chain, but in most cases this information is not communicated to consumers and other actors in the chain. The sociological surveys found that information on food ethics was desired by many interviewees. Therefore, the authors of this book have put together a catalogue of recommendations and policy options that hopefully can assist in the development of ethical traceability and informed food choice. The recommendations and policy options address four different groups of actors: the EC; civil society and NGOs; stakeholders in the food sector; and researchers.

The European Commission

The promotion of a wide and inclusive dialogue among relevant stakeholders about ethical traceability reveals the needs and wishes of stakeholders and consumers. On the basis of such dialogue, traceability regulations can be developed as a means of making ethical issues relating to food production practices visibile/transparent to consumers. As such, traceability regulations can function as a political instrument to guarantee minimum ethical standards in the food sector, in conformity with general consumer concerns. Furthermore, current traceability regulations could be strengthened to ensure food authenticity and guarantee truthful food labelling (avoiding misleading and deceptive information). In improving the regulations, it will be paramount to continue to argue in the WTO and Codex Alimentarius for the inclusion of ethical issues in food standards. On a less formal basis, the EC can encourage food companies to add ethical issues to existing traceability schemes and to support ethical capacity-building among small and medium-sized enterprises (SMEs).

Civil Society (NGOs, Such as Consumer, Animal Welfare and Environmental Organizations)

It is essential to push the political system and the food sector towards using current traceability legislation and private assurance schemes not only for food safety reasons but also as a means of promoting communication on food ethics with consumers. In doing so, it can be emphasized that the political system can use traceability legislation as a means to ensure and monitor that minimum ethical standards (e.g. on animal welfare, the environment, working conditions, etc.) in the food sector are not violated. Lastly, NGOs can consider their own active role in establishing and monitoring traceability schemes and in certifying traceability schemes.

The Food Sector

There are examples of good practice in existing food assurance schemes that can be built on to develop and improve communication with consumers. Existing assurance and traceability schemes can be developed to include more ethical parameters, in conformity with consumer concerns. For instance, they can be elaborated as a means of documenting production practices and trading conditions. Likewise, it will be important to promote examples of good practice and to encourage those elements of the food sector that either represent examples of good practice in ethical traceability, or have the potential to do so. Finally, it is essential to give ethical consumers better assistance in making their choices in supermarkets, shops and food service outlets.

Research

This book has highlighted several issues that merit further research. These include: how to involve diverse voices and their specific needs (consumer and citizen concerns) by way of participatory methods, and how to assure feedback on consumer needs (consumer concerns) to producers. In general, we can say that further attention needs to be paid to communication strategies that can operationalize ethical traceability from both a consumer's and a citizen's point of view. In doing so it will be important to consider how consumers react to ethical traceability schemes in practice, and especially how the trustworthiness of such schemes is conceived. Another issue that deserves notice is the cost-benefit analysis of including ethical issues in traceability schemes and of communicating traceability to consumers: this information will be essential for the future development of such schemes.

Ethical Traceability as a Communication Tool

We conclude this book with the observation that ethical traceability has the potential to function as a communication strategy for empowerment and involvement in ethical aspects of food production. This is true both for actors in the food supply chain and for consumers. For actors in the food supply chain, ethical traceability and informed food choice can help define the 'value-laden' and ethical qualities of their products and thus contribute to the 'identity' of their products. For consumers, ethical traceability is paramount both for making informed food choices and for engaging in ethical issues related to food production.

A central question for future research concerns which communication models are most suitable when dealing with ethical traceability and informed food choice. Instead of focusing on one-way information provision to consumers, it is more appropriate to reflect on the content and impact of consumer concerns from the perspective of the two-way communication models that are available, in order to determine which communication models can be recommended in specific cases. Due to the shifting nature of ethical concerns, ethical traceability requires dynamic, evolving structures to accommodate diverse points of view, cultures and narratives. This calls for continual dialogue along the chain.

Annex
Two Political Speeches: Consumers' Informed Choice and Ethical Traceability

On September 20, 2006 two political speeches were presented at a conference entitled *Ethical Traceability in the Food Chain — On Consumers' Informed Choice*. Margaritis Schinas, Head of Cabinet to Markus Kyprianou, European Commissioner for Health and Consumer Protection, presented the work and views on consumer's informed choice of Directorate General Health and Consumer Protection (DG SANCO). Mariann Fischer Boel, European Commissioner for Agriculture and Rural Development, talked about the role of ethics in the CAP. Both speeches are presented here.

Consumers' Informed Choice

Margaritis Schinas

I am very pleased to be here today, to represent Commissioner Kyprianou, who regrets very much that he is unable to attend in person. He has, however, asked me to pass on his warm regards and best wishes for an absorbing and successful conference.

Today, I want to focus on two specific areas which I trust will be of interest to you. First I will address the general area of *food labelling* and second, provide some thoughts on the sensitive issue of *animal welfare*.

Food Labelling

Many of you are aware that the Commission has been considering how best to amend the rules on food labelling for some considerable time.

Indeed, the review of food labelling legislation is necessarily a long-term project — and one which has been included in the Commission's simplification programme.

The general framework of the current labelling legislation dates back to 1979. This lays down the labelling requirements that are applicable to all foodstuffs intended for the ultimate consumer, and to foodstuffs supplied to restaurants and mass caterers.

The objective of that legislation is *to enable consumers to make and informed choice and not to be misled on the characteristics of foodstuffs*.

It needs to be *renovated and updated*, as both consumers and industry are unsatisfied with the current situation.

The broad aim of the review is to find a *sensible and practical balance* between the polarized positions: requests for *more information* on labels, and the need to have *clear and meaningful* labels, which are easily understood by consumers.

More specifically:

- Consumers want the information on the label to be as complete as possible but at the same time consider there is too much information on present labels, which make them less comprehensible

C. Coff et al. (eds.) *Ethical Traceability and Communicating Food*,

- Manufacturers consider the present labelling requirements as a burden, but at the same time, they "voluntarily" add extra information, which cuts back on space for the ingredients list and nutrition labelling

This view may appear contradictory and therefore lead to controversial discussions about the required changes

Nutrition labelling of food is also an issue where there is some controversy and therefore the Commission has been considering how to change the current legislation for a number of years.

Currently, such labelling becomes compulsory when a nutrition claim is made in the labelling, presentation or advertising of a foodstuff. But many stakeholders believe that nutrition labelling should become mandatory because of the on-going debate about rising obesity.

However, such views need to be considered alongside research which indicates that whilst more consumers are keen to have nutrition labels, there is evidence that the majority of consumers do not actually make use of them.

The situation has become more complex in recent years with the development by governments and industry, of front of pack ('signposting') schemes. The intention is to make nutrition information easier for the consumer to find and use.

The Commission's view is that such schemes should be in conformity with existing labelling legislation, and should not contradict or undermine efforts to improve existing EU rules.

Responses to the recent DG SANCO consultation (Labelling: competitiveness, consumer information and better regulation for the EU) — which closed at the end of June — will be used towards developing the revisions of the general food and nutrition labelling legislation.

The Commission services are now in the process of examining these contributions.

The member states will of course be actively involved in the process of the review, especially through the expert group on labelling.

The objective of the Commission as regards the review of the labelling legislation is to address and provide appropriate replies to the range of concerns through a proposal to renovate and modernize that legislation. This is scheduled for 2007.

Food Claims

I should also mention another important development — the new *Nutrition and Health Claims Legislation* on which agreement was recently reached. The aim here is to ensure that the claims that food companies make on packets regarding alleged nutritional or health benefits are *valid, and evidence based.*

This marks a major step towards making life easier for consumers to make *healthy choices* when selecting which food to buy.

Traceability

Before I move on to animal welfare, a few words about traceability. This is an obligation for all food business operators as laid down in the General Food Law Regulation of 2002.

The objective is to have in place a mechanism through which products can be traced back and forward through the food chain when food safety problems are identified.

Food and feed business operators need to be able to:

- Identify from whom and to whom a product has been supplied (one step back and one step forward)
- Have systems and procedures in place that allow for this information to be made available to the relevant authorities

These general requirements are valid for the whole food chain, apply to all EU producers and comprise foodstuffs, feed and food producing animals.

Labelling and traceability are different tools with different objectives. However, traceability could also allow for ensuring that information provided on labels is reliable. In that sense, there are complementarities between traceability and labelling where the latter can provide information on how the food was produced.

Animal Welfare

Animal welfare is an issue of great importance to European citizens. Indeed the sheer volume of mail Commissioner Kyprianou receives on this issue is a good measure of how it captures the public imagination.

Opinion surveys on this issue have revealed that a majority of European consumers are concerned about the welfare conditions of animals kept for food production.

Furthermore citizens cite animal welfare standards as an indicator of other product attributes — such as food safety, quality and healthiness. Many consumers equate good animal welfare standards with good food standards in general, and higher food quality.

And moreover, in addition to wanting to see more welfare friendly products on the shelves, citizens have expressed a readiness and willingness to pay for such products.

So far, so good. However, in a recent Eurobarometer survey a majority of citizens stated that the relevant and *appropriate information* to identify products produced under acceptable animal welfare conditions *is often missing* or inadequate.

The same survey also revealed that consumers' knowledge on the *way animals are farmed* is often insufficient to make a truly informed choice.

The commission has a strong commitment to improve animal welfare standards. In January this year the Community Action Plan on the Protection and Welfare of Animals was adopted.

The Action Plan identifies five areas of action to address the following:

1. To simplify and clarify current legislation facilitating its enforcement and to consider provisions for those species currently not addressed in EU legislation
2. To ensure that full regard is given to animal welfare across related policy fields (such as agriculture, research and food safety) and to contribute to the ongoing evolution of animal welfare as a scientific discipline
3. To improve marketing, labelling and communication strategies at European level enabling consumers to make better informed purchasing decisions;
4. To study and develop objective measurable indicators of animal welfare, allowing in particular the validation of farming systems applying higher welfare standards;
5. To promote animal welfare both within Europe and beyond, pressing for its acceptance at WTO level, and to analyze the potential trade impact on developing countries.

The Action Plan also suggests the possible creation of a specific *information platform* on animal welfare to facilitate the dialogue and exchange of information and experiences between all parties involved — producer, retailers and consumers.

The idea is currently under consideration and could pave the way towards a proper European labelling system for welfare friendly products.

Of course, whilst addressing animal welfare from the citizens' point of view the Commission will seek to ensure the concerns of the agriculture industry are also properly addressed.

Some parties voice the concern that high European welfare standards add to the pressure on EU producers from cheap imports. We firmly believe that promoting our high standards *can help counter the pressure* faced by EU producers from competition from third countries.

Europe is clearly leading the world in the promotion of animal welfare.

EU producers need to capitalize on the market advantages that spring from this. Our high animal welfare standards need to be considered and promoted as *an asset and an opportunity*, rather than viewed as an economic burden.

Clearly there is much work to be done to match consumers' aspirations with purchasing actions.

The Commission will continue to strive for better conditions for animals; increased marketing opportunities for European welfare friendly production; and better informed consumers.

Thank you.

'Just Deserts': Ethics, Quality and Traceability in EU Agricultural and Food Policy

Mariann Fischer Boel

Ladies and gentlemen,

First, I would like to thank the organizers of this conference for inviting me to speak on such an interesting subject. When I heard that I would have to talk about ethics, my first instinct was to try to telephone my good friend Aristotle. Unfortunately, he no longer works in the European Commission, so today I will have to take a less philosophical, more businesslike approach to the subject!

Traceability is not a new concept in European agriculture. Neither is ethics, if under this term we understand good treatment of farm animals and the environment, among other things.

But these two concepts have been gaining heavily in importance in the last few years as our agricultural policy places an ever-greater emphasis on two concerns: food safety and food quality.

I don't intend to say much about food safety today — important though that is – – because my colleague from the Commission's Health and Consumer Protection department is better-placed to do so.

Food Quality

Instead, I would like to focus on 'quality' (understood in a broad sense), highlighting its importance and situating issues of ethics and ethical traceability within it.

When applied to food and drink, the term 'quality' means different things to different people. It can certainly carry 'ethical' connotations — telling consumers about production methods and a product's relationship to animal welfare and the environment. It can refer to geographical origin. And of course, it can refer to that specific tang of a good cheese, or the way in which a good red wine goes down so smoothly.

It's essential for us to know what qualities consumers are looking for in food, and which ones will persuade them to pay higher prices. This is because of the changing international environment in which the agri-food industry now operates.

C. Coff et al. (eds.) *Ethical Traceability and Communicating Food*,

The barriers to international agricultural trade are falling fast, and there are many very powerful players out there who will fight tooth and claw for a larger share in world markets — especially bulk markets.

And let no one think that the challenge is less immediate now that the Doha Round of WTO trade talks is on hold. The forces of globalization were transforming our world long before the Doha Round got underway, and they will continue their work, whatever may happen around negotiating tables in Geneva.

I am confident that many farmers in the European Union will continue to compete effectively in bulk markets for the foreseeable future. But not all. Those that cannot win the battle on this ground must fight on different ground — and for many, that ground is high-quality production.

In the world of food and drink, 'Europe' and 'quality' are two words which have had a close association for many decades. Millions of consumers around the world could write out a long list naming the best of Europe's cheeses, wines, meats, *et cetera*, which are familiar to them from long and pleasant experience. In many markets, our products are also well known because they come from farming systems with high ethical standards — in other words, which take good care of the land and animals.

When we play our cards right, we can get a premium for these 'qualities'. But we cannot afford to rest on our laurels. We must keep up our efforts at producing the right goods with the right qualities. And of course we must make sure that potential buyers keep getting accurate messages about what we are offering them — through traceability systems, where appropriate — so that they will pay more.

In this battle, it is our farmers and the food and drink industry that are on the front line. We have to support them with the right policies. We do indeed have policies to promote quality and make it more identifiable, and I would like to bring you up to date on how they are developing.

European Union's Common Agricultural Policy

First, I should make it clear that the whole framework for high-quality agricultural production and high 'ethical' standards has changed. This is because of *fundamental improvements to the European Union's Common Agricultural Policy* (CAP).

In the past, critics of the CAP often claimed that it placed a heavy emphasis on quantity. There is some truth in this. One of the early aims of the CAP was higher productivity, to help prevent food shortages of the kind that still occurred after the Second World War. Under these circumstances, there was less emphasis on considerations which might now be described as 'ethical': environmental considerations, for example.

But circumstances have changed, and so has the CAP. In 2003 we agreed a new approach to agricultural policy which is bringing both product quality and ethics closer to the centre stage.

The core of the 2003 reforms is a new type of support payment to farmers, which is no longer linked to production. We call it the Single Farm Payment.

In order to receive this money, farmers do not have to farm a given product. Instead, they must meet high standards of environmentally friendly land management, animal welfare and public health. (We refer to this system as 'cross-compliance'.)

Obviously, this is a very strong incentive for more 'ethical' farming. Another effect of the new system is that farmers have much greater freedom to focus on product quality. In the past, they spent much of their time calculating which subsidy combination would pay the most. Now, they can spend this time on coming up with the new products which buyers want, and for which they will pay more.

In the same context, the reform has provided for the possibility for Member States to use up to 10% of their expenditure in direct payment for additional payments to be granted for specific types of farming which are important for the protection of the environment or for improving the quality and the marketing of agricultural products. Not many Member States, and not to the full extent, have made use of this possibility of introducing closer links between support and environmental and quality standards. Nevertheless this underlines even more the direction towards which CAP is going.

Within the framework set by the new-look CAP, one of the tools for supporting ethical farming and high-quality production is **rural development** policy — the so-called 'second pillar' of the CAP.

As I have already outlined, all farmers in the European Union must meet given standards of care for their land and their animals in order to receive their Single Farm Payment. But in addition to that, our rural development policy offers many possibilities for making extra payments to farmers who go beyond these standards.

There are also financial incentives for farmers to get involved in European Union or national schemes which improve product quality and production processes, or which certify product quality for consumers.

And from next year, when new rural development rules come into effect, there will also be funding to help the food chain co-operate in developing new products, processes and technologies.

Organic Farming and Food Production

Strongly linked to rural policy is our policy on **organic farming and food production**.

Different people buy organic products for different reasons. Some say that they prefer the taste. Others have a motivation which they would describe in terms of 'ethics': they like to buy products which cut down the use of chemicals.

The sector has generally been making strong progress, posting annual growth of about 30% since 1998. But I believe that we can give it a firmer basis for further growth, and so the European Commission is drafting new rules for organic production, to replace those which became law in 1991.

Labelling is one area of the rules that needs to change. I believe that the range of labels in the organic sector sometimes makes it difficult for the consumer to understand exactly which standards have been met by a given product.

I therefore think that every organic product in the European Union should carry either the EU organic logo or the standardized text fragment 'EU-ORGANIC' accompanied, where the case may be, by the national organic logos.

Geographical Indications

Having just spoken about labelling in the organic sector, of course I also have to talk about *Geographical Indications*, or GIs. Many of you will know what these are: names of food and drink products which refer to a place, and thereby to specific qualities which are thought to be worth a premium.

We in the European Union think it is important to protect GIs. We believe that the inclusion of a place in the name of a food or drink product should mean something clear. We believe that consumers want and deserve a transparent link between the name of a product and the reality of that product. This is traceability at its most basic!

This is why we have a system for safeguarding GIs. The system is popular with our producers and consumers, and it enables many producers to make better profits.

Unfortunately, it is much less popular with some of our trade partners around the world. Some of them do not wish us to get the benefit of product names whose fame was built up in Europe over many years through hard work. Many of them recognize trademarks which compete directly with some of our GIs.

Resolving our differences over these issues is a very difficult task. But a very necessary task, if we want to give our farmers and food and drink producers the export opportunities which they deserve — especially at a time when demand for high-value products is growing fast in markets such as China and India.

This is one reason why it is frustrating that the Doha Round of WTO trade talks has been suspended. These talks are the only forum within which the European Union can hope to achieve greater international recognition and protection for GIs. Now that the talks are on hold, for the time being we have to rely on bilateral agreements — which give us very patchy results.

Nevertheless, in the meantime we are making improvements to our internal system for handling GIs. We are carrying out a review to see how well it works for us. And we have recently made it more fully compliant with WTO rules. This means, for example, that producers in third countries can now apply directly to the European Union to get a product name protected in our territory, instead of applying through their own government.

Ladies and gentlemen, I hope you can see from my remarks today that ethical food production, food quality and the traceability of the characteristics of food hold a prominent position in the European Union's policies.

Farmers and food producers who aim to secure higher prices by delivering higher standards and quality — whether in terms of ethics or more generally — deserve support.

We must have the right tools in place to help them deliver those standards and quality, and to make sure that the result of their efforts can be clearly recognized and traced back through the food chain.

I am committed to doing whatever I can to provide those tools. They are indispensable to the future of farming and food production in the European Union.

Thank you for your attention.

Index

W

Printed in the United Kingdom
by Lightning Source UK Ltd.
135313UK00006B/181/P

9 781402 085239